Timing and Time Perception: Procedures, Measures, and Applications

# Timing and Time Perception: Procedures, Measures, and Applications

*Edited by*

Argiro Vatakis
Fuat Balcı
Massimiliano Di Luca
Ángel Correa

BRILL

LEIDEN | BOSTON

Cover design: Maria Kostaki, photograph courtesy of Charles E. Zapolski

The Library of Congress Cataloging-in-Publication Data is available online at http://catalog.loc.gov
LC record available at http://lccn.loc.gov/2018001126

Typeface for the Latin, Greek, and Cyrillic scripts: "Brill". See and download: brill.com/brill-typeface.

ISBN 978-90-04-28019-9 (hardback)
ISBN 978-90-04-28020-5 (e-book)



This book is printed on acid-free paper and produced in a sustainable manner.

# Contents


Preface VII

List of Illustrations IX

1 **Circadian Timing: From Genetics to Behavior** 1
*Patricia V. Agostino, Ivana L. Bussi and Carlos S. Caldart*

2 **Prospective and Retrospective Timing Processes: Theories, Methods, and Findings** 32
*Richard A. Block, Simon Grondin and Dan Zakay*

3 **Assessing Duration Discrimination: Psychophysical Methods and Psychometric Function Analysis** 52
*Karin M. Bausenhart, Massimiliano Di Luca and Rolf Ulrich*

4 **Methodological Issues in the Study of Prospective Timing** 79
*Giovanna Mioni*

5 **Duration Bisection: A User's Guide** 98
*Trevor B. Penney and Xiaoqin Cheng*

6 **Temporal Decision-making: Common Procedures and Contemporary Approaches** 128
*David Freestone and Fuat Balcı*

7 **Towards a Process Model of Temporal Generalization** 149
*Florian Klapproth*

8 **Reaction Time Analysis for Interval Timing Research** 165
*Patrick Simen*

9 **Analysing Multi-person Timing in Music and Movement: Event Based Methods** 177
*Mark T. Elliott, Dominic Ward, Ryan Stables, Dagmar Fraser, Nori Jacoby and Alan M. Wing*

10 **Measuring Temporal Preparation** 216
*Mariagrazia Capizzi and Ángel Correa*

11 **Temporal Order and Synchrony Judgments: A Primer for Students** 233
*Maria Kostaki and Argiro Vatakis*

12 **Perceived Temporal Order and Simultaneity: Beyond Psychometric Functions** 263
*Miguel A. García-Pérez and Rocío Alcalá-Quintana*

13 **Collecting and Interpreting Judgments about Perceived Simultaneity: A Model-Fitting Tutorial** 295
*Kielan Yarrow*

14 **Using the Simon Effect in Simultaneity/Asynchrony Discrimination Tasks: Interest, Methods, and Limits** 326
*Anne Giersch, Patrick E. Poncelet, Céline Z. Duval and Laurence Lalanne*

15 **Tracking Time in the Infant Brain** 341
*Franziska Kopp*

# Preface

Between 2010 and 2014, the European-funded ESF-COST Action *Time In MentaL ActivitY: Theoretical, behavioral, bioimaging, and clinical perspectives* (*TIMELY*) aimed at creating an international network of researchers working on timing and time perception, but it also started to address questions and issues that had not been attended to-date. One recurring issue was consistently pointed out to us particularly by junior researchers and students: the difficulty in acquiring theoretical understanding and practical advices about the various methods utilized to measure timing in behavior. The underlying problem was the typical spread of information in academia, where senior researchers passed their theoretical and practical knowledge to junior researchers in their lab or junior researchers were given a multitude of papers and sources so that they can figure out on their own how to conduct a given timing task. The theoretical knowledge was publicly available, but the practical knowledge was confined to a handful of labs where it was passed on from researcher to researcher. Only by being taught by someone in the lab (or by extensive trial-and-error) one could achieve the hands-on practical knowledge to correctly apply these research methods, handle the problems one normally encounters in their use, and avoid their potential pitfalls. A consequence of the limited spread of practical knowledge also led students that learned a given method to end up using that method for the majority of their research for the rest of their career without being able to integrate new techniques into their research.

In TIMELY, we aimed to provide hands-on methodological training on a large array of timing research methods to a wide audience by organizing a training school on *Timing and Time Perception: Procedures, Measures, & Applications* that was held in Corfu, Greece, between the 4th and the 8th of February 2013. During this event, numerous of the principal experimental tasks and data analyses employed to measure some aspect of timing behavior were presented theoretically and practically, with both talk and hands-on sessions. In a series of presentations by leading experts in the field, junior researchers were taught how to conduct, for example, a motor synchronization task, what programming code and equipment to use for running an experiment that could answer a scientific question using this method, how to analyze and best display the data obtained, the common issues and problems associated with this method, etc. The result of such a successful training school was that a team of 40 junior researchers got familiarized with all the major methodologies of measuring timing behavior and how to analyze these measurements, a learning experience that took only four days instead of the many years it could have taken with the conventional route. The students could also experience first-hand how to use traditional timing methodologies for real world applications such as how to study synchrony between the movements of a therapist and his/her patient and how the degree of synchrony may determine the speed and effectiveness of ones' well-being.

Because of the success of the training school and the lack of literature that systematically covers methods and techniques on timing and time perception, we decided to compile a book composed of contributions where literature, theory, practical applications in the form of computer programs, and experimental protocols covered all aspects of learning. This book you are now holding in your hands and the computer code associated with it is available online at https://github.com/ArgiroVat/Timing-and-Time-Perception-Book.

The book is a collection of the most utilized and known methods on timing and time perception. Specifically, it covers methods and analysis on circadian timing, synchrony perception, reaction/response time, time estimation, and alternative methods for clinical and developmental research. Most of the chapters are accompanied by a description of the experimental protocols for a sample experiment, the programming code to perform the experiment, sample results, and code for the analysis of the data. This practical section of the book is open source and will be continuously updated, but it serves as a companion to the chapters in the book (thus, first refer to the book, which is also available in open access via Brill's website).

The depth at which the different chapters composing the book treats the arguments varies from very introductory to more advanced so as to cover the needs of both the junior and senior researcher. We believe this is a representative source of the current methods on timing and time perception and we hope that this will be the first step in future efforts to document experimental methods and analysis both in a theoretical and in a practical manner. We would like to thank the TIMELY network and ESF-COST funding for supporting this endeavor and our authors, who contributed to this book and patiently waited for its completion and publication. We would also like to thank two specific individuals: Charlie Zapolski, who graciously provided us with the photograph of the book cover and allowed us to modify it freely, and Mary Kostaki, who worked on the design of the cover for this book providing us with a wonderful result.

*Argiro Vatakis, Fuat Balcı, Massimiliano Di Luca and Ángel Correa*

# List of Illustrations

## Figures


1.1 The molecular circadian clock. Simplified model of the transcriptional/translational feedback loops that constitute the mammalian circadian clock 3

1.2 Circadian entrainment to the environment 5

1.3 In the limit of circadian entrainment. Actograms representing different light/dark conditions in a nocturnal rodent 9

1.4 Characteristics of circadian rhythms 13

1.5 Double-plotted actograms and their respective waveforms 17

1.6 Double-plotted actograms and their respective periodograms 19

1.7 Graphical representation of Cosinor and Rayleigh in a polar plot 22

2.1 Summary of the main methods utilized for studying time perception (Adapted from Grondin, 2010) 35

2.2 Psychometric function used for estimating the difference threshold with the constant method 38

2.3 The attentional-gate model of prospective timing (from Block & Zakay, 2008; Zakay & Block, 1997) 41

3.1 Relative frequency of responding with $R_2$ (i.e., judging the second presented duration c as longer than the first presented duration s ) as a function of comparison duration (open circles) 55

3.2 Relative frequency of responding with $R_2$ (i.e., judging the second presented duration $c$ as longer than the first presented duration $s$ ) as a function of comparison duration (open circles) 59

3.3 Observed (open circles) and monotonized (black X and solid line) relative frequency of responding with $R_2$ (i.e., judging the second presented duration $c$ as longer than the first presented duration $s$ ) as a function of comparison duration 61

3.4 Relative frequency of responding with $R_{c>s}$ (i.e., judging the comparison duration $c$ as longer than the standard duration $s$) and psychometric functions for a hypothetical 2AFC experiment 65

3.5 Relative frequency of responding with $R_{same}$ (i.e., judging $c$ and $s$ as equally long). The solid line shows the best fitting psychometric function 68

3.6 Relative frequency of responding with $R_{same}$ (i.e., judging $c$ and $s$ as equally long). The solid line shows the best fitting psychometric function 70

3.7 Relative frequency of responding with $R_{same}$ (i.e., judging $c$ and $s$ as equally long) 71

5.1 Mean proportion of 'long' responses by a single human participant trained on anchor durations of .8 s and 3.2 s 102
5.2 Fits of pseudo-logistic, sigmoidal, and Weibull functions to duration bisection data of a single subject 103
5.3 Group mean psychometric response functions from three long (L)/short (S) anchor duration ratios following normalization by the group function bisection point. Data replotted from the 'No Distracter' condition in Experiment 2 of Penney, Yim, and Ng (2014) 104
5.4 Data from a single subject illustrating performance early and late in a duration bisection test session that comprised 392 trials 113
6.1 The peak interval task. Top: A schematic of the task. Bottom: The PsychoPy builder screenshot of a fixed interval trial 132
6.2 Example average response curves for Fixed Interval (blue) and Peak Interval (red) trials for a short 30-minute session for a single participant 133
6.3 Example single trial raster plots for Fixed Interval (left) and Peak Interval (right) trials for a short 30-minute session for a single participant 135
6.4 The switch task. Top: A schematic of the task. Bottom: The PsychoPy builder screenshot of the trial 139
6.5 Example switch times from a participant on the switch procedure. The short duration was 2 s and the long duration was 3 s 140
6.6 Example performance curves for the switch task. The black optimal performance curve shows that the optimal switch time decreases as the coefficient of variation increases 141
6.7 Example ways of assessing optimality in participants. Notice the y-limits start at .90, humans are often very close to optimal on this task 142
6.8 The DRL Task. Top: A schematic of the DRL task. Bottom: A screenshot of the PsychoPy builder used to run the task 144
6.9 Example inter-response times for a single human participant on the DRL task. The left panel shows every inter-response time in a scatter plot, and the right panel shows the histogram 145
7.1 The relation between abs(s – t)/t, response time (RT), and stimulus duration (T) 151
7.2 Temporal generalization gradients obtained from Experiment 1 of the Klapproth and Müller (2008) study. Modified after Klapproth and Müller (2008) 153
7.3 Standardized response times, plotted against comparison stimulus duration divided by the corresponding standard. Modified after Klapproth and Müller (2008) 154
7.4 A model of subjective probability of same or different responses in temporal generalization, dependent on the duration of the comparison interval t 156

7.5 The development of subjective probability of same or different responses in temporal generalization, dependent on the mental representation of the standard duration s* 157
7.6 Temporal-generalization gradients obtained from Experiment 2 of the Klapproth and Wearden (2011) study. Modified after Klapproth and Wearden (2011) 158
7.7 Mean latencies of same and different responses in Experiment 2 of the Klapproth and Wearden (2011) study. Modified after Klapproth and Wearden (2011) 159
7.8 Synopsis of the effects of speeded duration judgment in temporal generalization on the timing parameters *t*, *s*, and *b* 162
9.1 Example of a simple, low-cost experimental setup for capturing finger tap responses from group timing experiments 181
9.2 A typical hardware configuration for cue generation and response capture using MatTAP (reproduced from Elliott et al., 2009b) 186
9.3 Key stages involved in onset detection 190
9.4 Example sensor output signal resulting from a single tap 191
9.5 Temporal waveform, log magnitude spectrogram and four onset detection functions (ODF) extracted from a violin recording 199
9.6 Flowchart of the algorithm used to align onsets between two sources (taken from Elliott et al., 2009b) 203
9.7 Typical pattern of asynchronies when participant is exhibiting drift – i.e. not synchronising with the cue 206
10.1 (A) Experimental design. The variable foreperiod paradigm used by Vallesi, Lozano, and Correa (2013) 219
10.2 Experimental design. The temporal orienting task used by Capizzi, Sanabria, and Correa (2012) 221
10.3 Experimental design. The rhythm-based temporal preparation task used by de la Rosa, Sanabria, Capizzi, and Correa (2012) 225
11.1 Percentage of hypothetical 'synchronous' responses from an SJ task as a function of the SOA (blue bell-shaped curve) and hypothetical 'visual first' responses from a TOJ task plotted as a function of the SOA (magenta sigmoid curve) 239
12.1 Sample psychometric functions for SJ2 data (a), SJ3 data (b), and two sets of TOJ data (c) 265
12.2 Sample distributions of arrival time 268
12.3 Effects of SOA on the distribution of arrival times for reference and test stimuli (top row) and on the distribution of arrival-time differences (bottom row) 270
12.4 Psychometric functions in SJ2, SJ3, and TOJ tasks in four different scenarios resulting from the combination of two cases for the distributions of arrival

times (top row) and two cases for the width of the (symmetric) central region in decision space (left column) 276
12.5 PSSs and SRs (or JNDs) are insufficient to assess performance differences across groups or experimental conditions 280
12.6 Model extension to cover the SJ4 task. The decision space (center panel) includes two regions for "uncertain order" (U) judgments flanking the region for S judgments, besides the outer regions for TF and RF judgments 282
12.7 SJ4 data and fitted psychometric functions 284
12.8 Single-presentation methods cannot distinguish decisional bias from sensory acceleration 287
12.9 Distinguishing decisional bias from sensory acceleration with a dual-presentation task and a ternary response format 290
13.1 Example SJ data. Each row shows data collected under different instructions (see main text) 299
13.2 Schematic of the four-parameter observer model outlined in the main text 301
13.3 Schematic of process for the generation of a discrepancy function 307
14.1 Illustration of the Simon effect observed in patients affected by schizophrenia for SOAs of less than 20 ms, which is opposite in direction to the one found with larger asynchronies and in the healthy population 330
14.2 Illustration of the priming task used to explore how participants follow events in time 332
14.3 Illustration of the priming paradigm used to explore the influence of asynchronous priming stimuli (i.e., frames with an undetectable SOA of 17 ms) in a temporal order judgment task 333

## Tables

3.1 Exemplary outcome of a psychophysical experiment with s = 500 ms and nine comparison levels $c_1, \ldots, c_9$ ranging from $c_1$ = 300 ms to $c_9$ = 700 ms 54
3.2 Exemplary outcome of a psychophysical experiment with an equality judgment task 67
3.3 Overview of advanced toolboxes and functions for psychometric function fitting 73
5.1 These publications present data from variants of the duration bisection task 117
9.1 Summary of average measures for four state-of-the-art onset detectors submitted to MIREX (2015) audio onset detection contest 201

11.1 An overview of the sensitivity measures obtained from the TOJ and SJ tasks 238
11.2 An overview of the differences between the studies that have directly compared the TOJ, SJ, and SJ-3 tasks. (A: Auditory, V: Visual, AV: Audiovisual) 248

CHAPTER 1

# Circadian Timing: From Genetics to Behavior

*Patricia V. Agostino, Ivana L. Bussi and Carlos S. Caldart*

## 1 General Aspects of Biological Timing

In order to adapt to a changing environment, most living organisms have established the ability to measure time by the development of biological timing systems. There are different time-scales in which these temporal systems operate, thus covering more than 12 orders of magnitude, from microseconds to days or even seasons. All of these biological 'clocks' use time-keeping mechanisms that provide organisms with temporal information crucial to fulfill the required biological processes (Buhusi & Meck, 2005; Golombek, Bussi, & Agostino, 2014). Most chapters in this book are focused on temporal processing in the millisecond-to-minutes range, which is crucial to many forms of cognitive processing. In the present chapter, we will cover different aspects of biological timing in the circadian (i.e., 'about a day', 24-h) range. Since circadian timing is ubiquitous and regulates many physiological and behavioral functions – including cognitive functions – the study of circadian rhythms has a deep impact in neuroscience, medicine and health research.

Despite early classical experiments such as De Mairan's 1729 demonstration of an endogenous rhythm in leaf movement of Mimosa plants, among many others, the notion of an inner biological clock remained quite elusive until the mid-twentieth century (Golombek & Rosenstein, 2010). Several lines of evidence revealed the presence of an endogenous daily biological clock, and the term circadian was introduced in the fifties to identify self-sustained rhythms under constant conditions.

At present, it is very well known that the circadian clock is an endogenous timing system that organizes the environmental oscillations every 24-h. The periodic changes in the light/dark cycle produced by the rotation of the Earth influenced the development of circadian rhythms in almost all living organisms. Hence, organisms are able to adapt to and anticipate the changes in lighting conditions, thus settling their physiological processes to specific times of the day. Some physiological and behavioral functions regulated by the circadian system include the sleep-wake cycle, body temperature, hormone release, and gene expression. For example, nearly half of all mammalian genes are rhythmically expressed in one or more tissues (Yan, Wang, Liu, & Shao, 2008).

 DOI 10.1163/9789004280205_002

The circadian system have three fundamental characteristics: (1) the presence of an endogenous oscillator, which generates rhythms with a period close to 24-h in the absence of external cues; (2) the capacity to be synchronized by rhythmic environmental stimuli; and (3) temperature compensation, which means that the biochemical reactions that take place in the circadian clock do not change their speed with temperature variations (Buhr & Takahashi, 2013). Additionally, the circadian system is composed by three main components that allow the regulation of physiology and behavior: (1) an input pathway sensing the environment, (2) an endogenous core oscillator, and (3) an output pathway that couples the core oscillator with the rest of the organism. In mammals, the main circadian oscillator (called the core or master oscillator) is located in the suprachiasmatic nuclei (SCN) of the ventral hypothalamus. The SCN synchronizes subordinate organ and tissue clocks (called peripheral circadian oscillators) using mainly autonomic pathways from the hypothalamus, driving neurochemical, endocrine, and metabolic signaling pathways (Albrecht, 2012).

## 2 Molecular Components of the Circadian System

The SCN neurons have the ability to generate autonomous circadian rhythms (Welsh, Logothetis, Meister, & Reppert, 1995). The intrinsic rhythmicity of each neuron is based on a molecular oscillation that involves negative feedback loops of transcriptional and translational processes (Figure 1.1). In the mammalian circadian system, a primary feedback loop is composed by the positive elements CLOCK and BMAL1, which heterodimerize (i.e., join themselves to form a complex of non-identical monomers) and promote the transcription of *Per* (*Period*) and *Cry* (*Cryptochrome*) genes (negative elements) by acting on specific biding sites (E-box sequences) located in their promoter region (Gekakis et al., 1998; Munoz, Brewer, & Baler, 2006; Yoo et al., 2005). After transcription and translation, PER and CRY proteins heterodimerize and translocate to the nucleus to inhibit their own transcription by acting on the dimer CLOCK/BMAL1 (Ye et al., 2014).

Each transcriptional/translational loop takes near to 24-h to complete. In addition to this primary loop, another negative feedback loop contributes to clock precision and robustness. In the second loop, the protein REV-ERBα moves to the nucleus to repress *Bmal1* expression; inversely, Rorα can stimulate its transcription, both binding to RORE (Retinoic acid-related Orphan receptor Response Element) sites on the *Bmal1* promoter (Guillaumond, Dardente, Giguere, & Cermakian, 2005; Preitner et al., 2002; Sato et al., 2004). These oscillations of negative and positive elements generate circadian rhythmicity,

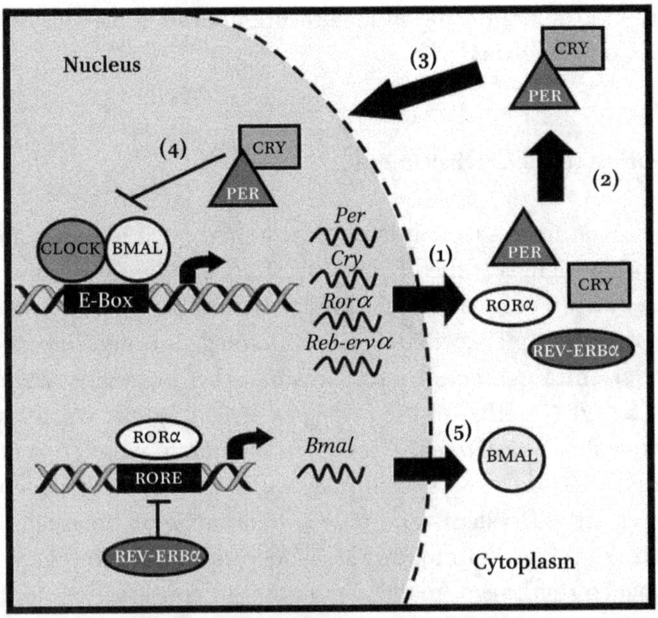


FIGURE 1.1 *The molecular circadian clock. Simplified model of the transcriptional/translational feedback loops that constitute the mammalian circadian clock. In the primary feedback loop, the positive elements CLOCK and BMAL1 initiate transcription of target genes containing E-box sequences (1), including* Period *(in mice,* Per1, Per2, *and* Per3) *and* Cryptochrome (Cry1 *and* Cry2). *Once in the cytoplasm, the resulting PER and CRY proteins heterodimerize (2) and translocate back to the nucleus (3). Negative feedback is achieved by PER:CRY heterodimers to repress their own transcription by acting on the CLOCK:BMAL1 complex (4). Another regulatory loop is induced by CLOCK:BMAL1 heterodimers activating transcription of* Rev-erbα *and* Rorα. *REV-ERBα and RORα subsequently compete to bind* RORE *elements present in* Bmal1 *promoter (5). Thus, the circadian oscillation of* Bmal1 *is both positively and negatively regulated by RORα and REV-ERBα.*

which regulates the circadian output pathway by driving downstream clock-controlled gene (CCG) expression. The regulation of CCGs is thought to be tissue-specific. Indeed, recent studies on mouse tissues indicate that roughly 50% of all genes oscillate with a circadian periodicity (Zhang, Lahens, Ballance, Hughes, & Hogenesch, 2014). Similar auto-regulatory feedback loops were also described for other organisms, including Drosophila, zebrafish, and cyanobacteria (Bell-Pedersen et al., 2005). Notably, the circuits that generate molecular oscillations of circadian clock genes are remarkably conserved among species, suggesting a possibly monophyletic origin of such mechanism (Dunlap, Loros, & DeCoursey, 2004).

## 3 Synchronization to the Environment

The circadian system in nature is normally exposed to a rhythmic 24-h environment, so that appropriate signals (called *Zeitgebers*, from German *Zeit*, 'time'; *geben*, 'to give'), such as light, temperature, or food, synchronize its oscillation (Golombek & Rosenstein, 2010). This effect in Chronobiology is called *entrainment*. Therefore, under entrained conditions, circadian rhythms adjust their endogenous period (called $\tau$) to that of the *Zeitgeber* (called T). It is worth to mention that the concept of *entrainment* – the phenomenon in which two or more independent rhythmic processes synchronize with each other – has been described in a wide variety of systems and over a wide range of timescales: some examples include fireflies illuminating in synchrony, human individuals adjusting their speech rhythms to match each other in conversation, foot tapping to the beat of a song, and attention to rhythmic patterns in the milliseconds scale (Jones, 1976; Clayton, Sager, & Will, 2005). Circadian rhythms of living organisms are entrained to rhythmic environmental cues, such as the 24-h light/dark alternation or daily temperature cycles. Locomotor activity rhythms are one of the most studied circadian outputs, and are frequently represented as *actograms* (Figure 1.2), where horizontal lines represent consecutive days and black bars describe the locomotor activity episode. Each horizontal line in Figure 1.2 shows one experimental day along the vertical axis, and the hours of the day are represented in the x axis. Further information about actograms is provided in Section 6.3.

In mammals, the most powerful synchronizer is the daily light/dark cycle. Under normal 24-h light/dark conditions, circadian rhythms are entrained with a period equal to 24-h (Figure 1.2A).

Light stimulates a group of photosensitive retinal ganglion cells (pRGC) that express the photopigment melanopsin (Panda et al., 2002) and project to the SCN through the retinohypothalamic tract (Morin, 2013). The effect of light on SCN neurons leads to an increase in the intracellular concentrations of $Ca^{2+}$, which initiates a signal transduction cascade that ultimately results in a phase shift of the circadian clock (Golombek, Agostino, Plano, & Ferreyra, 2004; Golombek et al., 2003; Golombek & Rosenstein, 2010; Morin & Allen, 2006). The circadian clock differs in its temporal responsiveness to external stimuli. For example, in nocturnal rodents, exposure to light synchronizes circadian rhythms by inducing phase delays during the early night and phase advances

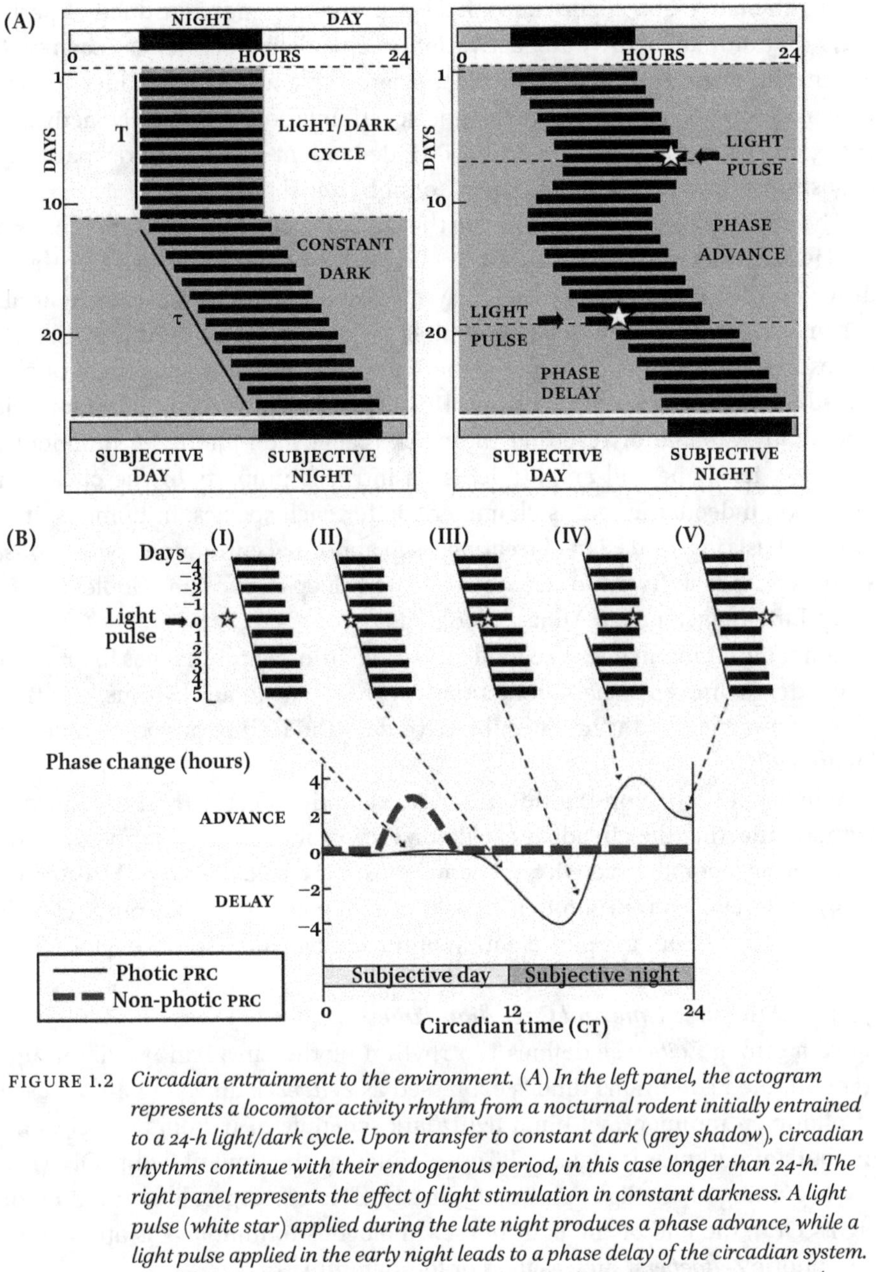


FIGURE 1.2 *Circadian entrainment to the environment. (A) In the left panel, the actogram represents a locomotor activity rhythm from a nocturnal rodent initially entrained to a 24-h light/dark cycle. Upon transfer to constant dark (grey shadow), circadian rhythms continue with their endogenous period, in this case longer than 24-h. The right panel represents the effect of light stimulation in constant darkness. A light pulse (white star) applied during the late night produces a phase advance, while a light pulse applied in the early night leads to a phase delay of the circadian system. (B) Phase response curves (PRC) of circadian rhythms. Actograms represent the response of locomotor activity rhythms to a light pulse during (I) the subjective day, (II) and (III) the early subjective night, and (IV) and (V) the late subjective night, producing a typical photic PRC (solid line). By convention phase advances are plotted as positive values and phase delays as negative values. Photic PRCs are typically biphasic, with phase delays at the beginning of the subjective night and phase advances at the end of that period, while non-photic PRCs (dashed line) exhibit clear phase advances during the subjective day and little (if any) responses during the subjective night.*

during the late night (Figure 1.2A), led by divergent signal transduction pathways that ultimately drive the activation of circadian clock genes (Lowrey & Takahashi, 2000). For example, during the late night, when light induces phase advances of behavioral rhythms, photic stimulation specifically activates the guanylyl cyclase (GC)/cGMP/cGMP-dependent kinase (PKG) pathway (Agostino, Plano, & Golombek, 2007; Golombek et al., 2004).

The effect of light in synchronizing the circadian clock depends on the time of day, thus defining a *phase-response curve* (*PRC*) that indicates the times at which light can induce phase delays, phase advances, or no change at all (Figure 1.2B, Dunlap et al., 2004; Golombek & Rosenstein, 2010; Johnson, Elliott, & Foster, 2003). The PRC is one of the main tools to study sensitivity of the circadian system to either light (photic entrainment) or many other stimuli potentially capable of resetting the clock (called non-photic entrainment). The PRC, its shape and amplitude, is an intrinsic property of the circadian oscillator. Indeed, the PRC is characteristic for each species. In humans, it is also the basis for the design of treatments in abnormal entrainment situations, such as jet-lag, shiftwork or circadian-related sleep disruption (Gooley, 2008; Lewy, Emens, Jackman, & Yuhas, 2006).

Other entrainment cues like food intake, environmental changes in temperature, drugs or even social interactions can reset circadian rhythms, eliciting phase shifts mainly during the subjective day (the resting period in rodents; Figure 1.2B).

When photic or non-photic *Zeitgebers* directly affect a rhythmic output without affecting the circadian oscillator, for instance directly suppressing or stimulating locomotor activity, this mechanism is called *masking* (Mrosovsky, 1999). Different from entrainment, masking influences a circadian output (such as behavior) without the participation of the endogenous circadian clock.

### 3.1 Zeitgeber *Time and Circadian Time*

Since a strong *Zeitgeber* defines the rhythm of the circadian oscillator, under entrained conditions time is expressed as *Zeitgeber* time (ZT). In the case of photic entrainment, within a light/dark schedule of 12 hours of light and 12 hours of darkness (LD 12:12), ZT0 is defined as the time of 'lights ON' (i.e., the beginning of the light phase). Similarly, ZT12 corresponds to the time of 'lights OFF', the end of the light phase. Analogous terminology is applied for non-photic *Zeitgebers*, such as food or temperature.

When organisms are kept under constant conditions by avoiding contact with external time cues, they display so-called *freerunning* or *circadian* rhythms, no longer equal to 24-h. Therefore, time cannot be expressed in ZT but it is expressed in circadian time (CT) units. One circadian cycle is divided into 24 CTs, with one unit being defined as the division of the internal period length ($\tau$) by 24 hours. The period under *freerunning* conditions differs among species. For example, the *freerunning* period of the human circadian rhythm in body temperature is close to 25-h.

### 3.2 *In the Limit of Synchronization: T-cycles, Forced Desynchronization and Relative Coordination*

Under natural conditions of entrainment to a *Zeitgeber* of T=24-h, variations of the subject's endogenous period ($\tau$) may lead to different phases of entrainment, that are associated with *chronotypes*. In this sense, short endogenous periods $\tau$ often lead to early phases ('morning larks') and long periods $\tau$ correspond to late phases ('night owls'). Mutations that affect the intrinsic period $\tau$ result in large shifts of the entrainment phase. For example, patients with familial advanced sleep phase syndrome have a short endogenous period of $\tau$ = 23.3-h leading to a phase advance of more than 3-h (Jones et al., 1999). Thus, the phase of entrainment is a function of the period mismatch $\tau - T$ (Bordyugov et al., 2015).

Most organisms can only entrain to *Zeitgebers* cycling with a period close to 24-h. If the entraining period is too short or too long – thus, exceeding the range of entrainment – the circadian system cannot follow the *Zeitgeber* anymore. In the laboratory, circadian rhythms can be entrained to variable *Zeitgebers* with different T periods (called T-cycles) in order to determine the range of entrainment for each species (Aschoff, Daan, & Honma, 1982; Jud, Schmutz, Hampp, Oster, & Albrecht, 2005). To determine this range of entrainment, T-cycles of different periods T≠24-h are applied. For example, 22-h or 26-h T-cycles may be used to define the limits of stable entrainment. These limits differ from species to species and depend on the nature of the applied *Zeitgeber.* Notably, under T periods close to the limits of circadian entrainment, nocturnal rodents present two patterns of behavior, called forced desynchronization and relative coordination (Campuzano, Vilaplana, Cambras, & Diez-Noguera, 1998; Casiraghi, Oda, Chiesa, Friesen, & Golombek, 2012; de la Iglesia, Cambras, Schwartz, & Diez-Noguera, 2004; Golombek et al., 2013). Forced desynchronization (Figure 1.3A) occurs when a system of dual, weakly coupled circadian oscillators is subjected to a *Zeitgeber* that is outside the range of entrainment of one of these oscillators. For example, it has been described that when rats were maintained under light/dark cycles with period of 22- and 23-h, two circadian rhythms were simultaneously present in their motor activity, temperature, and feeding activity: one rhythm followed the external light/dark cycle, whereas the other rhythm was under *freerunning* with a period of around 25-h (Madrid, Lax, Vilaplana, Cambras, & Díez-Noguera, 1992;

Vilaplana, Cambras, & Diez-Noguera, 1997). This behavior has been found as the manifestation of two groups of oscillators within the SCN: a ventrolateral SCN capable of entraining to the external cycle and a dorsomedial SCN incapable of such kind of entrainment (de la Iglesia, Cambras, Schwartz, & Diez-Noguera, 2004). Therefore, under certain external conditions, the circadian system can be dissociated in such a way that an entrained and a non-entrained rhythm (black bars and white bars, respectively, in Figure 1.3A) could appear simultaneously. On the other hand, relative coordination (Figure 1.3B) reflects the resetting effects of environmental *Zeitgebers* that are strong enough to influence the biological clock but are not strong enough to synchronize it (Golombek et al., 2013).

### 3.3 *Circadian Desynchronization*

When the timing of the light/dark cycle is altered – for example, as a result of transmeridian travel or work schedules – it leads to a misalignment of the circadian system with the environment.

A well-known consequence of circadian alteration is *jet-lag disorder*, which results from rapidly crossing times zones – with the severity of symptoms typically depending on the number of times zones crossed, the travel frequency, and the direction of travel (east or west). Symptoms vary from insomnia and excessive sleepiness to more severe complications, including increased cancer risk, metabolic disorder, cognitive deficits, reduced fertility, and deregulated immune responses (Castanon-Cervantes et al., 2010; Golombek et al., 2013; Karatsoreos, Bhagat, Bloss, Morrison, & McEwen, 2011; Mahoney, 2010; Scheer, Hilton, Mantzoros, & Shea, 2009). In animal models, interference with the cGMP-related photic entrainment pathway has been proven to be a useful tool for accelerating synchronization to advances of the light/dark cycle, which may reduce jet-lag symptoms due to eastbound flights (Agostino et al., 2007; Plano, Agostino, de la Iglesia, & Golombek, 2012).

Another circadian alteration is produced by *shift work disorder*, which affects people who frequently rotate shifts or work at night. Under these conditions, the alternation between a normal sleep/wake cycle and being awake at night and sleeping during the day not only alters the light/dark schedule in relation to the endogenous circadian clock, but also modifies feeding patterns, social habits, etc. (Reid & Abbott, 2015; Zee & Goldstein, 2010). Sleep loss, in addition to circadian misalignment, contributes to decreased alertness during night work. Current treatment of shift work disorder includes strategies to achieve and maintain some degree of circadian alignment (e.g., photic phase shift protocols, avoidance of light during the day, melatonin administration, etc.), improve sleep (using hypnotics, melatonin, and behavioral

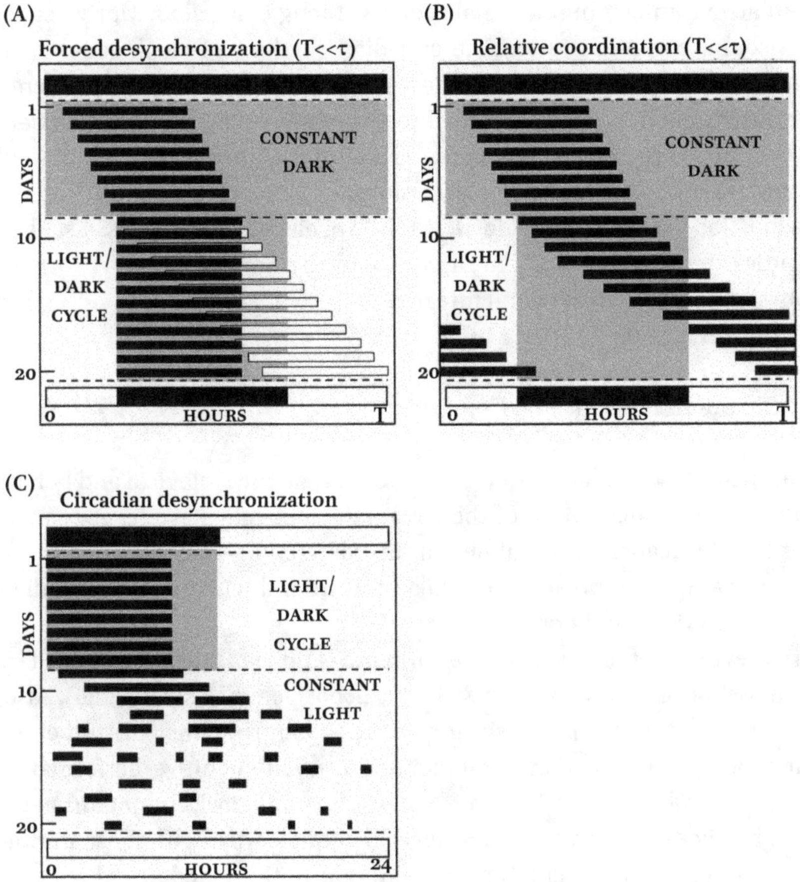


FIGURE 1.3 *In the limit of circadian entrainment. Actograms representing different light/dark conditions in a nocturnal rodent. (A) When a T-cycle is outside the synchronization range (in this example, T<<τ) it leads to forced desynchronization. A dissociation of the behavioral rhythm into two components can be observed, one that is synchronized to the light/dark schedule (black bars), and a second one (white bars) that runs in relative coordination. (B) Relative coordination (modulation of the circadian period). (C) Circadian desynchronization under constant light (LL) conditions.*

approaches), and facilitate alertness (using light, wake-promoting agents, and sleep scheduling).

Rodent models based on disruption of circadian rhythms provide useful information to study desynchronization due to modifications in the environment, such as abrupt changes in the light/dark cycle or nocturnal exposure to light. A commonly used model of circadian disruption is constant light (LL) exposure (Figure 1.3C), which induces period lengthening followed by circadian arrhythmicity in nocturnal rodents (Meng et al., 2010; Moriya et al., 2000). Also, LL conditions attenuate circadian rhythms of clock genes such as Per2 (Bussi, Levin, Golombek, & Agostino, 2014; Sudo et al., 2003). Altered circadian rhythms in rodents also lead to depressive-like symptoms, anxiety and anhedonia, expressed by reduced sucrose intake, altered patterns of food intake and high plasma levels of corticosterone (Fonken et al., 2009; Fonken et al., 2010; Tapia-Osorio, Salgado-Delgado, Angeles-Castellanos, & Escobar, 2013), indicating that entrainment of circadian rhythms is necessary for efficient physiological and behavioral integrity.

## 4 Circadian Influence on Cognition

As mentioned, the circadian system regulates several physiological and behavioral functions, and disruption of the circadian clock may have severe consequences for physical and mental health. In this section, we will overview an important example of circadian modulation: the influence of the circadian system on cognitive functions.

There is evidence that cognitive performance and learning is influenced by circadian oscillators (Eckel-Mahan & Storm, 2009; Gerstner & Yin, 2010). Also, regularly-timed cognitive processes impact circadian rhythms (Gritton et al., 2012), indicating a bidirectional interaction between cognitive performance and circadian processes. Furthermore, at least for some tasks in nocturnal rodents, night-phase performance can never be equaled by light-phase performance regardless of the strength of entrainment to the schedule, which may have deep implications for shiftwork therapies (Gritton et al., 2012). In addition, desynchrony between internal and environmental time has been associated with impaired cognitive function in animals (Devan et al., 2001; Gibson, Wang, Tjho, Khattar, & Kriegsfeld, 2010; Loh et al., 2010) and humans (Cho, 2001; Cho, Ennaceur, Cole, & Suh, 2000; Folkard, 1996; Santhi, Horowitz, Duffy, & Czeisler, 2007). Although it is well established that cognitive abilities vary as a function of daytime, there is still a widespread view that the circadian clock has a singular role in cognition related to sleep timing, and that the main factor for cognitive maintenance is the quality and duration of sleep. However, there is also evidence that indicates a circadian control of cognition beyond sleep timing, as revealed by forced desynchronization protocols (Kyriacou & Hastings, 2010; Wright, Hull, & Czeisler, 2002).

A fundamental component of cognition is the perception of the passage of time. In particular, temporal processing within the hundreds of milliseconds and the seconds-to-minutes range – known as interval timing – is crucial for many complex behaviors, such as speech comprehension, working memory, and decision making (Buhusi & Meck, 2005; Agostino & Cheng, 2016). Several evidences suggest that interval timing is also influenced by the circadian system. Indeed, time-of-day effects have been observed for the timing of both auditory and visual signals in the seconds-to-minutes range (Aschoff, 1985; Chandrashekaran et al., 1991; Meck, 1991; Pati & Gupta, 1994). For example, several studies have reported that time judgments in humans covary with normal circadian rhythms (Kuriyama et al., 2005; Lustig & Meck, 2001). Consistent with this finding, a circadian rhythm in interval timing was documented in control participants, but it was found to be disrupted in shift-workers (Pati & Gupta, 1994). Moreover, rats exhibit circadian variations in time perception similar to those that have been demonstrated in humans (Shurtleff, Raslear, & Simmons, 1990). Furthermore, sleep deprivation influences diurnal variation of interval timing in humans (Soshi et al., 2010). In Drosophila melanogaster, timing of short intervals is disrupted in circadian mutants for each of the three allelic *per* mutations, *per$^s$*, *per$^l$*, and *per$^o$* (Kyriacou & Hall, 1980). In mice, significant differences in the estimation of 24-sec intervals at different times of day were reported (Agostino, do Nascimento, Bussi, Eguia, & Golombek, 2011). These differences were maintained under constant dark conditions, but impaired in mice under constant light, which abolish circadian rhythmicity. Moreover, short time estimation in animals subjected to a 6-h advance of the light/dark cycle was transiently affected, indicating that temporal desynchronization of the circadian system is able to negatively affect interval timing. It has also been described that dopamine levels in mice present 24-h rhythms in the dorsal striatum, with lower levels during the day and peaking during the night (Bussi et al., 2014). Moreover, higher dorsal striatal dopamine levels during the night coincide with better performance on interval timing (i.e., peak location closer to the target time, higher peak amplitude and reduced peak width) in the nocturnal phase of the light/dark cycle in mice (Agostino et al., 2011; Bussi et al., 2014). Both interval timing and dopamine oscillation – as well as dopamine synthesis and turnover – in the dorsal striatum are affected by inducing circadian disruption under constant light conditions. In addition, circadian regulatory elements have been found in the promoter region of components involved in dopamine metabolism, like dopamine transporter (DAT), dopamine receptor 1A (DRD1A), tyrosine hydroxylase (TH), and monoamine oxidase (MAO), demonstrating that the expression of these components is under circadian regulation (Hampp et al., 2008). It was recently reported impaired learning in a timing task in pinealectomized rats (Bussi, Levin, Golombek, & Agostino, 2015). Furthermore, melatonin depletion increased striatal dopamine availability – which was reversed by external melatonin administration – indicating that this hormone may modulate interval timing on a circadian base.

## 5 Different Approaches to Measuring Circadian Rhythms

Several outputs are used to assess the overall status of the circadian system. These variables (called circadian marker rhythms) can be easily measurable over long periods of time, preferably using non-invasive methods. The most widely used circadian marker rhythms in several organisms are those of activity, temperature, hormones, and clock gene expression (LeSauter & Silver, 1998). In humans, the main variables used are actigraphy, core body temperature (CBT), and plasma or salivary melatonin (Bonmati-Carrion et al., 2014; Klerman, Gershengorn, Duffy, & Kronauer, 2002). In nocturnal rodents, wheel-running activity is the most commonly used and reliable method for assessing the output of the circadian clock. Other measures that can be used include general activity, food-bin approaches, drinking, sleep, and body temperature.

Bioluminiscence has become a powerful tool for studying circadian rhythms. Exogenous luciferase genes, under the control of promoters conferring circadian regulation, have been introduced into a wide variety of organisms, including cyanobacteria (Synechococcus), plants (Arabidopsis), insects (Drosophila), nematodes (C. elegans) and rodents (Welsh & Kay, 2005; Goya, Romanowski, Caldart, Bénard, & Golombek, 2016). These transgenic organisms exhibit robust circadian rhythms of luminescence, useful as a longitudinal assay of rhythmicity for a wide range of recent genetic and biochemical studies of circadian clock mechanisms.

## 6 Circadian Data Analysis

### 6.1 *Period, Amplitude and Phase*

Like other periodic patterns, circadian rhythms can be described using mainly three parameters: a) *period* ($\tau$), defined as the time in which a cycle is completed, b) *amplitude*, defined as the difference between the peak (or trough) and the mean value of the variable under study, and c) *phase* or *phase angle* ($\varphi$), the timing of a reference point in the cycle relative to a fixed event (Figure 1.4). There are several conventions to define the phase of a rhythm; however, the most commonly used is when the amplitude reaches its maximum, which is called *acrophase* in the Cosinor analysis (see below; Dunlap et al., 2004). Consequently, every time that we mention the phase of a rhythmic signal, it will be referring to its acrophase. Indeed, the algorithms that will be described in the following sections will also compute the acrophase (time of peak) of the signal.

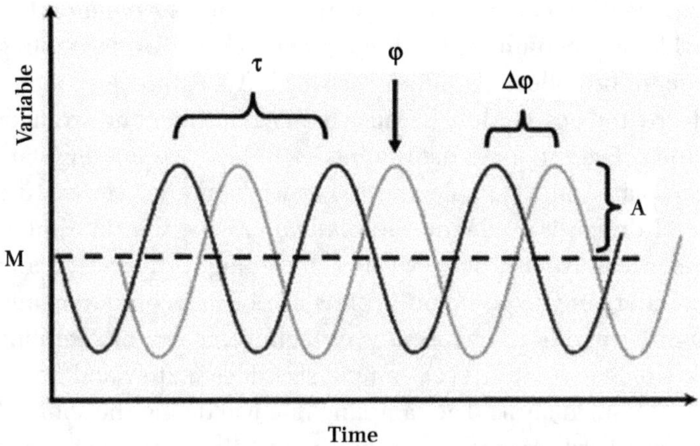


FIGURE 1.4 *Characteristics of circadian rhythms. Period (τ) is defined as the time to complete a cycle. It is commonly measured from peak to peak, but it could equally be measured from trough to trough or from any specified phase marker. The amplitude (A) of the rhythm is defined as one-half the peak-to-trough distance, from mesor (M) to peak or M to trough. Phase or phase angle (ϕ) is the timing of a reference point in the cycle (e.g., the peak) relative to a fixed event. The relationship of two mutually synchronized or entrained rhythms is expressed in terms of the phase angle difference (Δϕ).*

These parameters (period, amplitude, and phase) can describe any periodic pattern, but are normally used to characterize a sinusoidal wave. In many cases, however, this is not the real situation (i.e., locomotor activity may behave as a square-like rhythm). For this reason, most of the algorithms that are usually used to compute period and phase have been tested for different waveforms and are still valid for data analysis. Even though in this chapter we will only analyze the mentioned parameters, it is worth noting that according to the type of experimental data there are many other parameters that can be useful to analyze (e.g., sinusoidal dampening, fragmentation, coupling of the activity and *Zeitgeiber*, phase angle, etc.).

### 6.2 *Sampling and Signal Conditioning*

Before a complete overview of the methods used for analyzing circadian rhythms, it is important to refresh some basic concepts of signal analysis, such as sampling and signal conditioning (filtering and detrending). This information will allow us to know beforehand some of the limitations of our analytic methods.

A fundamental aspect of data analysis in temporal series is *sampling*. The first step is to decide between uniform sampling and non-uniform sampling (i.e., when data cannot be collected uniformly spaced). When possible, a uniform sample is clearly the best option, because it allows a wider and stronger set of statistical tools. Even if some data points were lost, the missing data could still be interpolated in a reliable and easier way. Another key feature of the experimental design is to define the *sampling rate* – the time interval between two consecutive samples, or *bin* – according to the needs and objectives of the undergoing experiment. In this age of fast computers and large hard drives, the trend is to choose a very high frequency of sampling and disregard the storage size; however, sample size does matter and it is a key factor for further data analysis. The sampling rate will define the shortest cycle that can be measured. According to the Nyquist theorem, the sample rate must be at least twice the higher frequency (i.e., 1/ τ) that it is aimed to study (Levine, Funes, Dowse, & Hall, 2002). For example, if we aim only to find the circadian frequency (24-h), we could just sample every 12 hours. However, in this case we cannot compute a phase or an accurate period. On the other hand, if the sampling rate were once per hour, then it would be possible to evaluate periodicities down to two hours (but not shorter), since a minimum of two points is required to describe a cycle. For this reason, the precision of the experimental approach is limited by the sampling procedure. However, a very high sampling rate would implicate low *sample size* – defining sample size as the length of the integrated time of the measure. In this case, when a very low sample size is used, the time for data processing and visualization greatly increases. According to the experimental output (e.g., locomotor activity, gene or protein oscillation, etc.), a compromise is reached between accuracy and speed. The usual sample size for circadian analysis of locomotor activity data varies between 5 to 30 minutes. This sample size will allow an accurate determination of circadian period and phase, as well as the study of the transient behavior of these parameters (including period and phase shift).

Sometimes there are some elements in the signal that may interfere with the calculation of circadian parameters. In this case, the raw data often need to be filtered for further study; this processing of the raw signal before the analysis is known as *signal conditioning.* Two of these interfering elements will be addressed in this section: the presence of a shifting temporal baseline (i.e., linear trend) and the presence of a high frequency noise.

In the first case (the presence of a shifting temporal baseline), the two types of linear effects that can alter the subsequent data analyses involve:

(1) a monotonically decrease or increase in the signal at a constant rate, thus producing a linear trend in the data, and (2) the rhythmic component of a signal can be obscured by high baseline activity. The first type of linear effect may have linear as well as non-linear components. This trend can be caused, for instance, by the natural aging of invertebrate models and it is common in their locomotor activity recordings. The second type of linear effect, a high baseline, could occur in many cases (e.g., data from imaging experiments). Even though the rhythm is quite strong in the circadian range, it cannot be computed before the conditioning. Removing this kind of trend is quite common and is highly recommended. It can be simply done by removing a regression line, to obtain a signal with slope and mean of 0. When the trend is non-linear, this detrend is a bit more complex; the most common approach is to use a moving-average window of fixed length to obtain the non-linear trend over time. Once we have obtained the trend, we can remove it with an amplitude correction by dividing the original signal by the trend, or by subtracting the signal. Thus, we keep the natural baseline of the signal, but we remove the non-linear trend. Detailed examples for detrending and normalizing a signal for circadian analysis can be found in Levine et al., 2002.

In the second case (the presence of high frequency noise), the approach is to use digital filters. These filters diminish the spectral contribution of a certain wavelength to the signal while keeping the others intact. The filters that reduce the high frequency components are called *low-pass* or *smooth*. We will mention a simple filter, *moving-average smoothing*, but there are other types of filters more complex, with different advantages and disadvantages. A filter can be defined as a function that transforms a time series into another. The moving-average filter is one of the most simple and used in the circadian field. It is used to analyze a set of data by creating a series of averages of different subsets of the full data set. Specifically, this filter averages a fixed number of points – it is better to have an odd number of points – in order to find the mean value of the initial subset of data. Then the subset is modified by moving one point forward in time, and the process is repeated until we reach the end of the data series. The resulting plot line connecting all the averages is the moving average: a set of numbers, each of which is the average of the corresponding subset of a larger set of data. In this case we give each point the same weight, but sometimes it is recommended weighting differently each point mainly with a Gaussian distribution (Levine et al., 2002) to give the central point of the window a higher weight. An example of moving average can be found in the online Appendix section (smoothma.m file; see book's GitHub repository).

### 6.3 *Actograms*

As previously shown in Figures 1.2 and 1.3, actograms are a useful tool to represent circadian locomotor activity rhythms in several species. The actogram is usually performed as a 'double plot' graph, where two cycles are represented in the same horizontal line. Specifically, double-plotted actograms show the second cycle on the right side of each line, as well as at the start of the following horizontal line, and so on. If the actogram is usually set at a fixed period of 24-h (for a solar day) in the x-axis, then the double plot representation will have 48-h duration (Figure 1.5, left panel). The cycle is divided into a time fraction with main activity episodes (called *alpha*) and other with major rest episodes (called *rho*). Sometimes, scattered activity can be observed in the *rho* time because the animal interrupts its sleep for a short time (Jud et al., 2005). Double-plotting is especially useful to visualize non-24-h rhythms or very long series.

One of the reasons that the actogram became a key representation in Chronobiology is that it allows a fast and reliable visual analysis. With a simple observation we can realize the changes of period and phase. The actogram was first introduced by Colin Pittendrigh at the beginning of the 1960s (Pittendrigh, 1965), well before any computational method for circadian data analysis was established. By that time, data were plot and measured by hand. Currently, there are many computational tools (e.g., a circadian plugin for *Imagej* and the software *El Temps*) that still maintain the concept of visual analysis ('paper and ruler' analysis, in which the actogram is printed and analyzed with common measuring instruments).

The main downfall of this approach is the variability of the observation between users and the lack of rigorous quantitative analysis. Nevertheless, this kind of analysis is still heavily used. A quick example of this visual inspection in the actogram can be seen in Figure 1.6, where locomotor records with different periods, as well as phase shifts, can be easily identified.

An example of a double-plotted actogram containing real data can be found in the online Appendix, by running the function Actogram (see the files SCRIPT.m and ScriptFilter.m for further details; see book's GitHub repository).

### 6.4 *Waveform*

Periodic data of known period could be divided into sections, with the length of these sections matching the underlying period. Overlaying these sections (cycles) will produce a clear *waveform*, with peak and trough (Figure 1.5, right panel). In other words, a waveform can be thought as an average of all the lines of the actogram. The length of the waveform is given by the length of the cycle, which can be empirically fixed (e.g., the length of a light/dark cycle) or previously calculated by any method. Even though it can be used in many cases, this technique is more common with evenly sampled series (uniform sampling) containing a good number of complete cycles (10 or higher). The observation and analysis of the average wave yields important characteristics of the rhythm, such as the duration of the activity time (*alpha*) and the rest time (*rho*). To define these different states of activity we need to find the beginning of the main activity episode, called *activity onset*, and the end or *activity offset*, which in many cases is a quite easy task. However, when these points are more difficult to find, as for instance due to fragmentation of activity episodes, is better to set an objective criterion. One approach is to find the median or the average of activity within the waveform cycle. Thus, when the activity goes over this threshold for a certain amount of time, we can define a statistical criterion for the *activity onset*. Similarly, when the activity falls under this threshold for the same amount of time, we can define the *activity offset*. The time difference between activity offset and onset is used to calculate *alpha* (see Figure 1.5).

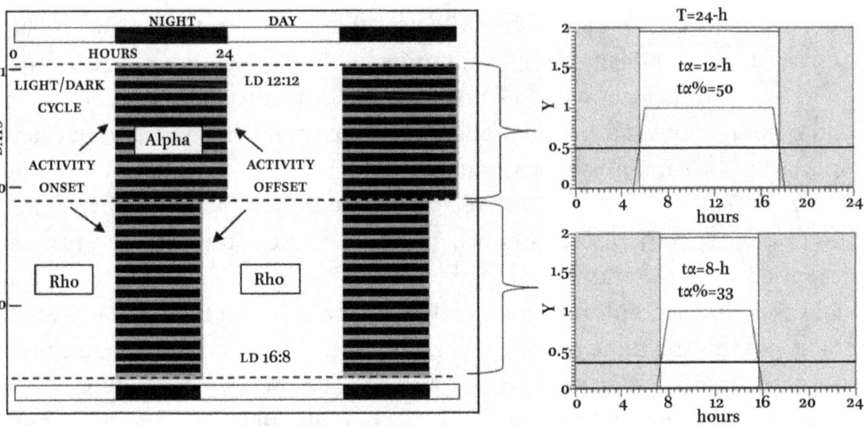


FIGURE 1.5 *Double-plotted actograms and their respective waveforms. To better visualize behavioral rhythms, actograms (left panel) are often double-plotted by aligning two consecutive days horizontally (e.g., day 1 left and day 2 right). In this example, two different light/dark (LD) conditions with a period of 24-h are shown: a LD 12-12 cycle (i.e., 12-h of light and 12-h of darkness), and a LD 16-8 cycle (i.e., 16-h of light and 8-h of darkness). Activity onset, activity offset,* alpha *and* rho *are indicated in the actogram. The resulting waveforms (right panel) for these two LD conditions display the same period (24-h). In each waveform, the length of the activity portion* (alpha, *white background) is measured between the activity onset and the activity offset, being* tα*=12-h for the upper waveform and* tα*=8-h for the bottom waveform. The percent of this length, according to the total length of the cycle* (alpha percentage), *is 50% and 33%, respectively. The length of the rest time* rho (tρ) *and its percentage* (tρ%) *can be obtain as the complement of* tα *and* tα% *respectively.*

Once the limits of *alpha* and *rho* phases are defined, their length (tα and tρ) is easily measured by taking into account that T = tα + tρ. The most usual way to represent these times is relative to the length of T (for *alpha* tα% = 100 tα / T). For example, for T = 24-h and tα = 12-h, then tα% = 50, meaning that half the total period corresponds to the activity phase. Along with the length of *alpha* and *rho* phases, it can be also calculated the area under the curve for both time lapses (Aα and Aρ) and to express them as percentages of the total area (Aα% and Aρ%). Finally, with both the length and amount of activity in both states, it can be computed the *degree of agglomeration* (Aα% / tα%). When this degree is near 1, the distribution of activity is uniform (i.e., we have a weak variation in the rhythm), while when this degree is higher than 1, most of the activity is concentrated in the alpha phase (Madrid & de Lama, 2006).

### 6.5 *Periodogram*

A key step in the analysis of circadian data is to accurately calculate the underlying period. There are several methods for determining period, which range from graphic approaches, such as the actogram – mentioned in Section 6.3 – to more mathematically complex approaches, such as Fourier-transform and non-Fourier-transform based algorithms (Zielinski, Moore, Troup, Halliday, & Millar, 2014). The latter methods allow a level of significance for period calculation, as well as an unbiased estimation.

A *periodogram* is a function relating periodic components of a time series to their spectral power. Many algorithms for period estimation had been described in the literature, all of them with a varying degree of complexity and different assumptions. Nevertheless the calculation method, all periodograms possess the same graphic representation, which makes them quite easy to understand (Figure 1.6). Several authors (Refinetti, 1993; Zielinski et al., 2014) compare the different methods for periodogram analysis (e.g., Fourier, Enright, Lomb-Scargle, etc.) and describe the best option for different cases. Each method has its own strength and weakness. In this section, we will mention two widely used methods: Lomb-Scargle (LS, very useful in short series and non-uniform sampling) and the Chi-square method, also known as Sokolove-Bushell (SB).

The most popular and conceptually simpler method is the Chi-square periodogram – its name corresponds to the statistical distribution that it uses. This method consists in splitting the data in sections of a fixed length. When this fixed length matches the period of the series, the overlay of all sections will give a clear waveform (with a strong variation) as the repeating patterns of the series coincide. However, if the length of sections does not coincide, then the

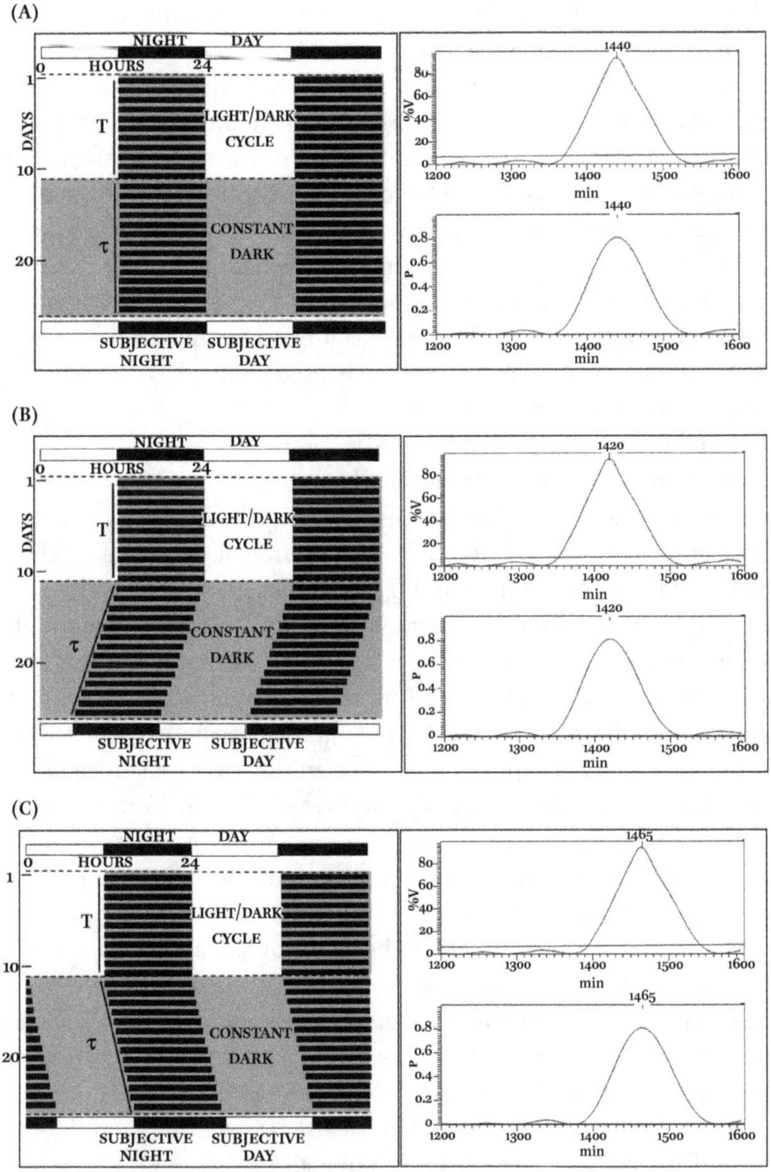


FIGURE 1.6 *Double-plotted actograms and their respective periodograms. (A) T = τ, with a period of 1440 minutes (24-h). (B) When τ < T, the activity seems to drift left. In this example, the endogenous period is τ = 1420 minutes. (C) When T < τ, the activity drifts to the right. In this example, the endogenous period is τ = 1465 min. In all cases, the peridogram represents the constant dark condition. Upper panel: Sokolov-Bushel (SB) periodogram. Lower panel: Lomb-Scargle (LS) periodogram. It is also possible to have an idea of the resulting period by the actogram's visual inspection. By using the proper tool, the endogenous period can be calculated as the slope of the change in τ ('paper and ruler' method).*

peaks and troughs will not match, and overlaying the sections together will result in a small-amplitude signal with a non-significant variation. To analyze data with unknown period, the algorithm evaluates a wide range of length for different periods. When the maximum correlation is found – and it is statistically significant – we have found the main period of the series that gave the highest amplitude of the waveform. One of the best features of this method is its independence from the waveform of the rhythmic signal, because it only requires a repeating pattern, while other methods, such as Fourier or LS, assume a sinusoidal waveform. The main downfall of the SB method is that the step size between periods that can be tested is limited by the sample size and the duration of the series. This method is widely used in long time series (more than 10 cycles) with a high sampling rate.

Another general approach is based in the curve-fitting criterion. This method assumes an *a-priori* periodic signal (mainly sinusoidal) and optimizes the signal parameters to best fit the data. In this case, we plot the correlation coefficient after the fit with different periods. This approach allows using shorter series and even non-uniform sampling. The Fourier method is the most basic approach of this set of algorithms. As we use a sinusoidal signal to fit, the method is especially sensitive to this waveform; however, it is not as good to fit other types of rhythmic signals. The Lomb-Scargle periodogram consists of a spectral analysis (a modification of the Fourier periodogram). This method creates a spectrum representing the significance of each frequency with a normalized power, and is usually the best choice in short series and non-uniform sampling. This method combines both cosine and sine fit to find periodicity regardless of the data phase. Another interesting feature is that it reduces the subsequent sub-harmonics of the main period (if the period is T, the peaks in 2T, 3T, etc.) and, therefore, the periodogram analysis and representation are 'clean'.

There are many other methods to calculate the underlying period, such as MESA (Maximum Entropy Spectral Analysis) or FFT-NLLS (Fast Fourier Transform Non-linear Least Square algorithm). MESA is very useful because it can resolve multiple periods that are relatively close (Dowse, 2013). It is beyond the objective of the present chapter to describe all of them, but it is important to recognize which method is the best choice according to the specific experimental data. A comparison of the most used algorithms for determining period can be found in Zielinski et al. (2014).

An example containing the periodogram calculation of real data can be found in the online Appendix, by running the functions chisquare (Chi-square or Sokolov-Bushel periodogram) and lsssigx (a simplify application of the fastlomb.m that gives the complete spectrum of the Lomb-Scargle periodogram). See the files SCRIPT.m and ScriptFilter.m for further details (see book's GitHub repository).

### 6.6 *Cosinor*

As it was already shown in Figure 1.4, a simple rhythm can be described as a sinusoidal wave. By taking into account all the rhythm parameters, the signal can be mathematically expressed as:

$$y = M + A\cos\left(\omega t - \varphi\right); \tag{1}$$

where M is the mean of the adjusted data (Midline Estimating Statistic Of Rhythm, MESOR), φ is the acrophase, A is the amplitude (difference between φ and M) and t corresponds to time. Finally, ω = 2π/T is the angular frequency, being T the rhythm period. The mathematical procedure that fits the best curve – best fit according to the less-square method – is known as Cosinor. The expression of $y$ leads to a nonlinear optimization to find the values of M, A and φ. The equation (1), by using the Euler's equation through the formula cos(α – β) = cos(α)cos(β) + sin(α)sin(β), is transformed to:

$$y = M + \alpha\cos\left(\omega t\right) + b\sin\left(\omega t\right); \tag{2}$$

where $a$ [A cos (φ)] and $b$ [A sin (φ)] are known as the Euler coefficients, and are used to calculate the values of A and φ in the original equation (1) as follows:

$$A = \sqrt{a^2 + b^2};\ \varphi = \tan^{-1}\left(b / a\right) \tag{3}$$

If the signal was uniform sampled, the calculation is even simpler:

$$M = \frac{1}{n}\sum_{i=1}^{n} y; a = \frac{2}{n}\sum_{i=1}^{n} y_i \cos\left(\omega t_i\right); b = \frac{2}{n}\sum_{i=1}^{n} y_i \sin\left(\omega t_i\right) \tag{4}$$

The Cosinor method is mainly used to analyze a rhythmic time series with a known period (T), which can be obtained from the periodogram analysis (Cornelissen, 2014; Diez-Noguera, 2013). The Cosinor analysis will allow to obtain the phase (φ) and the amplitude (A) of the rhythm. A polar graph is usually performed to plot the results (see Figure 1.7 and online Appendix). In this graph, a circle represents the period T, with a vector that origins in the center of the circle. This vector has a size that depends on the amplitude (A) and its direction depends on the time of the acrophase (φ). Additionally, it is

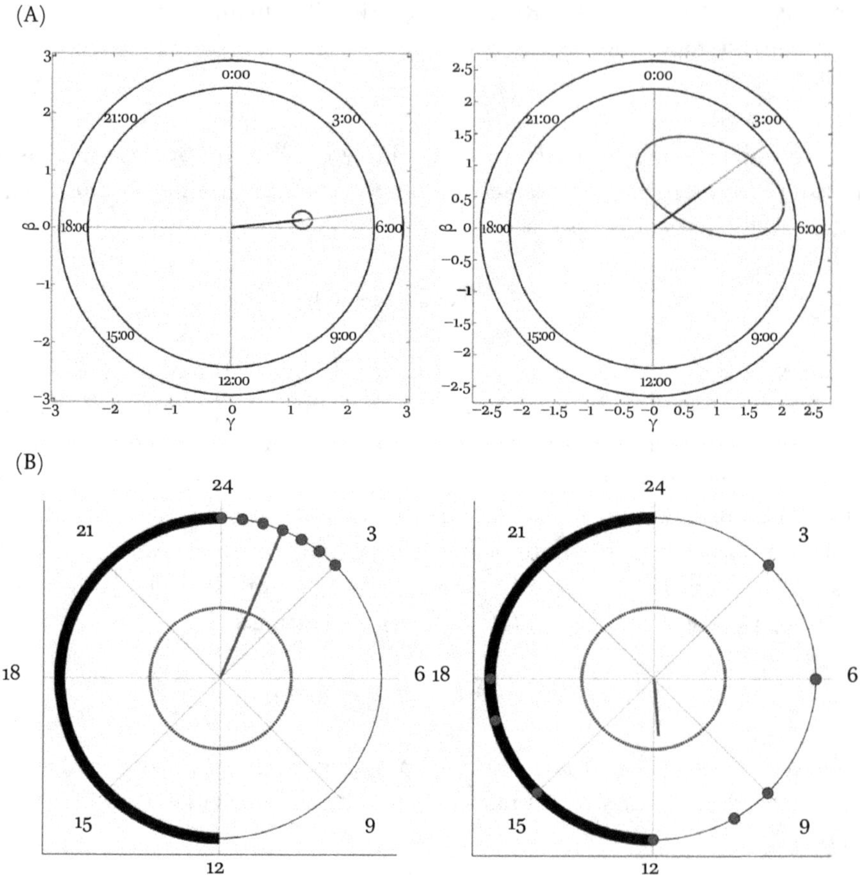

FIGURE 1.7 *Graphical representation of Cosinor and Rayleigh in a polar plot. (A) Cosinor analysis of two data sets. This is illustrated by the elliptical 95% confidence interval (CI) for the amplitude-acrophase pair. The 95% CI is located at the endpoint of the vector. In the left panel, the phase can be found and is significant as it pass the null amplitude hypothesis (Mesor = 0.019; Amplitude = 1.22; Acrophase = −1.46 radians = 5.57 hours; $p < 0.05$). In the right panel, the data does not pass this test (Mesor = 0.12; Amplitude = 1.10; Acrophase = −0.92 radians = 3.51 hours; $p > 0.05$). (B) Rayleigh test of two data sets. In the left panel there is a significant unimodal aggrupation ($p$ = 2e-4), while and in the right panel there is a non-significant data set ($p$ = 0.38). The significance can be clearly observed as the resultant vector goes beyond the significance circle in the center of the left panel graph, while it does not go beyond the circle in the right panel.*

usually represented in the same graph the confidence interval (CI) for each parameter, forming an ellipse around the further point of the vector. When the ellipse includes the vector's originating point, then the amplitude (and thus the rhythm) is statistically non-significant (i.e., A is non-different from zero). This is caused when the fitted curve represents less than 60% of the total variation of the signal (40% residual variation of the less-squared fitting). In a similar way, when more than one signal is represented in the same graph, a superposition of the CI indicates that the two signals do not have a significant difference.

The Cosinor method is widely used to find the amplitude (as criterion for significant rhythmicity), but more importantly, the acrophase of the time series. However, as many other curve-fitting methods, is strongly dependent on the signal's waveform, and responds much better (both in accuracy and significance) with smooth sinusoidal data than signals with abrupt changes or nonsymmetrical shapes (Cornelissen, 2014).

The online Appendix includes examples of Cosinor analysis of locomotor activity series. To analyze this kind of data, first select the time period and then run the function cosinor (see the files SCRIPT.m and ScriptFilter.m for further details in book's GitHub repository).

### 6.7 *Circulars Statistics and Rayleigh*

Circular statistics refers to the development of statistical techniques for the use with data represented on an angular scale (Berens, 2009). Due to their circular nature, many circadian data can be analyzed with circular statistics. The phase of time series, when we study a rhythmic signal, is an example of circular data. In this case, there is no designated zero and, in contrast to a linear scale, high/low values are arbitrary. This idea of 'circular time' is especially strong in the case of analyzing repeating patterns with a fixed time range (i.e., a cycle of period T and how this pattern changes). For this reason, the rhythmic signal is naturally represented as a circular graph (as shown in the Cosinor's section). The rhythm's phase can be studied as a fraction of the complete cycle (in time, angle or radians).

There are many other tests for circular statistics as an alternative for Rayleigh, such as the Omnibus test (Zar, 1999). This test works well with unimodal, bimodal and multimodal data, but its use is more selective and is not as extensive as others (Berens, 2009).

In case that there is a set of signals (e.g., many individuals in the same experiment), each one will have its own phase. We can study phase similarities by representing all the phases in the same circular graph and test them for a significant clustering by means of the Rayleigh test (Figure 1.7). This test asks how large the vectorial sum of the individual vectors must be to indicate a non-uniform distribution, according to the Fisher circular distribution (Fisher, 1995). The Rayleigh test is especially strong for detecting a unimodal deviation from uniformity (Berens, 2009). The test estimates the module (r) of the sum vector of the unit vectors corresponding to the phases of each individual. The higher the value of r, the greater degree of phase homogeneity. The p-value of the data can be computed by obtaining the critical value from the distribution. By setting the significance threshold in p=0.05 the critical value of the vector can be obtained beforehand. In this way, we can approximate the critical value = r ≈ 1.6732268 / (1-n) 0.492018, where n is the number of cases used (Diez-Noguera, 2013).

Examples of circular statistics can be found in the online Appendix section (Rayleigh.m file; see book's GitHub repository).

## 7 Conclusions

The circadian system is ubiquitous, being essential for most living organisms' physiology. The interaction of several circadian clock genes and proteins generates oscillations in expression of output target genes which temporally regulate numerous molecular and cellular processes. A precise regulation of circadian rhythms is crucial to keep track with the environment, and perturbations in the circadian system are linked to numerous molecular dysfunctions and may result in severe pathologies.

Hence, a comprehensive knowledge regarding the mechanistic of the circadian system is necessary to develop new procedures to investigate pathologies associated with a deregulated clock.

In the online appendix section (see book's GitHub repository), we incorporate important tools for circadian data analysis, including codes and sample data. We include the codes for LS, SB, Cosinor, and Rayleigh analysis as well as a code to plot actograms. There are many other software – some of them freeware – that possess a wider range of circadian tools. We include links to some software too.

## References

Agostino, P.V., & R.K. Cheng (2016). Contributions of dopaminergic signaling to timing accuracy and precision. *Current Opinion in Behavioral Sciences, 8,* 153–160.

Agostino, P.V., M. do Nascimento, I.L. Bussi, M.C. Eguía, & D.A. Golombek (2011). Circadian modulation of interval timing in mice. *Brain Research, 1370,* 154–163.

Agostino, P.V., S.A. Plano, & D.A. Golombek (2007). Sildenafil accelerates reentrainment of circadian rhythms after advancing light schedules. *Proceedings of the National Academy of Science U S A, 104,* 9834–9839.

Albrecht, U. (2012). Timing to perfection: The biology of central and peripheral circadian clocks. *Neuron, 74,* 246–260.

Aschoff, J. (1985). On the perception of time during prolonged temporal isolation. *Human Neurobiology*, *4*, 41–52.

Aschoff, J., S. Daan, & K.I. Honma (1982). Zeitgeber, entrainment, and masking: Some unsettled questions. In Aschoff, J., S. Daan, G.A. Gross, *Vertebrate Circadian System (Structure and Physiology)*. Berlin: Springer-Verlag.

Bell-Pedersen, D., V.M. Cassone, D.J. Earnest, S.S. Golden, P.E. Hardin, T.L. Thomas, & M.J. Zoran (2005). Circadian rhythms from multiple oscillators: Lessons from diverse organisms. *Nature Reviews Genetics*, *6*(7), 544–556.

Berens, P. (2009). CircStat: A MATLAB Toolbox for Circular Statistics. *Journal of Statistical Software*, *31*(10), 1–21.

Bonmati-Carrion, M.A., B. Middleton, V. Revell, D.J. Skene, M.A. Rol, & J.A. Madrid (2014). Circadian phase assessment by ambulatory monitoring in humans: Correlation with dim light melatonin onset. *Chronobiology International*, *31*(1), 37–51.

Bordyugov, G., U. Abraham, A. Granada, P. Rose, K. Imkeller, A. Kramer, & H. Herzel (2015). Tuning the phase of circadian entrainment. *Journal of The Royal Society Interface*, *12*(108), 20150282.

Buhr, E.D., & J.S. Takahashi (2013). Molecular components of the mammalian circadian clock. *Handbook of Experimental Pharmacology*, *217*, 3–27.

Buhusi, C.V., & W.H. Meck (2005). What makes us tick? Functional and neural mechanism of interval timing. *Nature Reviews Neuroscience*, *6*(10), 755–765.

Bussi, I.L., G. Levin, D.A. Golombek, & P.V. Agostino (2015). Melatonin modulates interval timing in rats: Effect of pinealectomy. *International Journal of Comparative Psychology*, 28(1).

Bussi, I.L., G. Levín, D.A. Golombek, & P.V. Agostino (2014). Involvement of dopamine signaling in the circadian modulation of interval timing. *European Journal of Neuroscience*, *40*(1), 2299–2310.

Campuzano, A., J. Vilaplana, T. Cambras, & A. Diez-Noguera (1998). Dissociation of the rat motor activity rhythm under T cycles shorter than 24 hours. *Physiology & Behavior*, *63*(2), 171–176.

Casiraghi, L.P., G.A. Oda, J.J. Chiesa, W.O. Friesen, & D.A. Golombek (2012). Forced desynchronization of activity rhythms in a model of chronic jet lag in mice. *Journal of Biological Rhythms*, *27*(1), 59–69.

Castanon-Cervantes, O., M. Wu, J.C. Ehlen, K. Paul, K.L. Gamble, R.L. Johnson, R.C. Besing, M. Menaker, A.T. Gewirtz, & A.J. Davidson (2010). Dysregulation of inflammatory responses by chronic circadian disruption. *The Journal of Immunology*, *185*(10), 5796–5805.

Chandrashekaran, M.K., G. Marimuthu, R. Subbaraj, P. Kumarasamy, M.S. Ramkumar, & K. Sripathi (1991). Direct correlation between the circadian sleep-wakefulness rhythm and time estimation in humans under social and temporal isolation. *Journal of Biosciences*. *16*, 97–101.

Cho, K. (2001). Chronic 'jet lag' produces temporal lobe atrophy and spatial cognitive deficits. *Nature Neuroscience, 4,* 567–568.

Cho, K., A. Ennaceur, J.C. Cole, & C.K. Suh (2000). Chronic jet lag produces cognitive deficits. *The Journal of Neuroscience, 20*(6), 1–5.

Clayton, M., R. Sager, & U. Will (2005). In time with the music: The concept of entrainment and its significance for ethnomusicology. *European meetings in ethnomusicology, 11,* 1–82.

Cornelissen, G. (2014). Cosinor-based rhythmometry. *Theoretical Biology and Medical Modelling, 11*(1), 1.

de la Iglesia, H.O., T. Cambras, W.J. Schwartz, & A. Diez-Noguera (2004). Forced desynchronization of dual circadian oscillators within the rat suprachiasmatic nucleus. *Current Biology, 14*(9), 796–800.

Devan, B.D., E.H. Goad, H.L. Petri, E.A. Antoniadis, N.S. Hong, C.H. Ko, L. Leblanc, S.S. Lebovic, Q. Lo, M.R. Ralph, & R.J. McDonald (2001). Circadian phase-shifted rats show normal acquisition but impaired long-term retention of place information in the water task. *Neurobiology of learning and memory, 75*(1), 51–62.

Díez-Noguera, A. (2013). Methods for serial analysis of long time series in the study of biological rhythms. *Journal of circadian rhythms, 11*(1), 7.

Dowse, H.B. (2013). Maximum entropy spectral analysis for circadian rhythms: Theory, history and practice. *Journal of circadian rhythms, 11*(1), 6.

Dunlap, J.C., J.J. Loros, & P.J. DeCoursey (2004). *Chronobiology. Biological timekeeping.* Sunderland, MA: Sinauer Associates, Inc.

Eckel-Mahan, K.L., & D.R. Storm (2009). Circadian rhythms and memory: not so simple as cogs and gears. *EMBO Reports, 10,* 584–591.

Fisher, N.I. (1995). *Statistical analysis of circular data.* Cambridge, uk: Cambridge University Press.

Folkard, S. (1996). Effects on performance efficiency. In Coquhoun, W.P., G. Costa, S. Folkard, and P. Knauth (Eds.), *Shiftwork: Problems and Solution* (pp. 65–87). Frankfurt: Peter Lang.

Fonken, L.K., M.S. Finy, J.C. Walton, Z.M. Weil, J.L. Workman, J. Ross, & R.J. Nelson (2009). Influence of light at night on murine anxiety- and depressive-like responses. *Behavioural Brain Research, 205*(2), 349–354.

Fonken, L.K., J.L. Workman, J.C. Walton, Z.M. Weil, J.S. Morris, A. Haim, & R.J. Nelson (2010). Light at night increases body mass by shifting the time of food intake. *Proceedings of the National Academy of Science USA, 107*(43), 18664–18669.

Gerstner, J.R., & J.C. Yin (2010). Circadian rhythms and memory formation. *Nature Reviews Neuroscience, 11,* 577–588.

Gekakis, N., D. Staknis, H.B. Nguyen, F.C. Davis, L.D. Wilsbacher, D.P. King, J.S. Takahashi, & C.J. Weitz (1998). Role of the CLOCK protein in the mammalian circadian mechanism. *Science,* 280, 1564–1569.

Gibson, E.M., C. Wang, S. Tjho, N. Khattar, & L.J. Kriegsfeld (2010). Experimental 'jet lag' inhibits adult neurogenesis and produces long-term cognitive deficits in female hamsters. *PLoS One 5*, e15267.

Golombek, D.A., & R.E. Rosenstein (2010). Physiology of circadian entrainment. *Physiological reviews*, *90*, 1063–1102.

Golombek, D.A., P.V. Agostino, S.A. Plano, & G.A. Ferreyra (2004). Signaling in the mammalian circadian clock: The NO/cGMP pathway. *Neurochemistry International*, *45*, 929–936.

Golombek, D.A., I.L. Bussi, & P.V. Agostino (2014). Minutes, days and years: Molecular interactions among different scales of biological timing. *Philosophical Transactions of the Royal Society of London B: Biological Sciences*, *369*(1637), 20120465.

Golombek, D.A., L.P. Casiraghi, P.V. Agostino, N. Paladino, J.M. Duhart, S.A. Plano, & J.J. Chiesa (2013). The times they're a-changing: Effects of circadian desynchronization on physiology and disease. *Journal of Physiology-Paris*, *107*(4), 310–322.

Golombek, D.A., G.A. Ferreyra, P.V. Agostino, A.D. Murad, M.F. Rubio, G.A. Pizzio, M.E. Katz, L. Marpegan, & T.A. Bekinschtein (2003). From light to genes: Moving the hands of the circadian clock. *Frontiers of Bioscience, 8*, s285–293.

Gooley, J.J. (2008). Treatment of circadian rhythm sleep disorders with light. *Annual Academy of Medice Singapore*, *37*, 669–676.

Goya, M.E., A. Romanowski, C.S. Caldart, C.Y. Bénard, & D.A. Golombek (2016). Circadian rhythms identified in Caenorhabditis elegans by in vivo long-term monitoring of a bioluminescent reporter. *Proceedings of the National Academy of Science USA*, *113*(48), E7837–E7845.

Gritton, H.J., A. Kantorowski, M. Sarter, & T.M. Lee (2012). Bidirectional interactions between circadian entrainment and cognitive performance. *Learning & Memory*, *19*(3), 126–141.

Guillaumond, F., H. Dardente, V. Giguère, & N. Cermakian (2005). Differential control of Bmal1 circadian transcription by REV-ERB and ROR nuclear receptors. *Journal of Biological Rhythms*, *20*, 391–403.

Hampp, G., J.A. Ripperger, T. Houben, I. Schmutz, C. Blex, S. Perreau-Lenz, I. Brunk, R. Spanagel, G. Ahnert-Hilger, J.H. Meijer, & U. Albrecht (2008). Regulation of monoamine oxidase A by circadian-clock components implies clock influence on mood. *Current Biology*, *18*(9), 678–683.

Johnson, C.H., J.A. Elliott, & R. Foster (2003). Entrainment of circadian programs. *Chronobiology International*, *20*, 741–774.

Jones, M.R. (1976). Time, our lost dimension: Toward a new theory of perception, attention, and memory. *Psychological Review*, *83*(5), 323.

Jones, C.R., S.S. Campbell, S.E. Zone, F. Cooper, A. DeSano, P.J. Murphy, B. Jones, L. Czajkowski, & L.J. Ptč (1999). Familial advanced sleep-phase syndrome: A short-period circadian rhythm variant in humans. *Nature Medicine*, *5*(9), 1062–1065.

Jud, C., I. Schmutz, G. Hampp, H. Oster, & U. Albrecht (2005). A guideline for analyzing circadian wheel-running behavior in rodents under different lighting conditions. *Biology Proceeding Online, 7,* 101–116.

Karatsoreos, I.N., S. Bhagat, E.B. Bloss, J.H. Morrison, & B.S. McEwen (2011). Disruption of circadian clocks has ramifications for metabolism, brain, and behavior. *Proceedings of the National Academy of Science USA, 108,* 1657–1662.

Klerman, E.B., H.B. Gershengorn, J.F. Duffy, & R.E. Kronauer (2002). Comparisons of the variability of three markers of the human circadian pacemaker. *Journal of Biological Rhythms, 17*(2), 181–193.

Kuriyama, K., M. Uchiyama, H. Suzuki, H. Tagaya, A. Ozaki, S. Aritake, K. Shibui, T. Xin, L. Lan, Y. Kamei, & K. Takahashi (2005). Diurnal fluctuation of time perception under 30-h sustained wakefulness. *Neuroscience Research, 53,* 123–128.

Kyriacou, C.P., & J. Hall (1980). Circadian rhythm mutations in Drosophila melanogaster affect short-term fluctuations in the male's courtship song. *Proceedings of the National Academy of Science U S A, 77,* 6729–6733.

Kyriacou, C.P., & M.H. Hastings (2010). Circadian clocks: Genes, sleep, and cognition. *Trends in Cognitive Sciences, 14,* 259–267.

LeSauter, J., & R. Silver (1998). Output signals of the SCN. *Chronobiology International, 15*(5), 535–550.

Levine, J.D., P. Funes, H.B. Dowse, & J.C. Hall (2002). Signal analysis of behavioral and molecular cycles. *BMC Neuroscience, 3*(1), 1.

Lewy, A.J., J. Emens, A. Jackman, & K. Yuhas (2006). Circadian uses of melatonin in humans. *Chronobiology International, 23*(1–2), 403–412.

Loh, D.H., J. Navarro, A. Hagopian, L.M. Wang, T. Deboer, & C.S. Colwell (2010). Rapid changes in the light/dark cycle disrupt memory of conditioned fear in mice. *PLoS One, 5,* e12546.

Lowrey, P.L., & J.S. Takahashi (2000). Genetics of the mammalian circadian system: Photic entrainment, circadian pacemaker mechanisms, and posttranslational regulation. *Annual Review of Genetics, 34*(1), 533–562.

Lustig, C., & W.H. Meck (2001). Paying attention to time as one gets older. *Psychological Science,* 12(6), 478–484.

Madrid, J.A., & A.R. de Lama (2006). Cronobiología básica y clínica. Editec@ red. Madrid, España.

Madrid, J.A., P. Lax, J. Vilaplana, T. Cambras, & A. Diez-Noguera (1992). Presence of two differentiated circadian components in the eating and motor behaviour in the young rat. *Journal of Interdisciplinary Cycle Research, 23,* 211–212.

Mahoney, M.M. (2010). Shift work, jet lag, and female reproduction. *International Journal of Endocrinology, 2010,* 813764.

Meck, W.H. (1991). Modality-specific circadian rhythmicities influence mechanisms of attention and memory for interval timing. *Learning and Motivation*, *22*(1), 153–179.

Meng, Q.J., E.S. Maywood, D.A. Bechtold, W.Q. Lu, J. Li, J.E. Gibbs, S.M. Dupré, J.E. Chesham, F. Rajamohan, J. Knafels, B. Sneed, L.E. Zawadzke, J.F. Ohren, K.M. Walton, T.T. Wager, M.H. Hastings, & A.S. Loudon (2010). Entrainment of disrupted circadian behavior through inhibition of casein kinase 1 (CK1) enzymes. *Proceedings of the National Academy of Sciences USA*, *107*(34), 15240–15245.

Morin, L. (2013). Neuroanatomy of the extended circadian rhythm system. *Experimental Neurology*, *243*, 4–20.

Morin, L.P., & C.N. Allen (2006). The circadian visual system. *Brain Research Reviews*, *51*(1), 1–60.

Moriya, T., Y. Yoshinobu, Y. Kouzu, A. Katoh, H. Gomi, M. Ikeda, T. Yoshioka, S. Itohara, & S. Shibata (2000). Involvement of glial fibrillary acidic protein (GFAP) expressed in astroglial cells in circadian rhythm under constant lighting conditions in mice. *Journal of Neuroscience Research*, *60*(2), 212–218.

Mrosovsky, N. (1999). Masking: History, definitions, and measurement. *Chronobiology International*, *16*(4), 415–429.

Muñoz, E., M. Brewer, & R. Baler (2006). Modulation of BMAL/CLOCK/E-Box complex activity by a CT-rich cis-acting element. *Molecular and Cellular Endocrinology*, *252*(1), 74–81.

Panda, S., T.K. Sato, A.M. Castrucci, M.D. Rollag, W.J. DeGrip, J.B. Hogenesch, I. Provencio, & S.A. Kay (2002). Melanopsin (Opn4) requirement for normal light-induced circadian phase shifting. *Science*, *298*, 2213–2216.

Pati, A.K., & S. Gupta (1994). Time estimation circadian rhythm in shift workers and diurnally active humans. *Journal of Biosciences*, *19*(3), 325–330.

Pittendrigh, C.S. (1965). On the mechanism of the entrainment of a circadian rhythm by light cycles. In Aschoff, J., *Circadian clocks*. Amsterdam: Elsevier.

Plano, S.A., P.V. Agostino, H.O. de la Iglesia, & D.A. Golombek (2012). cGMP-phosphodiesterase inhibition enhances photic responses and synchronization of the biological circadian clock in rodents. *PLoS One*, 7(5), e37121.

Preitner, N., F. Damiola, L. Lopez-Molina, J. Zakany, D. Duboule, U. Albrecht, & U. Schibler (2002). The orphan nuclear receptor REV-ERBalpha controls circadian transcription within the positive limb of the mammalian circadian oscillator. *Cell*, *110*, 251–260.

Refinetti, R. (1993). Laboratory instrumentation and computing: Comparison of six methods for the determination of the period of circadian rhythms. *Physiology & Behavior*, *54*(5), 869–875.

Reid, K.J. & S.M. Abbott (2015). Jet lag and shift work disorder. *Sleep Medicine Clinics*, 10(4), 523–535.

Santhi, N., T.S. Horowitz, J.F. Duffy, & C.A. Czeisler (2007). Acute sleep deprivation and circadian misalignment associated with transition onto the first night of work impairs visual selective attention. *PLoS One, 2*, e1233.

Sato, T.K., S. Panda, L.J. Miraglia, T.M. Reyes, R.D. Rudic, P. McNamara, K.A. Naik, G.A. FitzGerald, S.A. Kay, & J.B. Hogenesch (2004). A functional genomics strategy reveals Rora as a component of the mammalian circadian clock. *Neuron, 43*, 527–537.

Scheer, F.A., M.F. Hilton, C.S. Mantzoros, & S.A. Shea (2009). Adverse metabolic and cardiovascular consequences of circadian misalignment. *Proceedings of the National Academy of Science USA, 106*, 4453–4458.

Shurtleff, D., T.G. Raslear, & L. Simmons (1990). Circadian variations in time perception in rats. *Physiology & Behavior, 47*(5), 931–939.

Soshi, T., K. Kuriyama, S. Aritake, M. Enomoto, A. Hida, M. Tamura, Y. Kim, & K. Mishima (2010). Sleep deprivation influences diurnal variation of human time perception with prefrontal activity change: A functional near-infrared spectroscopy study. *PLoS One, 5*, e8395.

Sudo, M., K. Sasahara, T. Moriya, M. Akiyama, T. Hamada, & S. Shibata (2003). Constant light housing attenuates circadian rhythms of mPer2 mRNA and mPER2 protein expression in the suprachiasmatic nucleus of mice. *Neuroscience, 121*(2), 493–499.

Tapia-Osorio, A., R. Salgado-Delgado, M. Angeles-Castellanos, & C. Escobar (2013). Disruption of circadian rhythms due to chronic constant light leads to depressive and anxiety-like behaviors in the rat. *Behavioural Brain Research, 252*, 1–9.

Vilaplana, J., T. Cambras, & A. Diez-Noguera (1997). Dissociation of motor activity circadian rhythm in rats after exposure to LD cycles of 4-h period. *American Journal of Physiology-Regulatory, Integrative and Comparative Physiology, 272*(1), R95–R102.

Welsh, D.K., & S.A. Kay (2005). Bioluminescence imaging in living organisms. *Current Opinion in Biotechnology, 16*(1), 73–78.

Welsh, D.K., D.E. Logothetis, M. Meister, & S.M. Reppert (1995). Individual neurons dissociated from rat suprachiasmatic nucleus express independently phased circadian firing rhythms. *Neuron, 14*, 697–706.

Wright, K.P. Jr., J.T. Hull, & C.A. Czeisler (2002). Relationship between alertness, performance, and body temperature in humans. *American Journal of Physiology-Regulatory, Integrative and Comparative Physiology, 283*(6), R1370–1377.

Yan, J., H. Wang, Y. Liu, & C. Shao (2008). Analysis of gene regulatory networks in the mammalian circadian rhythm. *PLoS Comput Biol, 4*(10), e1000193.

Ye, R., C.P. Selby, Y.Y. Chiou, I. Ozkan-Daglivan, S. Gaddameedhi, & A. Sancar (2014). Dual modes of CLOCK:BMAL inhibition mediates by Cryptochrome and Period proteins in the mammalian circadian clock. *Genes & development, 28*(18), 1989–1998.

Yoo, S.H., C.H. Ko, P.L. Lowrey, E.D. Buhr, E.J. Song, S. Chang, O.J. Yoo, S. Yamazaki, C. Lee, & J.S. Takahashi (2005). A noncanonical E-box enhancer drives mouse

Period2 circadian oscillations in vivo. *Proceedings of the National Academy of Sciences* USA, *102*, 2608–2613.

Zar, J.H. (1999). *Biostatistical Analysis.* 4th edition. Prentice Hall, Upper Saddle River, New Jersey.

Zee, P.C., & C.A. Goldstein (2010). Treatment of shift work disorder and jet lag. *Current Treatment Options in Neurology*, *12*(5), 396–411.

Zhang, R., N.F. Lahens, H.I. Balance, M.E. Hughes, & J.B. Hogenesch (2014). A circadian gene expression atlas in mammals: Implications for biology and medicine. *Proceedings of the National Academy of Sciences* USA, *111*, 16219–16224.

Zielinski, T., A.M. Moore, E. Troup, K.J. Halliday & A. J Millar (2014). Strengths and limitations of period estimation methods for circadian data. *PLoS One*, *9*(5), e96462.

CHAPTER 2

# Prospective and Retrospective Timing Processes: Theories, Methods, and Findings

*Richard A. Block, Simon Grondin and Dan Zakay*

## 1 Introduction

Although one could date the history of the study of psychological time back to antiquity, with early hominids' experiences and Aristotle's philosophical speculations, that is a matter for historians. The experimental study of time estimation has a history that apparently started with research published in 1868 by Vierordt (Lejeune & Wearden 2009). This was almost two decades before the widely mentioned beginning of psychological research (Wundt 1886). Also, in the early times of what later became psychology, several interesting theories on timing and time perception were proposed (e.g., Hooke, 1682, as cited in Hintzman 2003; Guyau, 1888; see Michon, Pouthas, & Jackson, 1988). Beginning especially with theoretical reviews by the philosopher-turned-psychologist James (1890) and continuing through Fraisse in his famous books, *Psychologie du temps* (1957/1963) and *Psychologie du rythme* (1974), the psychology of timing and time perception have blossomed. This is evident in a recent edited volume, also with the title *Psychology of time* (Grondin 2008b). For a more in-depth historical review, see Hancock and Block (2012), and for a summary of recent review articles, see Block and Grondin (2014).

In the present chapter, we mainly focus on time perception and time estimation (see Block & Hancock 2013, for an annotated bibliography). The database PsycINFO distinguishes these terms in a slightly overlapping way: The keyword *time perception* is defined as "perception of duration, simultaneity, or succession in the passage of time." The keyword *time estimation* is defined as "estimation of duration or passage of time." The MEDLINE, or PubMed, database simply uses the keyword *time perception,* which is defined as "the ability to estimate periods of time lapsed or duration of time."

In the experimental literature, which we review later, Hicks, Miller, and Kinsbourne (1976) conducted a seminal study. Instead of using James's (1890) descriptions (time in passing and in retrospect), they used the terms *prospective* and *retrospective.* In the prospective paradigm, a person is told that time estimation is relevant and important. In the retrospective paradigm, a person

RICHARD A. BLOCK ET AL., 2018 | DOI 10.1163/9789004280205_003 is misdirected from attending to time, such as with a cover story or a cover task. In everyday life, a time estimate becomes relevant and important when an environmental situation makes demands to respond in a temporally accurate way. Retrospective duration estimates are used in situations such as those involving remembering, eyewitness testimony, and so on.

Researchers have revealed several spans that they think might involve different processes: 100 ms (Block 1979) to about 1.3 s (Grondin 2010; Grondin, Laflamme, & Mioni, 2015), 3 s (Pöppel 1978), or 7 s (James 1890), and of course even longer. We distinguish between these. We start with brief temporal events, and then we review research on longer episodes. We first detail the importance of different time scales and review procedures, methods, and measures. Then we discuss prospective and retrospective duration judgment processes and findings. We conclude by reviewing research on several related issues, such as temporal illusions and what is usually called *prospective memory*.

## 2 Durations of Temporal Experiences

If a series of identical visual stimuli occurs at brief interstimulus intervals (<100 ms), according to many old experiments, some interesting phenomena occur (see reviews in Block 1979; Patterson 1990). These phenomena were found decades ago and labeled as 'the psychological moment'; all events occurring within this period would be processed as co-temporal (but see Elliott & Giersch 2016).

However, the experience of duration can be approached differently. It is now known that for the discrimination of auditory brief intervals, the Weber fraction is not constant. When intervals to-be discriminated are longer than circa 1.3 s (Grondin 2012) the fraction increases—i.e., the discrimination is more difficult (for reviews, see Gibbon, Malapani, Dale, & Gallistel, 1997; Grondin 2014), just as if a crucial cognitive capacity would be exceeded with long intervals (Grondin et al. 2015). Indeed, there is a possibility to avoid this increase of the Weber fraction. It has been shown that for discriminating brief auditory intervals, it is worth adopting an explicit counting strategy when intervals are longer than 1.2 s (Grondin, Meilleur-Wells, & Lachance, 1999).

At longer interstimulus intervals, another phenomenon is experienced, as revealed by several experiments. According to Pöppel (1997), there is a low-frequency mechanism binding successive events—a kind of temporal integration—into perceptual or action units when these events occur no more than 3 s apart. This phenomenon is sometimes referred to as 'the subjective present'.

Although these phenomena might result from relatively early analyses of sensory information, other phenomena have been reported as the "flow of time" at longer interstimulus intervals in any of several sensory modalities (visual, auditory, or tactual), even with non-identical stimuli (e.g., Gruber & Block 2013). In other words, some of these temporal experiences appear to be multimodal in origin. These experiments suggest that these temporal phenomena might be a result of sensory persistence, dynamic modal completion, and possibly also conceptual processes.

A longer time-related phenomenon also occurs—possibly up to James's (1890) and other reviewers' identification of 5–7 s as yet another critical period. These might be a result of what are usually now called working memory processes.

## 3 Prospective Timing Methods and Processes

Several methodological procedures are used to study the processes involved in timing and time perception. In prospective timing, the participants know in advance that the targeted interval will have to be estimated (such as reproduced). Some authors report four classical methods. In addition to verbal estimation and the method of production, which involve chronometric units, research has been conducted by using other methods, such as the reproduction method and the comparison method (Bindra & Waksberg 1956; Grondin 2010, 2014; Wallace & Rabin 1960; Zakay 1993). Figure 2.1 summarizes the main methods for studying time perception.

### 3.1 *Estimates Relying on Conventional (Learned) Time Units*

In the case of verbal estimations, participants provide a numerical estimation of the duration, using temporal units (seconds or minutes), of a stimulus or a series of stimuli previously presented, such as a flash, sound, series of words or pictures, and so on. Verbal estimates tend to be variable, and they are not suitable for studies involving young children who have not yet learned what seconds or minutes mean. Productions are also based on the use of chronometric units. In the production method, participants are asked to produce, for instance, by tapping twice on the spacebar of a keyboard, to mark the beginning and end of an interval lasting several seconds (Mioni, Stablum, Prunetti, & Grondin, 2016).

### 3.2 *Estimates Involving Duration Comparisons*

Young children must learn how to make time estimates using conventional verbal units. However, researchers use other methods to assess time perception. With young children, the reproduction method is often used: A child is shown a stimulus lasting for several seconds, and then is asked to press a button for the

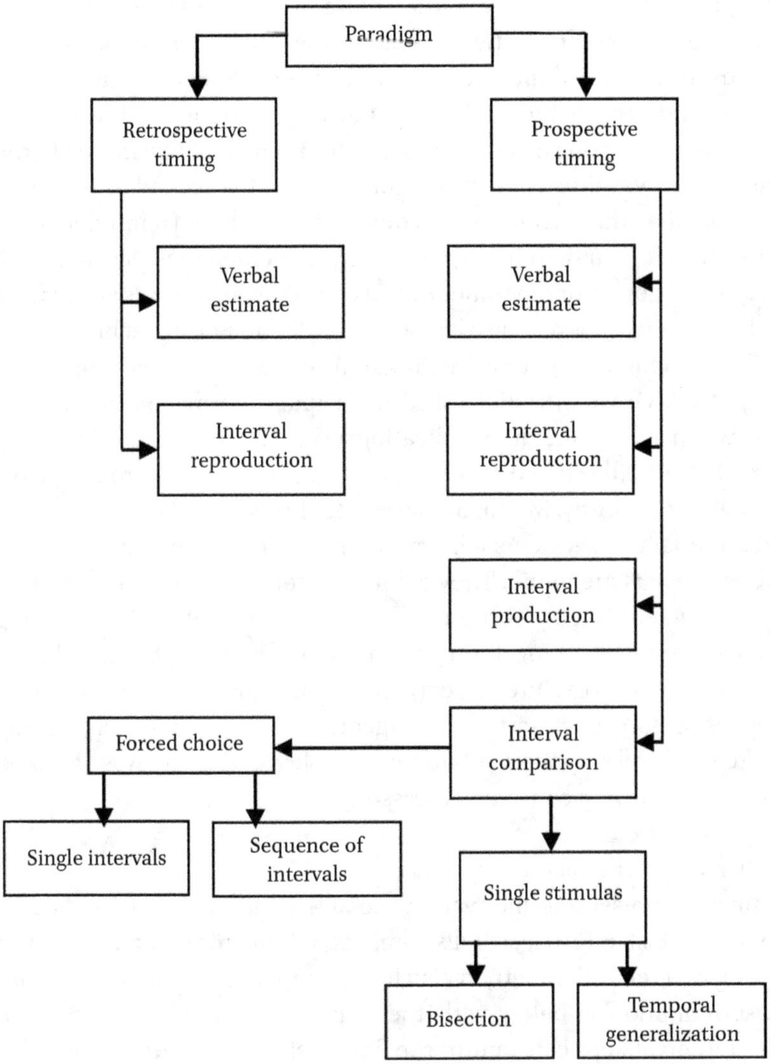


FIGURE 2.1 *Summary of the main methods utilized for studying time perception (Adapted from Grondin 2010)*

same amount of time that the stimulus appeared (Zakay, 1992a). The wording of these instructions should be adjusted for the age of the child. Also, research using nonhuman animals must use other methods, such as the peak-procedure method, which we do not review here (see, for example, Church 2003; Chapter 6, this volume).

The method of comparison is a procedure that involves comparing two durations. Using this method, participants are asked to judge the relative duration of intervals presented successively, one standard interval and one comparison interval. Each to-be compared interval could be marked by continuous sounds or continuous flashes (filled intervals) or by brief sensory signals marking the beginning and end of each interval (empty intervals). Participants are asked to indicate, such as by pressing an appropriate button, which of two intervals, the first or the second, was shorter or the longer one; or if the second interval was shorter or longer than the first one. In the comparison method, a reminder method indicates that the standard interval is always presented first (Macmillan & Creelman 1991): the order of the standard and comparison intervals varies from trial to trial. Other methods may also be used (e.g., Macmillan & Creelman 1991).

Using the comparison method, with durations of many seconds retrospectively judged, Ornstein (1969) asked participants to make a mark on a second line when given a first line, indicating a comparison of durations. Several researchers have discussed this method (e.g., Block 1974; Grondin 1993; Rammsayer 2014). Recently, Mioni, Stablum, McClintock, and Grondin (2014) emphasized the fact that when slight variations are introduced in reproductions, different results are found. They compared conditions in which participants were instructed either: (a) press only at the end of the interval after having been presented a signal marking the beginning of the interval to-be reproduced, (b) press to start and stop the interval, or (c) press continuously during the interval. The highest accuracy (closest to the target time) was obtained when using keypresses to start and stop the reproduction, but less variability was obtained with the method involving continuous pressing.

### 3.2.1 Time-Order Effects

In psychophysics, presenting intervals successively induces what is called a time-order error (Eisler, Eisler, & Hellström, 2008; Hellström 1985). Moreover, using a roving or a reminder method also has an impact on duration discrimination: Discrimination is better with the reminder method—that is, with the standard interval kept constant in the first position (Grondin & McAuley 2009; Hellström & Rammsayer 2004). The time-order effect is sometimes referred to as a Type-A effect, and the fact of having better performances in the standard-comparison order than in the comparison-standard order is sometimes referred to as a Type-B effect (Bausenhart, Dyjas, & Ulrich, 2015), or standard position effect (Hellström & Rammsayer 2015). Time-order errors tend to be negative for relatively short durations—that is, the second duration is judged as longer than the first—but tend to be positive for relatively longer durations, especially those judged in retrospect—that is, the first duration is judged longer than the second (Block 1985).

To account for the standard position effect, Hellström (1979) proposed a sensation-weighting model. According to this model, in a discrimination task a person compares the scaled subjective difference between the first and the second stimuli (see Hellström & Rammsayer 2015). Dyjas, Bausenhart, and Ulrich (2014) proposed that the standard position effect can be explained by an internal reference model, which stipulates that discrimination is not based on the internal representation of the standard and comparison intervals, but rather on the internal reference that is built up dynamically from trial to trial during the experiment. In this model, in each experimental trial, participants compare the updated internal reference to the second stimulus.

### 3.3 *Other Timing Methods*

Another way to investigate processes of prospective timing is to return to Fechner's traditional methods. Kuroda and Hasuo (2014) described ways of using method of limits and of adjustments, in addition of the constant method. For instance, a participant can be presented the standard and comparison intervals, and adjust the comparison interval to make it equal to the standard (method of adjustment). Over a series of trials, the mean adjusted value of the comparison intervals would provide a point of subjective equality (PSE) with the standard, and the variability (standard deviation) of the series of adjusted values could be interpreted as a just noticeable difference (JND; Hasuo, Nakajima, & Ueda, 2011).

A strict form of the method of limits is not used in time perception studies, but the adaptive procedure could be seen as a variation of this method as it may involve ascending trials and descending trials from specific points above or below threshold. Basically, after each trial, the level of difficult in a discrimination task is adjusted after each trial. With an adaptive procedure, an experimenter should decide what the magnitude of the changes are after a correct and after an incorrect response, when a series of trials ends, how many series of trials are necessary, and how the threshold is operationally defined and calculated (Macmillan & Creelman 1991). In the classical adaptive method called the *staircase procedure*, the steps up and down of the comparison stimulus are changed by a fixed amount, and a series of trials could end after a certain number of changes or a certain number of trials. There are several other adaptive procedures, including parameter estimation by sequential testing and the procedures based on a Bayesian procedure or maximum likelihood (Shen 2013). The adaptive procedure usually provides a good approximation of a threshold value with a reasonable number of trials and is often used in duration discrimination studies (e.g., Rammsayer 2014).

With the constant method, a series of comparison intervals is selected and one of these, in a random order from trial to trial, is presented after the standard. After multiple presentations of each comparison intervals, a psychometric function can be drawn, plotting the probability of responding that the comparison interval is longer than the standard on the *Y*-axis, as a function the length of the comparison intervals on the *X*-axis (Grondin 2008a, 2010). On a psychometric function, the PSE is given by the value on the *X*-axis corresponding 50% on the *Y*-axis (see Figure 2.2). It is classical to define the difference threshold as the difference, divided by 2, between the *X* values corresponding to 75% and 25% on the *Y*-axis. These two percentages correspond to mid-point between perfect discrimination (0 or 100%) and random responses (50%). Another frequent way of expressing the difference threshold is to estimate one standard deviation on the psychometric function (Killeen & Weiss 1987).

The use of the constant method and of a psychometric function raises the question of the model adopted for drawing the function through the data points.

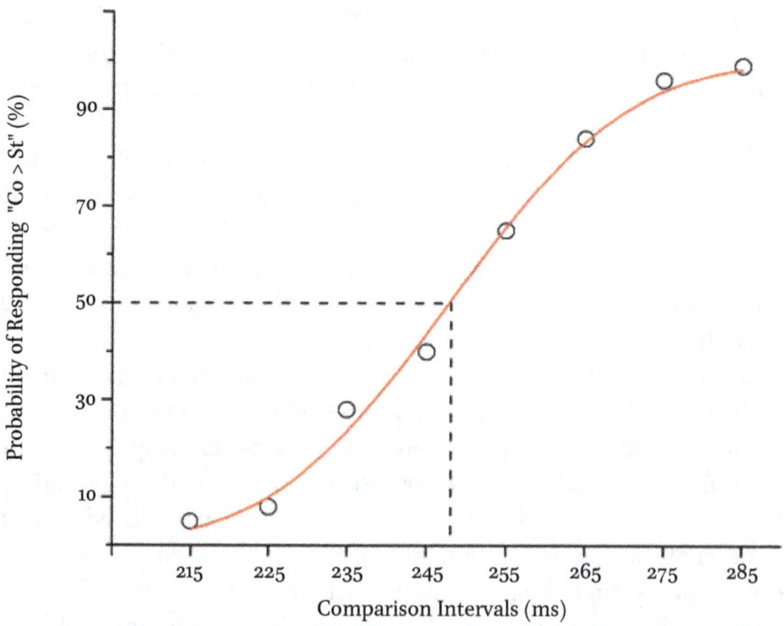


FIGURE 2.2 *Psychometric function used for estimating the difference threshold with the constant method. "Co > St" means that the comparison interval is judged as longer than the standard (250 ms). The model for fitting the data points in this example is the cumulative normal distribution, and the dashes indicate the point of subjective equality (Adapted from Grondin 2008a)*

In psychology, it is typical to assume that a phenomenon is distributed normally, and, thus, a Gaussian model is adopted (cumulative normal distribution; see for instance Laflamme, Zakay, Gamache, & Grondin, 2015). Among the other models, Macmillan and Creelman (1991) proposed the logistic and the Weibull functions, and the reader will also find the use of a pseudo-logistic function in the time perception literature (Grondin 2001a; Killeen, Fetterman, & Bizo, 1997).

In a variant of the constant method, called the *single-stimulus method*, a participant makes a judgment after each interval presentation. This involves assigning the interval to one of two categories, short or long; there is no presentation of the standard on each trial (see for instance Grondin, Laflamme, & Gontier, 2014, Experiment 3; Kuroda, Grondin, Miyazaki, Ogata, & Tobimatsu, 2016). A classical method in the animal timing literature, called the *bisection method*, is a widely used variant of this method in human timing studies (Mioni, Meligrana, Grondin, Perini, Bartolomei, & Stablum, 2015a; Mioni, Zakay, & Grondin, 2015b; Penney, Gibbon, & Meck, 2008; Chapter 4, this volume). With this method, a series of intervals are selected and the shortest and the longest intervals (referred to as standards) of this series are first presented several times. In the following trials, each interval of this series is presented several times, in a random order, and on each trial the interval should be categorized as closer to one of the two standards. As for the constant method, a psychometric function can be drawn from these data. There are many other methods in the study of animal timing, including temporal generalization, which is also used in the human timing literature (Wearden & Lejeune 2006). With this method, the mid-point interval in a series of intervals is first presented several times and in subsequent trials, participants indicate whether the interval presented is or is not of the same length as the standard.

Depending on the specific field of interest, specific variants of methods can be used. Researchers interested in motor behavior and timing have used a production method, which is indeed most often a task where a participant first listen to a series of isochronous sounds and eventually tries to synchronize finger tapping with the sound; at some point, there is no more sound but the participant continues to tap at the same pace. The dependent variable of interest is the variability of the inter-tap intervals in the continuous phase, and sometimes the mean inter-tap intervals. It is possible to distinguish the part of variance, in the overall observed variance, belonging to the motor system (implementation of the tap) and the part belonging to the internal timekeeping system (Wing & Kristofferson 1973).

Finally, researchers more interested in rhythm perception than interval timing often use duration discrimination methods, but instead of presenting single intervals (a single standard or a single comparison interval), multiple successive standard or comparison intervals are presented. Increasing the number of intervals generally leads to improved discrimination (for a review, see ten Hoopen, Miyauchi, & Nakajima, 2008). Beyond four presentations, in the auditory modality, there is not much improvement to expect (ten Hoopen, van den Berg, Memelink, Bocanegra, & Boon, 2011); and, in the visual modality, the gain with multiple presentations depends on the duration under investigation and on presenting or not the standard and comparison intervals in continuity (Grondin 2001a).

## 4 Prospective Timing Processes

Much evidence reveals that attention to time is involved in prospective timing. Block (1990) criticized this concept because of the descriptive nature of the term, without any experimental or theoretical work on underlying processes. Later, Block (2003), for example, came to clarify and accept the term. He and others suggested that it involves what is now called *recursive reminding* (e.g., Hintzman 2004). If a person is asked to indicate when an experimenter-specified duration has elapsed, he or she often thinks: "Is it time now?" Every time that the person has that thought, it retrieves a time-dated memory of the previous occurrence. How often that happens depends on secondary-task attentional demands: If a person has to perform another cognitive task, the recursive reminding process is interrupted. Thus, duration productions lengthen, for example.

However, this explanation does not completely reveal the reason why cognitive load affects prospective duration judgments. The attentional-gate model (AGM; see, for example, Zakay & Block 1997) provides that kind of explanation. In this model, attention to time affects prospective duration judgments. A classical explanation of prospective timing is based on an internal-clock device made of a pacemaker emitting pulses, and of a counter accumulating these pulses, the perceived duration being proportional to the number of pulses accumulated. The AGM proposed by Zakay and Block (Figure 2.3; see Block & Zakay 2008; Zakay & Block 1997, for a description) determines the number of pulses, with more pulses being accumulated when more attention is allocated to the passage of time, and less to a nontemporal secondary task (in the case of a dual-task paradigm).

Perhaps both ideas are needed, recursive reminding to explain what it means when a person attends to time, and attentional gating to explain dual-task interference effects. Additional evidence on prospective duration judgments

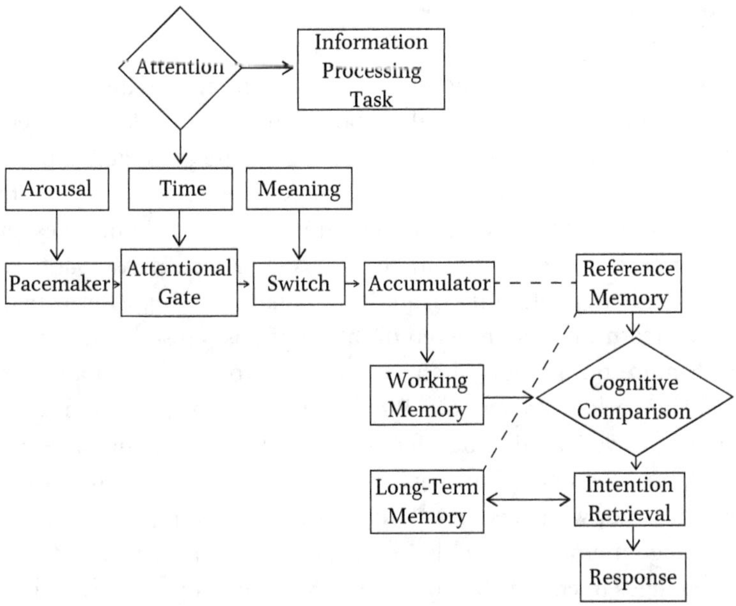


FIGURE 2.3 *The attentional-gate model of prospective timing (from Block & Zakay 2008; Zakay & Block 1997)*

comes from many experiments that have been reviewed in two major meta-analyses (Block, Hancock, & Zakay, 2010; Block & Zakay 1997) of studies using durations greater than 3 s. These findings reveal that variables involving cognitive load affect prospective timing. Mainly, these are response demands (active vs. passive), attentional demands (divided or selective vs. unitary), and processing difficulty (difficult vs. easy). With any of these cognitive demands,-prospective duration judgments made with the production method lengthened and verbal estimates decreased, for example.

Zakay (1992b, 2015)) reported research on temporal relevance and temporal uncertainty as variables that affect prospective timing. If an interval is relevant to a person's current concern or if the person is uncertain about when an interval will end, prospective temporal productions lengthen. This model supports a model of prospective timing that emphasizes attention. For-example, waiting intervals are perceived as longer than same clock-time intervals without waiting, because while waiting time is a major concern and the end of the waiting is not certain. Zakay (2005) showed that when timing is done concurrently with a nontemporal task, the duration of the interval is perceived to be longer when a person is instructed to treat timing as a primary rather than as a secondary task.

## 5 Retrospective Timing Processes

In another duration judgment paradigm, a participant is not informed in advance that time has to be estimated. Experiments using this retrospective paradigm are far less numerous than those using the prospective paradigm, probably because only one duration estimate may be obtained before the participant is aware that time judgments are of interest. Sometimes, at the onset a cover story or cover task might be used to lead participants to think that duration is not relevant (Grondin & Laflamme 2015). Retrospective judgments concern the remembered duration of past episodes, and they are based much more on memory than on attention to time. Meta-analytic findings (Block, Hancock, & Zakay, 2010; Block & Zakay 1997) reveal that variables involving cognitive load do not affect retrospective judgments. The important variables are segmentation (e.g., high-priority events or contextual interruptions) and other variables that affect memory encoding and retrieval. A contextual change memory model is implicated: If a person is able to remember more changes in context, retrospective duration judgments lengthen (Block & Reed 1978).

Two main methods have been most frequently used to study retrospective timing: the method of reproduction and the method of verbal estimation. When intervals under investigation are not too long, it is reasonable to consider the possibility to ask a participant to reproduce the duration. In such a case, this means that during the encoding phase (the interval to be reproduced), the participant is not informed that the duration (of the activity for instance) will have to be reproduced. For very long intervals, it is not realistic to adopt the reproduction method. Using verbal estimation is more practical. The problem that occurs with the verbal estimation is that people tend to round up their judgments to the nearest second if intervals are relatively brief, to the nearest 5 or 10 seconds if intervals last at least one minute, and to the nearest half-minute, or even minute, if the intervals last several minutes. One feature of retrospective judgments is that people should not be aware of the need to judge duration. Consequently, after completing a single trial, participants are aware of the importance of the duration of an activity or event. One way of collecting more than one judgment per participant is to ask to complete a series of activities before asking for duration judgments. Such a strategy was adopted by Boltz (1995), which involved a series of brief melodies (<15 s) to be learned before reproducing their duration (Experiment 1) and the presentation of long films (Experiment 2). Brown and Stubbs (1988) used a multiple-activity approach. They presented musical excerpts lasting 96 to 570 s. Boltz presented brief videotape sequences. Grondin and Plourde (2007) asked participants to complete five cognitive tasks (e.g., counting backward, recalling names of animals in alphabetical order) lasting 2–8 minutes, before asking for a verbal estimate of the duration of each of these tasks.

Researchers using retrospective timing are usually interested in the accuracy of judgments, not by the variability of estimates as is often the case for prospective timing. Limiting an investigation to a single trial actually makes it difficult to study the variability issue. A way to approach the problem is to ask not only for a verbal estimate of the activity, but also for a window within which the duration is certainly included. For example, a researcher may ask for the maximum and minimum duration of the activity (Bisson & Grondin 2013; Grondin & Plourde 2007; Tobin, Bisson, & Grondin, 2010; Tobin & Grondin 2012). The difference between the maximum and minimum could be interpreted as an uncertainty window, of a measure of variability.

## 6 Temporal Illusions

Like many other perceptual processes, time perception suffers from illusions that result from the processes underlying prospective and retrospective timing. Temporal illusions occur when the perception of duration of an interval do not faithfully represent the objective (clock-time) duration of that interval.

A well-known illusion is the filled-duration illusion: Empty intervals are perceived to be shorter or longer than filled intervals, which objectively endure exactly the same clock time. The direction of the illusion depends on the paradigm. If timing is done prospectively, then empty intervals, which do not demand many attentional resources for information processing, will be perceived as longer than filled intervals, which demand many attentional resources (e.g., Wearden, Norton, Martin, & Montford-Bebb, 2007). The opposite is the case regarding retrospective estimates. In this case, empty intervals are perceived as shorter than same clock time compared to filled intervals.

Thus, the perceived duration of an interval depends on cognitive load during the interval (Block et al. 2010.) This contradicts the essential characteristics of physical time. For example, it is often said that time flies when a person is having fun. The reason is that while having fun, attention is focused on the "fun" aspects (e.g., an attractive film) and not on time. Thus, the same clock-time interval will be perceived as longer when a person suffers instead of having fun. This is, of course, another illusion.

The "watched-pot" phenomenon is another temporal illusion. When a person waits for something to occur (when the water in a pot will boil), the main concern is time (when will this happen), and as a result most attentional resources are focused on time. This results in a longer temporal experience as compared to a regular interval (Block, George, & Reed, 1980).

An everyday situation in which people experience this illusion is when they have to wait for someone to come or some event to happen. Waiting intervals are perceived as longer than same clock-time intervals, without waiting. There are many more temporal illusions, most of them are prospective and reflect the dependency on attentional resources. The AGM provides a good explanation for that.

Temporal-illusions are an example for the dependency of subjective timing processes on measurement methods, interval durations, and other factors. For example, Hasuo, Nakajima, Tomimatsu, Grondin, and Udea (2014) found that for short intervals (40–520 ms) the illusion was more likely to occur with magnitude estimation than with the method of adjustment. The magnitude of the illusion increased as the interval duration lengthened. Taking a somewhat philosophical perspective, one may ask if not all temporal experience, including the flow of time itself, is actually an illusion (Gruber & Block 2013). Subjective durations are rarely identical with objective durations.

## 7 Temporal Dating of Memories

For centuries, dating to Aristotle, philosophers have thought that time is intimately related to memory. Theorists such as James (1890) and Hooke (1705/1969; see Hintzman 2003) proposed ideas on time and memory in the brain. These earliest ideas were reviewed by Hintzman (2003) and others (Block & Zakay 2008). Now, researchers have revealed how such temporal contiguity—proximity in time—functions (Hintzman 2016). Hintzman concluded that there is little or no evidence of temporal organization by contiguity per se. He said that research does not support the hypothesis that memory is not organized by time or by the principle of contiguity. Then he also elaborated on the present status of these ideas. In our view, time perception is mainly automatic only in some kinds of judgments that do not require voluntary attention. Automatic encoding might apply only to temporal dating of events, not necessarily to the dual-task interference seen in prospective duration estimation. Prospective duration estimates are based on attentional processes, as the AGM suggests. Temporal dating of specific memories is based on other processes, which involve long-term memory.

## 8 Prospective Memory: Timing the Future

Another interesting topic, which relates to processes involved in prospective and retrospective duration judgments, concerns prospective memory, which has also been called *prospective remembering*, or *timing the future*. These tasks involve intending to perform a time-based or an event-based action at some future time or occasion (Block & Zakay 2006; Graf & Grondin 2006; Labelle, Graf, Grondin, & Gagné-Roy, 2009; Mioni & Stablum 2014). For example, what role does temporal memory of past events serve in remembering to perform future actions? If people think about something that might likely occur (perhaps with what is called *stimulus-independent mentation*, or *daydreaming*), what time-related processes are involved?

Time-based and event-based prospective remembering processes are similar to those implicated in prospective and retrospective duration judgments, respectively. In timed-based tasks, participants are asked to make a response when they think that an experimenter-specified duration has elapsed; this is similar to a prospective duration estimate. In event-based tasks, people are asked to respond when a particular event (such as presentation of an animal word, or in everyday life, meeting a specific person) is encountered.

Time-based prospective memory tasks are similar to those involved in laboratory studies on prospective duration judgments, such as using the method of production. While is performing another task, the person must signal when a specified time interval (such as 30 s) has ended. The findings reveal that, such as in prospective duration judgments, attentional demands of the nontemporal task affect these processes. In everyday life, time-based tasks are often performed successfully by using reminder methods, such as calendars or electronic notes.

Event-based prospective memory tasks rely on the intent to respond in a certain way when a situation is encountered. Meeting a person or seeing a particular kind of event occur may lead to automatic retrieval of the intent. Of course, this is also somewhat likely to lead to failures to remember the previous intent.

## 9 Summary

In this chapter, we reviewed the history, methods, and current status of psychological research on timing and time perception, focusing mainly on prospective and retrospective judgments of time. Time perception varies as a function of duration, or interstimulus interval. Different phenomena arguably occur at about 100 ms, 1.3 s, 3 s, 5–7 s, and longer. Various methods are used, including those that require knowledge of conventional time units as well as others that do not. The latter have been used in studies of nonhuman animals and young children. Prospective time judgments, which involve a situation in which the estimation of durations is relevant and important, are affected by other information-processing demands. Retrospective duration judgments, on the other hand, depend mainly on retrieval of memories from the time period, and those are mainly affected by contextual encoding and retrieval processes. Other kinds of time-related phenomena occur in prospective memory situations, for example. Time-based remembering processes are similar to those of prospective timing, especially using the production method. Event-based remembering relies on memory processes similar to those involved in retrospective duration judgments. We did not review evidence on the many brain areas and processes involved in psychological time, mainly because the evidence is still unclear; future researchers should focus on that issue in order to deepen the understanding of these processes.

## References

Bausenhart, K.M., O. Dyjas, & R. Ulrich (2015). Effects of stimulus order on discrimination sensitivity for short and long durations. *Attention, Perception, & Psychophysics, 77*, 1033–1043.

Bindra, D., & H. Waksberg (1956). Methods and terminology in the studies of time estimation. *Psychological Bulletin, 53*, 155–159.

Bisson, N., & S. Grondin (2013). Time estimates of Internet surfing and video gaming. *Timing & Time Perception, 1*, 39–64.

Block, R.A. (1974). Memory and the experience of duration in retrospect. *Memory & Cognition, 2*, 153–160.

Block, R.A. (1979). Time and consciousness. In Underwood, G. & R. Stevens (Eds.), *Aspects of consciousness: Vol. 1. Psychological issues* (pp. 179–217). London: Academic Press.

Block, R.A. (1985). Contextual coding in memory: Studies of remembered duration. In Michon, J.A. & J.L. Jackson (Eds.), *Time, mind, and behavior* (pp. 169–178). Berlin: Springer-Verlag.

Block, R.A. (1990). Models of psychological time. In Block, R.A. (Ed.), *Cognitive models of psychological time* (pp. 1–35). Hillsdale, NJ: Erlbaum.

Block, R.A. (2003). Psychological timing without a timer: The roles of attention and memory. In Helfrich, H. (Ed.), *Time and mind II: Information processing perspectives* (pp. 41–59). Göttingen, Germany: Hogrefe & Huber.

Block, R.A., & S. Grondin (2014). A review of recent reviews on the psychology of time. *Frontiers in Psychology: Perception Science, 5*, 1–3.

Block, R.A., & P.A. Hancock (2013). Time perception. *Oxford Bibliographies*. [0123.xml?rskey=rykwK8&result=146].

Block, R.A., & M.A. Reed (1978). Remembered duration: Evidence for a contextual-change hypothesis. *Journal of Experimental Psychology: Human Learning and Memory, 4*, 656–665.

Block, R.A., & D. Zakay (1997). Prospective and retrospective duration judgments: A meta-analytic review. *Psychonomic Bulletin & Review, 4*, 184–197.

Block, R.A., & D. Zakay (2006). Prospective remembering involves time estimation and memory processes. In Glicksohn, J. & M.S. Myslobodsky (Eds.), *Timing the future: The case for a time-based prospective memory* (pp. 25–49). River Edge, NJ: World Scientific.

Block, R.A., & D. Zakay (2008). Timing and remembering the past, the present, and the future. In Grondin, S. (Ed.), *Psychology of time* (pp. 367–394). Bingley, England: Emerald.

Block, R.A., E.J. George, & M.A. Reed (1980). A watched pot sometimes boils: A study of duration experience. *Acta Psychologica*, 46, 81–94.

Block, R.A., P.A. Hancock, & D. Zakay (2010). How cognitive load affects duration judgments: A meta-analytic review. *Acta Psychologica, 134*, 330–343.

Boltz, M.G. (1995). Effects of event structure on retrospective duration judgments. *Perception & Psychophysics, 57*, 1080–1096.

Brown, S.W., & D.A. Stubbs (1988). The psychophysics of retrospective and prospective timing. *Perception, 17*, 297–310.

Church, R.M. (2003). A concise introduction to scalar timing theory. In Meck, W.H. (Ed.), *Functional and neural mechanisms of interval timing* (pp. 3–22). Boca Raton, FL: CRC Press.

Dyjas, O., K. Bausenhart, & R. Ulrich (2014). Effects of stimulus order on duration discrimination sensitivity are under attentional control. *Journal of Experimental Psychology: Human Perception and Performance, 40*, 292–307.

Eisler, H., A. Eisler, & A. Hellström (2008). Psychophysical issues in the study of time perception. In Grondin, S. (Ed.). *Psychology of time* (pp. 75–110). Bingley, UK: Emerald Group Publishing.

Elliott, M.A., & A. Giersch (2016). What happens in a moment. *Frontiers in Psychology: Perception Science, 7*(6), 1–8.

Fraisse, P. (1967/1973). Psychologie du temps [Psychology of time] (2nd ed.). Paris: Presses Universitaires de France.

Fraisse, P. (1974). Psychologie du rythme [Psychology of rhythm]. Vendome, France: Presses Universitaires de France.

Gibbon, J., C. Malapani, C.L. Dale, & C. Gallistel (1997). Toward a neurobiology of temporal cognition: Advances and challenges. *Current Opinion in Neurobiology, 7*, 170–184.

Graf, P., & S. Grondin (2006). Time perception and time-based prospective memory. In Glicksohn, J. & M.S. Myslobodsky (Eds.), *Timing the future: The case for a time-based prospective memory* (pp. 1–24). River Edge, NJ: World Scientific.

Grondin, S. (1993). Duration discrimination of empty and filled intervals marked by auditory and visual signals. *Perception & Psychophysics, 54,* 383–394.

Grondin, S. (2001a). Discriminating time intervals presented in sequences marked by visual signals. *Perception & Psychophysics, 63,* 1214–1228.

Grondin, S. (2001b). From physical time to the first and second moments of psychological time. *Psychological Bulletin, 127,* 22–44.

Grondin, S. (2008a). Methods for studying psychological time. In Grondin, S. (Ed.). *Psychology of time* (pp. 51–74). Bingley, UK: Emerald.

Grondin, S. (2008b). *Psychology of time.* Bingley, UK: Emerald.

Grondin, S. (2010). Timing and time perception: A review of recent behavioral and neuroscience findings and theoretical directions. *Attention, Perception & Psychophysics, 72,* 561–582.

Grondin, S. (2012). Violation of the scalar property for time perception between 1 and 2 seconds: Evidence from interval discrimination, reproduction, and categorization. *Journal of Experimental Psychology: Human Perception and Performance, 38,* 880–890.

Grondin, S. (2014). About the (non)scalar property for time perception. In Merchant, H. & V. de Lafuente (Eds.), *Advances in Experimental Medicine and Biology. Neurobiology of interval timing* (vol. 829, pp. 17–32). New York: Springer.

Grondin, S., & V. Laflamme (2015). Stevens's law for time: A direct comparison of prospective and retrospective judgments. *Attention, Perception, & Psychophysics, 77,* 1044–1051.

Grondin, S., & J.D. McAuley (2009). Duration discrimination in crossmodal sequences. *Perception, 38,* 1542–1559.

Grondin, S., & M. Plourde (2007). Judging multi-minute intervals retrospectively. *Quarterly Journal of Experimental Psychology, 60,* 1303–1312.

Grondin, S., G. Meilleur-Wells, & R. Lachance (1999). When to start explicit counting in time-intervals discrimination task: A critical point in the timing process of humans. *Journal of Experimental Psychology: Human Perception and Performance, 25,* 993–1004.

Grondin, S., V. Laflamme, & E. Gontier (2014). Effect on perceived duration and sensitivity to time when observing disgusted faces and disgusting mutilation pictures. *Attention, Perception & Psychophysics, 76,* 1522–1534.

Grondin, S., V. Laflamme, & G. Mioni (2015). Do not count too slowly: Evidence for a temporal limitation in short-term memory. *Psychonomic Bulletin & Review, 22,* 863–868.

Gruber, R.P., & R.A. Block (2013). The flow of time as a perceptual illusion. *Journal of Mind and Behavior, 34,* 91–100.

Guyau, A. (1890). *Le genese de l'idee au temps.* Felix Alcan (ed. A Fouilee).

Hancock, P.A., & R.A. Block (2012). The psychology of time: A view backward and forward. *American Journal of Psychology, 125,* 267–274.

Hasuo, E., Y. Nakajima, & K. Ueda (2011). Does filled duration illusion occur for very short time intervals? *Acoustical Science & Technology, 32*, 82–85.

Hasuo, E., Y. Nakajima, E. Tomimatsu, S. Grondin, & K. Ueda (2014). The occurrence of the filled duration illusion: A comparison of the method of adjustment with the method of magnitude estimation. *Acta Psychologica, 147*, 111–121.

Hellström, Å. (1979). Time errors and differential sensation weighting. *Journal of Experimental Psychology: Human Perception and Performance, 5*, 460–477.

Hellström, Å. (1985). The time-order error and its relatives: Mirrors of cognitive processes in comparing. *Psychological Bulletin, 97*, 35–61.

Hellström, Å., & T.H. Rammsayer (2004). Effects of time-order, interstimulus interval, and feedback in duration discrimination of noise bursts in the 50- and 1000-ms ranges. *Acta Psychologica, 116*, 1–20.

Hellström, Å., & T.H. Rammsayer (2015). Time-order errors and standard-position effects in duration discrimination: An experimental study and an analysis by the sensation-weighting model. *Attention, Perception, & Psychophysics, 77*, 2409–2423.

Hicks, R.E., G.W. Miller, & M. Kinsbourne (1976). Prospective and retrospective judgments of time as a function of amount of information processed. *American Journal of Psychology, 89*, 719–730.

Hintzman, D.L. (2003). Robert Hooke's model of memory. *Psychonomic Bulletin & Review, 10*, 3–14.

Hintzman, D.L. (2004). Judgment of frequency versus recognition confidence: Repetition and recursive reminding. *Memory & Cognition, 32*, 336–350.

Hintzman, D.L. (2016). Is memory organized by temporal contiguity? *Memory & Cognition*, 44, 365–375.

Hooke, R. (1969). *The posthumous works of Robert Hooke: With a New Introduction by Richard S. Westfall.* New York: Johnson Reprint Corp. (Original work published 1705)

ten Hoopen, G., R. Miyauchi, & Y. Nakajima (2008). Time-based illusions in the auditory mode. In Grondin, S. (Ed.). *Psychology of time* (pp. 139–187). Bingley, UK: Emerald.

ten Hoopen, G., S. van den Berg, J. Memelink, B. Bocanegra, & R. Boon (2011). Multiple-look effects on temporal discrimination within sound sequences. *Attention, Perception, & Psychophysics, 73*, 2249–2269.

James, W. (1890). *The principles of psychology.* New York: Holt.

Killeen, P.R., & N.A. Weiss (1987). Optimal timing and the Weber function. *Psychological Review, 94*, 455–468.

Killeen, P.R., J.G. Fetterman, & L.A. Bizo (1997). Time's causes. In Bradshaw, C.M. & E. Szabadi (Eds.), *Time and behavior: Psychological and neurobehavioral analyses* (pp. 79–131). Amsterdam: Elsevier.

Kuroda, T., & E. Hasuo (2014). The very first step to start psychophysical experiments. *Acoustical Science & Technology, 35*, 1–9.

Kuroda, T., S. Grondin, M. Miyazaki, K. Ogata, & S. Tobimatsu (2016). The kappa effect with only two visual markers. *Multisensory Research, 29*(8), 703–725.

Labelle, M.-A., P. Graf, S. Grondin, & L. Gagné-Roy (2009). Time-related processes in time-based prospective memory and in time-interval production. *European Journal of Cognitive Psychology, 21,* 501–521.

Laflamme, V., D. Zakay, P.-L. Gamache, & S. Grondin (2015). Foreperiod and range effects on time interval categorization. *Attention, Perception, & Psychophysics, 77,* 1507–1514.

Lejeune, H., & J. Wearden (2009). Vierordt's the experimental study of the time sense (1868) and its legacy. *European Journal of Cognitive Psychology, 21,* 941–960.

Macmillan, N.A., & C.D. Creelman (1991). *Detection theory: A user's guide.* New York: Cambridge University Press.

Michon, J.A., V. Pouthas, & J.L. Jackson (1988). Jean-Marie Guyau: Life and ideas. In Michon, J.A., V. Pouthas & J.L. Jackson (Eds.), *Guyau and the idea of time* (pp. 19–36). Amsterdam: North-Holland.

Mioni, G., & F. Stablum (2014). Monitoring behaviour in a time-based prospective memory task: The involvement of executive functions and time perception. *Memory, 22,* 536–552.

Mioni, G., F. Stablum, S.M. McClintock, & S. Grondin (2014). Different methods for reproducing time, different results. *Attention, Perception, & Psychophysics, 76,* 675–681.

Mioni, G., L. Meligrana, S. Grondin, F. Perini, L. Bartolomei, & F. Stablum (2015a). The effects of the facial expression of emotions on time perception in patients with Parkinson's disease. *Journal of the International Neuropsychological Society, 21.*

Mioni, G., D. Zakay, & S. Grondin (2015b). Faster is briefer: The symbolic meaning of speed influences time perception. *Psychonomic Bulletin & Review, 22,* 1285–1291.

Mioni, G, F Stablum., E Prunetti., & S Grondin. (2016). Time perception in anxious and depressed patients: A comparison between time reproduction and time production tasks. *Journal of Affective Disorders, 196,* 154–163.

Ornstein, R.E. (1969). *On the experience of time.* Harmondsworth, England: Penguin.

Patterson, R. (1990). Perceptual moment models revisited. In Block, R.A. (Ed.), *Cognitive models of psychological time* (pp. 85–100). Hillsdale, nj: Erlbaum.

Penney, T.B., J. Gibbon, & W.H. Meck (2008). Categorical scaling of duration bisection in pigeons (Columba livia), mice (Mus musculus), and humans (Homo sapiens). *Psychological Science, 19,* 1103–1109.

Pöppel, E. (1978). Time perception. In Held, R., H.W. Leibowitz & H.-L. Teuber (Eds.), *Handbook of sensory physiology* (Vol. 8, pp. 713–729). Berlin: Springer-Verlag.

Pöppel, E. (1997). A hierarchical model of temporal perception. *Trends in Cognitive Sciences, 1,* 56–61.

Rammsayer, T. (2014). The effects of type of interval, sensory modality, base duration, and psychophysical task on the discrimination of brief time intervals. *Attention, Perception, & Psychophysics, 76,* 1185–1196.

Shen, Y. (2013). Comparing adaptive procedures for estimating the psychometric function for an auditory gap detection task. *Attention, Perception, & Psychophysics, 75,* 771–780.

Tobin, S., & S. Grondin (2012). Time perception is enhanced by task duration knowledge: Evidence from experienced swimmers. *Memory & Cognition, 40,* 1339–1351.

Tobin, S., N. Bisson, & S. Grondin (2010). An ecological approach to prospective and retrospective timing of long durations: A study involving gamers. *PLoS ONE, 5*(2), e9271.

Wallace, M., & A. Rabin (1960). Temporal experience. *Psychological Bulletin, 57,* 213–235.

Wearden, J.H., & H. Lejeune (2006). Temporal generalization in humans. *Journal of Experimental Psychology: Animal Behavior Processes, 18,* 134–144.

Wearden, J.H., R. Norton, S. Martin, & O. Montford-Bebb (2007). Internal clock processes and the filled-duration illusion. *Journal of Experimental Psychology: Human Perception and Performance, 33,* 716–729.

Wing, A.M., & A.B. Kristofferson (1973). Response delays and the timing of discrete motor responses. *Perception & Psychophysics, 14,* 5–12.

Wundt, W. (1886). *Eléments en psychologie physiologique.* Paris: Alcan.

Zakay, D. (1992a). The role of attention in children's time perception. *Journal of Experimental Child Psychology, 54,* 355–371.

Zakay, D. (1992b). On prospective time estimation, temporal relevance and temporal uncertainty. In Macar, F., V. Pouthas, & W.J. Friedman (Eds.), *Time, action and cognition: Towards bridging the gap* (pp. 109–117). New York: Kluwer Academic/Plenum.

Zakay, D. (1993). Time estimation methods: Do they influence prospective duration estimates? *Perception, 22,* 91–101.

Zakay, D. (2005). Timing is a natural secondary task. *Proceedings of the annual meeting of the International Society for Psychophysics* (Fechner Day), Traverse City, USA.

Zakay, D. (2015). The temporal-relevance temporal-uncertainty model of prospective duration judgment. *Consciousness and Cognition, 38,* 182–190.

Zakay, D., & R.A. Block (1997). Temporal cognition. *Current Directions in Psychological Science, 6,* 12–16.

CHAPTER 3

# Assessing Duration Discrimination: Psychophysical Methods and Psychometric Function Analysis

*Karin M. Bausenhart, Massimiliano Di Luca and Rolf Ulrich*

## 1 Introduction

An important aspect in timing and time perception research is investigating the ability to perceive and compare temporal intervals, that is, the study of duration discrimination (Bindra & Waksberg 1956; Grondin 2010; Matthews & Meck 2016). Just as in every perceptual domain, a central problem in this field is how the relation between physical stimulus input (e.g., a tone lasting for 500 ms) and the sensation evoked by this input (the perceived duration of this stimulus) can be quantified. The scientific study of this relation is called psychophysics (Fechner 1889; Gescheider 1997).

One fundamental issue in psychophysics is the measurement of the difference threshold (just noticeable difference, *JND*; difference limen, *DL*), or in other terms, discrimination sensitivity. It is often loosely defined as the minimal physical difference between two stimuli (e.g., a 500 ms vs. a 550 ms interval) that a participant can just notice. A second important concept in psychophysics concerns the magnitude of the sensation evoked by a given stimulus. Typically, this sensation magnitude is determined by identifying the physical magnitude of a stimulus that is judged to be equal to the magnitude of another stimulus defined as the standard stimulus. For example, one might pinpoint that an auditorily presented temporal interval must be 480 ms to appear as having the same duration as a visually presented standard interval of 500 ms duration. This point along the duration dimension is termed the point of subjective equality (*PSE*), and just as in the example above, it often does not correspond to the point of objective equality (*POE*), which indexes physical equality with the standard stimulus.

Although these definitions appear simple, the experimental determination of these indices of discrimination performance can be quite cumbersome. For example, *PSE* can be influenced by perceptual and decisional biases, and this may even depend on the specific procedures employed for data collection. For example, when a participant is asked to compare the duration of two successively presented identical intervals, there may be a general tendency (i.e., bias) to judge the second presented duration as longer than the first presented one. Another major problem is that discrimination performance randomly fluctuates from moment to moment. For example, sometimes a given physical difference between two stimuli is perceived, while sometimes this difference is not perceived. To overcome such obstacles, 19th century researchers already invented various psychophysical tools for measuring discrimination performance, such as the method of constant stimuli (Hegelmaier 1852; Renz & Wolf 1856).

In this chapter, we review several of these tools and methods that are especially useful for measuring duration discrimination performance. Numerical examples are provided to illustrate these psychophysical procedures. In the first section, we introduce the standard psychometric function for comparative judgments and its associated parameters. We discuss various experimental paradigms, which are typically used to collect such data for assessing discrimination performance. In the second section, we present data collection and analysis methods based on equality judgments. For each type of judgment, we introduce several parametric and non-parametric procedures for computing indices of discrimination performance from these data, including exemplary Matlab scripts implementing these procedures (see book's GitHub repository). In the final conclusion, we briefly review several advanced toolboxes available for assessing discrimination performance.

## 2 Comparative Judgments

Several of the experimental paradigms, which are typically employed in timing research, involve comparative judgments. Specifically, these judgments require that participants decide whether a given stimulus duration is longer or shorter than a certain target duration. For example, in the so-called reminder task, the participant receives two successive durations in each experimental trial. One of the two durations is the target duration that is kept constant across a block of trials. This duration is traditionally called the *standard* or reference duration $s$ (Guilford 1954; Woodworth & Schlosberg 1954). The other duration varies randomly from trial to trial and is usually called the *comparison* or test duration $c$.

In most experiments, several different comparison durations are used, some larger than $s$ and some smaller than $s$. Typically, between 6 and 12 different values of $c$ are arranged symmetrically around $s$. It is convenient to index these comparison levels as $c_1,\ldots,c_k$ from the smallest to the largest.[1] These selected comparison levels are presented several times (usually 10 to 20 repetitions) during the course of a single experiment in a random sequence. The order of $s$ and $c$ may be either constant (fixed stimulus order), for example, in each trial $s$ is presented first, or it may vary randomly from trial to trial (random stimulus order). In the following, we will introduce some typical experimental paradigms employing either fixed or random stimulus order and describe several methods for analyzing the data emerging from these paradigms.

### 2.1 *Fixed Order of Standard and Comparison Stimuli*

Presumably the most elementary psychophysical approach uses a fixed order of $s$ and $c$ (e.g., Luce & Galanter 1963). For example, in the classical *reminder task*, $s$ precedes $c$ in every trial. Participants are typically asked whether the first or second stimulus appears longer, and consequently select the response $R_1$ or $R_2$, respectively. It is important to note that participants have to choose one of the two response alternatives in every trial – if a judgment cannot be made with certainty, the subject is asked to choose the alternative that seems most appropriate or simply to guess an alternative. After each trial, the experimenter simply records whether the participant responded with $R_1$ or $R_2$.

Table 3.1 contains an outcome example of such a psychophysical experiment comprising $k = 9$ comparison durations centered symmetrically around $s = 500$ ms. For these data, the relative frequency $f_i$ of responding with $R_2$ as a

TABLE 3.1 *Exemplary outcome of a psychophysical experiment with $s = 500$ ms and nine comparison levels $c_1,\ldots,c_9$ ranging from $c_1 = 300$ ms to $c_9 = 700$ ms. Row $n_i$ shows how many responses per comparison level $c_i$ were recorded during the course of the experiment. For most c levels, data of 15 trials were available, yet for some levels fewer data were recorded – for example, because the participant forgot to respond or occasionally pressed the wrong response key on the keyboard. The rows $n_{1,i}$ and $n_{2,i}$ show the number of $R_1$ and $R_2$ responses, respectively. The row $f_i$ contains the relative frequency of $R_2$ responses per c level.*

| | $c_1$<br>300 | $c_2$<br>350 | $c_3$<br>400 | $c_4$<br>450 | $c_5$<br>500 | $c_6$<br>550 | $c_7$<br>600 | $c_8$<br>650 | $c_9$<br>700 |
|---|---|---|---|---|---|---|---|---|---|
| $n_i$ | 15 | 15 | 14 | 15 | 12 | 15 | 15 | 15 | 14 |
| $n_{1,i}$ | 14 | 11 | 12 | 5 | 3 | 0 | 1 | 0 | 0 |
| $n_{2,i}$ | 1 | 4 | 2 | 10 | 9 | 15 | 14 | 15 | 14 |
| $f_i$ | 0.07 | 0.27 | 0.14 | 0.67 | 0.75 | 1.00 | 0.93 | 1.00 | 1.00 |

1 Usually the two extreme values in the range, $c_1$ and $c_k$, are selected in such a way that the comparisons cover the full range of the psychometric function from 0 to 1. Weber fractions

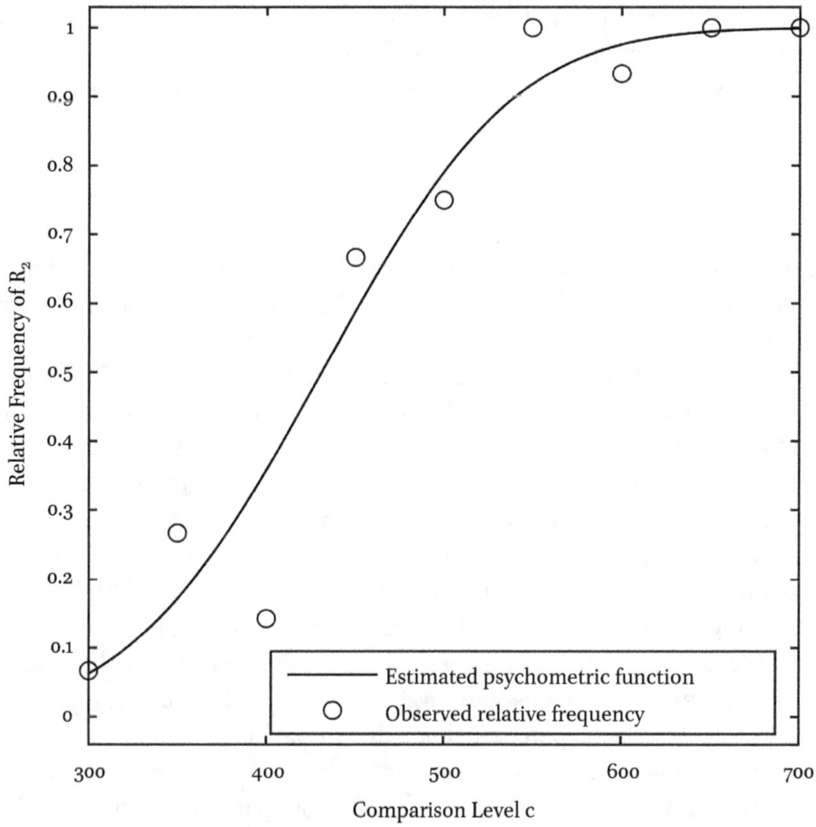


FIGURE 3.1 *Relative frequency of responding with $R_2$ (i.e., judging the second presented duration c as longer than the first presented duration s) as a function of comparison duration (open circles), for the example data given in Table 3.1. The solid line shows the best fitting psychometric function derived by means of probit analysis.*

function of comparison level $c_i$ is depicted in Figure 3.1. Apart from the statistical noise involved in such data, one would expect that this relative frequency increases with increasing duration of $c$.

### 2.1.1 Probit Analysis

In order to enable a more comprehensive analysis of the data emerging from such an experiment, one typically fits a psychometric function $\Psi(c)$ to the relative frequencies of $R_2$ responses per $c$ level (e.g., Luce & Galanter 1963).

---

may help to select these values. For example, assume that $s = 500$ ms and the participant is asked to discriminate auditory intervals, for which the Weber fraction typically amounts to approximately 0.1 (Rammsayer 2010; Rammsayer & Ulrich 2012). As a rule of thumb, $c_1$ may be selected as $s \cdot (1 - 4 \cdot 0.1)$ and $c_k$ as $s \cdot (1 + 4 \cdot 0.1)$. For $s = 500$ ms, this would yield $c_1 = 300$ ms and $c_k = 700$ ms.

For example, the cumulative density function (cdf) of a normal distribution has been often used as a mathematical model for $\Psi(c)$. This function increases monotonically from 0 to 1 with increasing values of $c$ and can be expressed as

$$\Psi(c) = \Phi\left(\frac{c-\mu}{\sigma}\right), \tag{1}$$

where $\Phi$ denotes the cdf of a standard normal distribution, $\mu$ is the location parameter, and $\sigma$ represents the slope of $\Psi$. This approach of modeling the psychometric function is also called probit analysis (Finney 1952).[2]

The parameter $\mu$ denotes the level of $c$ at which the probability of responding with $R_2$ is equal to 0.5, that is, at this level the two responses $R_1$ and $R_2$ are equally likely. This level is often called the *PSE*, because it denotes the duration of $c$, which is judged to have the same duration as $s$. The *PSE* needs not to be equal to $s$. For example, the *PSE* is often smaller than $s$ because participants usually tend to overestimate the second duration compared to the first one, a phenomenon termed the time-order error (Eisler, Eisler, & Hellström, 2008; Köhler 1923). In general, the difference between objective physical equality and subjective equality has been termed *constant error* (*CE*) and has been defined as $CE = PSE - s$ in the psychophysical literature. Shifts of the *PSE* away from the *POE* may reflect a perceptual or a decisional bias.

A second parameter of major importance that can be computed from a psychometric function is the *DL* or *JND*. This parameter indexes the discrimination sensitivity of a participant, with smaller values of *DL* indicating a higher level of sensitivity. The *DL* is related to the steepness of the psychometric function. It is typically defined as half its interquartile range, that is, $DL = (c_{0.75} - c_{0.25})/2$, where $c_{0.75}$ and $c_{0.25}$ represent the stimulus levels at which the response $R_2$ is elicited with probability 0.75 and 0.25, respectively (Luce & Galanter 1963). Consequently, *DL* indexes the duration difference between $s$ and $c$, which enables the subject to identify $c$ as being either shorter or longer than $s$ with an accuracy level of 75%. For the function embodied in Equation 1, the *DL* is given by

$$DL = \sigma \cdot z_{0.75} \tag{2}$$

2 Other functional families than the normal distribution are often used to model the psychometric function, such as the logistic or the Weibull function. However, the logistic and the probit model produce virtually the same results (Lord, Novick, & Birnbaum, 1968, p. 399).

where $z_{0.75}$ is the 75% percentile of the standard normal distribution, i.e., $z_{0.75} \approx 0.6745$.[3]

An especially efficient method for estimating the parameters *PSE* and *DL* is Fisher's maximum-likelihood procedure. In brief, one uses Equation 1 to compute the likelihood of the observed data,

$$L(Data \mid \mu,\sigma) = \Pi_{i=1}^{k} \Psi(c_i)^{n_{2,i}} \times \left[1-\Psi(c_i)\right]^{n_{1,i}} \tag{3}$$

where $n_{1,i}$ and $n_{2,i}$ denote the frequencies of observed $R_1$ and $R_2$ responses at each comparison level (compare Table 3.1). The maximum likelihood estimates of $\mu$ and $\sigma$ are those numerical values that maximize this likelihood function. The maximum of this function can be found numerically using a computer, a procedure known as numerical optimization.

A simple Matlab (R2016b) script ("MLEPsyProbit.m") for performing this optimization is available (see book's GitHub repository). It finds the parameters $\mu$ and $\sigma$ at which the function $L(Data \mid \mu,\sigma)$ has its extremum. This script requires as input the vectors $\vec{c} = (c_1,\ldots,c_k)$, $\vec{n}_1 = (n_{1,1},\ldots,n_{1,k})$, and $\vec{n}_2 = (n_{2,1},\ldots,n_{2,k})$ and provides the maximum-likelihood estimates of *PSE* and *DL* together with their standard errors and their corresponding 95% confidence intervals as outputs. This script computes the standard errors from the observed Fisher information. Applying the script to the data in Table 3.1, one obtains *PSE* = 430.9 ms, *Se* = 13.3 ms with a 95%-confidence interval of *CI* = [404.9, 457.0], and *DL* = 57.6 ms, *Se* = 8.8 ms with *CI* = [40.5, 74.8]. On the basis of the *PSE* result, the script computes *CE* = −69.1 ms, *Se* = 13.3 ms with *CI* = [−95.1, −43.0]. The *CE* indicates a systematic overestimation of the comparisons relative to the standard duration *s* = 500 ms, which might be attributed, for example, to a negative time-order error. Figure 3.1 depicts the relative

3 Several researchers (Treutwein 1995; Treutwein & Strasburger 1999; Wichmann & Hill 2001) have suggested to include also lapse parameters in the estimation of psychometric functions to account for trials in which the participant commits stimulus-independent lapses due to phasic inattention or "finger errors". These events will result in scaled psychometric functions, which do not cover the full range from 0 to 1. Even though such processing failures are rare events, typically estimated to occur in between 0% and 5% of trials (Wichmann & Hill 2001), their presence can nonetheless distort the estimation of *DL*. Therefore, if empirical evidence suggests the presence of lapses, corresponding extended psychometric functions should be used for data analysis (Wichmann & Hill 2001, also see Table 3.3 for a list of tools available for performing such advanced analyses). Models comparison statistics can be used as a principled way of choosing the function with or without lapses.

response proportions $f_i$ from Table 3.1 and the resulting psychometric function derived by this probit analysis, which is a standard psychophysical approach for estimating *PSE*, *DL*, and *CE*.

### 2.1.2 Pseudo-Gaussian Function

Killeen, Fetterman, and Bizo (1997) proposed an alternative to Equation 1 that often provides an excellent fit to observed data (Allan & Gerhardt 2001; Birngruber, Schröter, & Ulrich, 2014; Grondin 2001). This approach takes Weber's law into account, according to which variability in perceived duration should linearly increase with physical duration. Specifically, let $S$ and $C$ represent the internal representations of the standard $s$ and the comparison $c$, respectively. In addition, assume that the internal difference $\Delta = C - S$ follows a normal distribution with mean $E[\Delta|c] = c - (\varepsilon + s)$, where the parameter $\varepsilon$ has the status of a constant error. If the standard deviation of the difference $\Delta$ follows Weber's law $\sigma_c = w \cdot c, w > 0$, then the psychometric function is given by the *Pseudo-Gaussian function*,

$$\Psi(c) = \Phi\left(\frac{c - (\varepsilon + s)}{w \cdot c}\right), \tag{4}$$

where $\Phi$ again denotes the cumulative density function of a standard normal variable, and the parameters are the constant error $\varepsilon$ and the Weber fraction $w$.[4] This Pseudo-Gaussian function is actually not a genuine psychometric function because it does not converge to 1. However, this deviation from 1 is negligible for realistic values of $w$. The supplementary Matlab script "MLEPSyPseudoGaussian.m" (see book's GitHub repository) provides maximum likelihood estimates of the parameters $\varepsilon$ and $w$. Applying this script to the data in Table 3.1 yields for $\varepsilon$ an estimate of −86.8 ms, $Se$ = 12.3 ms, $CI = [-110.8, -62.7]$ and for $w$ an estimate of 0.190, $Se = 0.026$, $CI = [0.138, 0.241]$.

Moreover, for this Pseudo-Gaussian function, it can be shown that the *PSE* is given by

$$PSE = s + \varepsilon \tag{5}$$

4 As a further extension, one may replace $\sigma_c = w \cdot c$ by the generalized Weber's law $\sigma_c = \sqrt{w_1 \cdot c^2 + w_2 \cdot c + w_3}$ (see Killeen et al. 1997). A similar model has been proposed by García-Pérez (2014). Also note that for $w_1 = w_2 = 0$ this extended model becomes a special case of the probit model discussed above.

and the DL by

$$DL = (s + \varepsilon) \cdot \frac{w \cdot z_{0.75}}{1 - (w \cdot z_{0.75})^2} \tag{6}$$

with $z_{0.75} \approx 0.6745$. Inserting the above estimates into these equations yields $PSE = 413.2$ ms and $DL = 53.7$ ms. It can be noticed that these estimates differ numerically from the ones of the standard approach embodied by Equation 1, which must be attributed to the different assumptions underlying both models.

Figure 3.2 depicts the relative response proportions $f_i$ from Table 3.1 and the psychometric function resulting from the Pseudo-Gaussian model. A potential

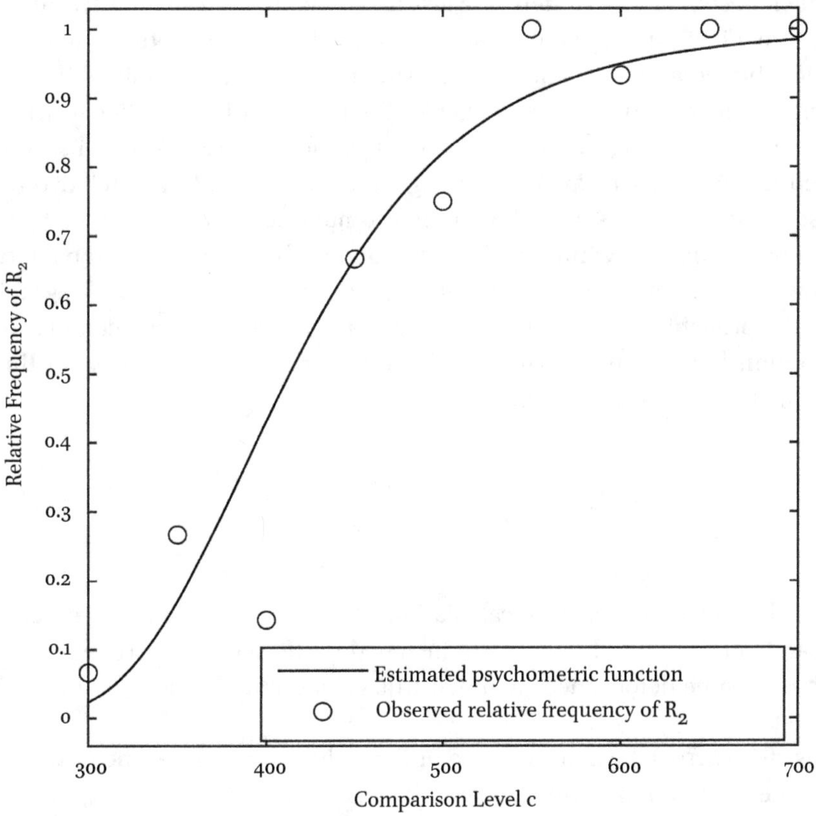


FIGURE 3.2 *Relative frequency of responding with $R_2$ (i.e., judging the second presented duration c as longer than the first presented duration s) as a function of comparison duration (open circles), for the example data given in Table 3.1. The solid line shows the best fitting psychometric function derived by means of the Pseudo-Gaussian model.*

drawback of this Pseudo-Gaussian model is that the estimate of DL is affected by the size of $\varepsilon$, that is, the constant error.

### 2.1.3 Spearman-Kärber Method

In addition to the parametric approaches discussed above, one can also use a nonparametric approach, the *Spearman-Kärber method* (Kärber 1931; Spearman 1908), for estimating the location and the spread of the psychometric function (Miller & Ulrich 2001; Sternberg, Knoll, & Zukofsky, 1982). This method has several advantages. In contrast to parametric approaches, the Spearman-Kärber method does not require specific assumptions about the functional family of the true underlying psychometric functions. Also, it allows for estimating higher-order moments as skewness and kurtosis, in addition to location and spread of the psychometric function. Moreover, this method is computationally efficient compared to others, because it does not require an iterative fitting procedure. Finally, parameter estimates obtained with this method are often even less biased and less variable than parameter estimates obtained by employing parametric approaches (Miller & Ulrich 2001; Ulrich & Miller 2004).

In the Spearman-Kärber method, the range of comparison stimuli is subdivided into bins, each ranging from $c_{i-1}$ to $c_i$, for $i = 1,\ldots,k$. The relative response frequencies $f_i$ associated with each stimulus level $c_i$ are assumed to be uniformly distributed within each corresponding bin. Thus, the probability density within each bin is estimated as $(f_i - f_{i-1})/(c_i - c_{i-1})$. The resulting histogram of probability densities approximates the continuous true cumulative distribution function underlying the data. Each $r$th raw moment $m'_r$ of this psychometric function can then be calculated as

$$m'_r = \frac{1}{r+1}\sum_{i=1}^{k+1}\frac{(f_i - f_{i-1}) \cdot \left(c_i^{r+1} - c_{i-1}^{r+1}\right)}{c_i - c_{i-1}}. \tag{7}$$

It must be noted that in this calculation, the values of the most extreme comparison levels $c_0$ and $c_{k+1}$ are not included in the actual experimental design but must be determined such that true values of $f_0 = 0$ and $f_{k+1} = 1$ can be assumed.

This step is crucial whenever $f_1 > 0$ or $f_k < 1$, that is, whenever the observed psychometric function is truncated (i.e., it does not start at 0 or reach 1). For example, this may be the case if the chosen range of comparison levels for testing was not broad enough to cover the whole range of the psychometric function. Similarly, lapses, finger errors, or simply binomial random error might cause such truncated psychometric functions. In this case, the specific values chosen for $c_0$ and $c_{k+1}$ will affect the CDF's raw moments, and consequentially the estimates of location, spread, etc. derived from these raw moments. Therefore, it is advised to interpret parameters obtained by means of the Spearman-Kärber method especially carefully whenever truncated psychometric functions are present. Furthermore, for computing higher raw moments (i.e., $r > 1$), it is necessary to monotonize the observed psychometric function before computing these moments with Equation 7 (see Ayer, Brunk, Ewing, Reid, & Silverman, 1955; Miller & Ulrich 2001, cf. also Figure 3.3).

From the raw moments, one can derive estimates of location, spread, skewness and kurtosis (Miller & Ulrich 2001). For example, the first raw moment $m_1'$ corresponds to the arithmetic mean and, thus, indexes the location of the psychometric function (i.e., it serves as an estimate of *PSE*). The standard deviation of the underlying CDF can be estimated with $\sigma = \sqrt{m_2' - (m_1')^2}$.

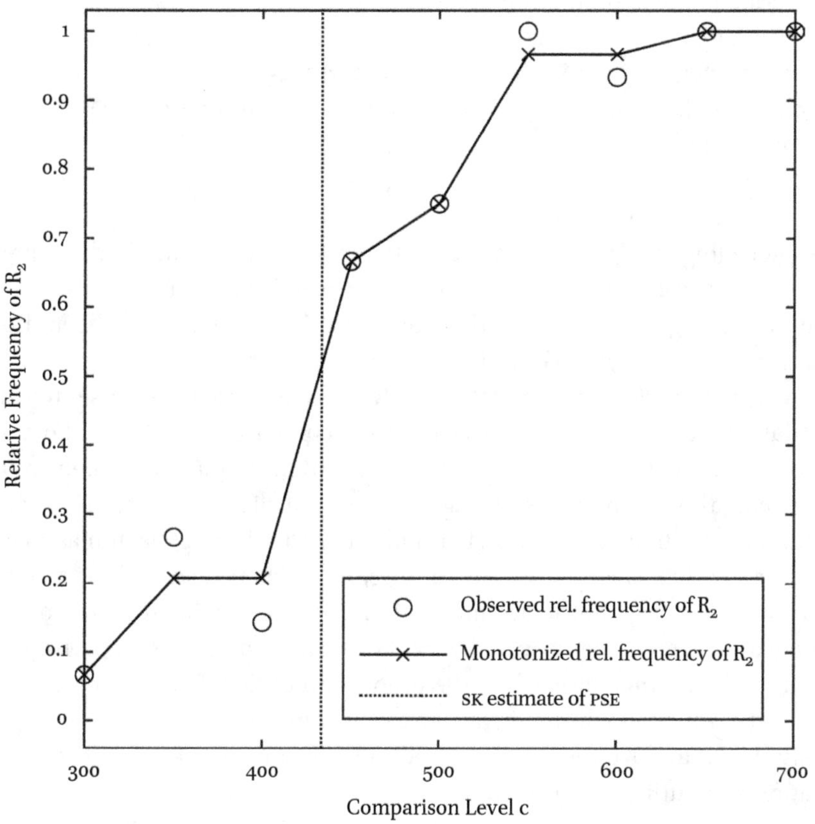


FIGURE 3.3 *Observed (open circles) and monotonized (black X and solid line) relative frequency of responding with $R_a$ (i.e., judging the second presented duration c as longer than the first presented duration s) as a function of comparison duration, for the example data given in Table 3.1. The dotted vertical line corresponds to the* PSE *estimate derived by the Spearman-Kärber method.*

For comparative purposes, a convenient estimate of DL can then be approximated by multiplying $\sigma$ by $z_{0.75} \approx 0.6745$.

The provided Matlab script "SpearmanKaerber.m" (see book's GitHub repository) monotonizes the observed psychometric function and then computes the Spearman-Kärber estimates of PSE, $\sigma$, and DL for the example data contained in Table 3.1 (see Figure 3.3). By default, the extreme values $c_0$ and $c_{k+1}$ are set such that $c_1 - c_0 = c_2 - c_1$, and $c_{k+1} - c_k = c_k - c_{k-1}$, that is, equidistance between the first 3 and the last 3 comparison levels is assumed. The script outputs the observed response frequencies $f_i$ and the monotonized response frequencies $\overline{f_i}$, as well as a vector containing estimates of PSE, $\sigma$, DL, and CE. For the example data given in Table 3.1, the corresponding estimates are PSE = 433.5 ms, $\sigma$ = 82.1 ms, DL = 55.4 ms, and CE = −66.5 ms. These parameter estimates correspond quite well with the estimates derived by the probit analysis described above. In addition, this function provides bootstrap estimates of these parameters based on 1000 replications, including standard errors and CIs. For example, for PSE: *Se* = 12.1 ms, CI = [408.3, 455.4], for $\sigma$: *Se* = 10.6 ms, CI = [58.4, 100.4], for DL: *Se* = 7.2 ms, CI = [39.4, 67.7], and for CE: *Se* = 12.1 ms, CI = [−91.7, −44.6].[5]

#### 2.1.4 Variants of Data Collection

In the preceding sections, it is assumed that in each trial a standard *s* is presented before the comparison duration *c* (i.e., reminder task). Especially in the domain of timing research, several variants of this basic task have been proposed (for an overview, see Grondin 2010).

First, in the *single-stimulus method* only the comparison is presented in each trial. The participant then classifies each comparison as either short or long, presumably against an internal standard that is quickly formed from experiencing the comparisons during the course of the experiment (Bausenhart, Bratzke, & Ulrich, 2016; Dyjas, Bausenhart, & Ulrich, 2012; Nachmias 2006; Woodworth & Schlosberg 1954). Sometimes, researchers also present a standard *s* for several times at the beginning of the experiment, in order to provide a more explicit reference for classifying the duration of each comparison as short or long. In either case, when the proportion of "long" responses is plotted against comparison duration, an ogive psychometric function will emerge. Estimating PSE and DL then can proceed in the same manner as in the standard approach outlined above.

5 Naturally, these bootstrapped values will randomly fluctuate with each execution of the provided Matlab function.

Second, a further methodological variant of the standard approach is the *bisection method*. Here, at the beginning of the experiment the shortest (i.e., $c_1$) and the longest (i.e., $c_k$) comparisons are presented several times as anchor stimuli. During the experiment, only comparisons are presented (as in the single-stimulus method) and the participant must classify each comparison as more similar to the short or to the long anchor duration (Allan & Gibbon 1991; Wearden, Rogers, & Thomas, 1997). The data analysis again proceeds as outlined above.

Third, in comparative judgments, researchers may allow for a *third response option* besides $R_1$ and $R_2$, i.e., an "uncertain" or "same" response (Woodworth & Schlosberg 1954, pp. 212–217). Historically, two response categories have been preferred over three response categories in psychophysics (Woodworth & Schlosberg 1954, p. 217). Nevertheless it is sometimes useful to employ three categories for theoretical reasons (e.g., Rammsayer & Ulrich 2001; Ulrich 1987) and more complex models of discrimination performance may be fitted to the data emerging from three-response categories to identify the relevant parameters indicating discrimination performance (García-Pérez 2014; García-Pérez & Alcalá-Quintana 2013).

Finally, all data collection variants as described above may be regarded as instances of the method of constant stimuli, in which the researcher preselects a range of comparison levels and typically presents each comparison level for a predetermined number of repetitions, with all trials presented in random order. This has sometimes been criticized as relatively inefficient, since many points along the psychometric function are sampled with an equal and large number of trials. Yet, some of these points, typically those demarcating threshold values as *PSE* and *DL*, are of especially high interest to the researcher, and an efficient data collection procedure might focus on assessing these points with high precision instead. Since the threshold values are of course not known in advance of testing, but depend on the participants' performance, comparison levels then cannot be specified in advance. Rather, the experimenter's decision about which comparison level should be presented in a given trial must depend on the participant's responses given in previous trials. There is a vast number of data collection schemes and analysis variants for such *adaptive testing* procedures (see Kaernbach 1991; Leek 2001; Treutwein 1995), although some caution is required when applying these procedures (e.g., García-Pérez 1998).

### 2.2 *Random Order of Standard and Comparison Stimuli*

In the methodological variants for data collection described in the preceding section, the temporal order of *s* and *c* is either the same in each experimental trial, or only $c$ is presented. In contrast, in the so-called *two-alternative forced-choice task* (2AFC, sometimes also two-interval forced-choice task or 2IFC), this order of the standard and comparison varies randomly from trial to trial. Thus, in each trial, stimulus order is either $\langle sc \rangle$ or $\langle cs \rangle$. Participants typically indicate whether the first or second stimulus appears longer by responding with $R_1$ or $R_2$, respectively.[6] Since the order of $s$ and $c$ varies randomly, the range of $c$ levels can be restricted to values $c \geq s$, but it is also possible to employ values ranging from $c_1 < s$ to $c_k > s$. In the latter case, $R_1$ and $R_2$ responses can be recoded as $c > s$ responses, for the stimulus orders $\langle cs \rangle$ and $\langle sc \rangle$, respectively. From these data, a psychometric function depicting the proportion of $c > s$ responses emerges. Given a sufficiently large range of $c$ values, and disregarding the possibility of lapses or finger errors, this function covers the full range from 0 to 1. Then, as a measure of discrimination sensitivity, DL is often estimated as half the interquartile range of this psychometric function (analogously to the procedure outlined above for the reminder task). In the former case, researchers often plot the proportion of correct responses (i.e., $c > s$ responses), resulting in a psychometric function restricted from 0.5 (i.e., guessing probability) to 1. An often-employed procedure to derive DL from such psychometric functions is to compute it as $DL = c_{0.75} - s$ (cf. Ulrich 2010; Ulrich & Vorberg 2009).

In both cases outlined above, however, the common practice of collapsing the raw data across the two orders of $s$ and $c$ can lead to loss of information and even to severe distortions in the estimated parameters of the psychometric function. To avoid such distortions, data from the two stimulus orders $\langle sc \rangle$ or $\langle cs \rangle$ should be plotted and analyzed separately (Ulrich 2010; Ulrich & Vorberg 2009). Consequently, two order-dependent psychometric functions emerge in the 2AFC design (cf. Figure 3.4). Specifically, let $S_1$ and $S_2$ denote the stimulus in the first or second position, respectively. Define $F_1(c) \equiv P(R_1 | \langle cs \rangle)$ and $F_2(c) \equiv P(R_2 \mid \langle sc \rangle)$ as the conditional probability with which the participant judges the comparison $c$ as the larger of the two stimuli when it was presented first or second, respectively. Note that the two conditional psychometric functions monotonically increase with $c$.

Importantly, these two conditional psychometric functions can differ in their location ("Type A order effect") and in their spread ("Type B order effect"). A prominent example for a Type A order effect is the typically observed negative time-order error, in which the duration of the first of two subsequently presented intervals is underestimated compared to the second one (Eisler et al. 2008; Hellström 1985; Köhler 1923). Specifically, this would correspond to

6 Participants are usually not aware that there is a constant standard, which appears first or second.

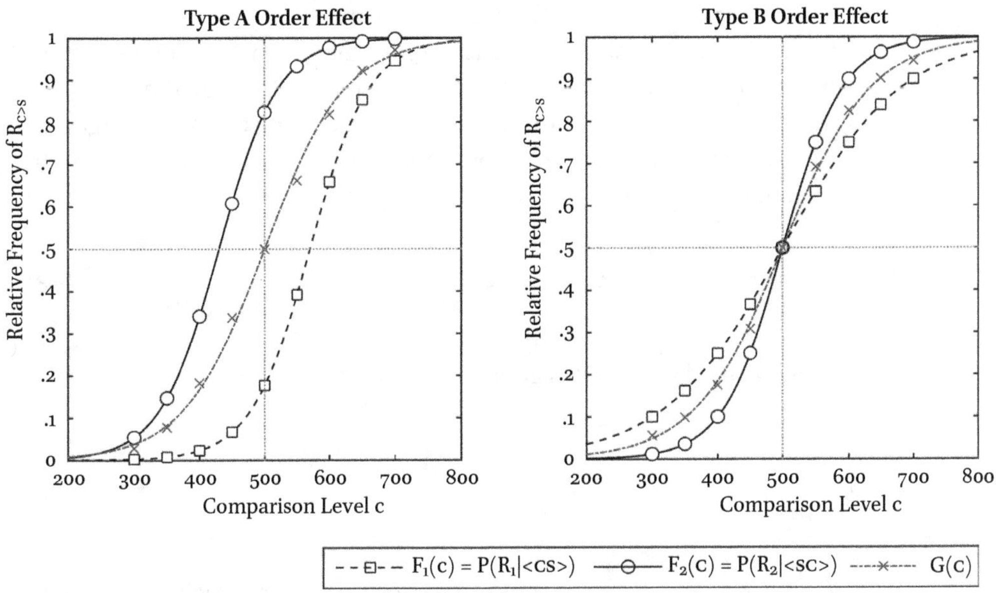


FIGURE 3.4 *Relative frequency of responding with* $R_{c>s}$ *(i.e., judging the comparison duration c as longer than the standard duration s) and psychometric functions for a hypothetical 2AFC experiment. Depicted are the order-conditional functions* $F_1(c)$ *(dashed line and squares) and* $F_2(c)$ *(solid line and circles) for stimulus orders* ⟨cs⟩ *and* ⟨sc⟩*, respectively. In addition, the psychometric function G(c) (grey dash-dotted line and x) corresponds to the observed response frequencies aggregated across presentation orders. The left panel depicts a type A order effect and the right panel a type B order effect. These effects will be concealed by the common practice of fitting a single psychometric function to the data aggregated across stimulus orders. Moreover, whenever a Type A order effect is present in the data, the aggregated psychometric function G(c) is less steep than either of the order-conditional functions (see left panel). Consequently, DLs derived from such aggregated functions will be overestimated.*

a lateral shift of the conditional psychometric functions away from the *POE*, at which $s = c$, such that the location of the conditional psychometric function for stimulus order ⟨*cs*⟩ is shifted to *POE* + $\gamma$ and the mean location of conditional psychometric function for stimulus order ⟨*sc*⟩ is shifted to *POE* – $\gamma$. An example of a Type B order effect, which is often observed in duration discrimination, is a shallower slope of the conditional psychometric function for stimulus order ⟨*cs*⟩ than for stimulus order ⟨*sc*⟩ (Bruno, Ayhan, & Johnston, 2012; Dyjas et al. 2012; Nachmias 2006). Consequently, such an effect indicates a higher discrimination sensitivity for two subsequent intervals, when the first of these intervals is a standard interval with constant duration, rather than when it is varied randomly from trial to trial.

The preceding explanation assumes that researchers choose the stimuli presented in a 2AFC task such that they vary only along a single stimulus dimension. In a duration discrimination task, for example, *s* and *c* would be identical in all respects, except for their duration. In this case, a restriction emerges for the estimated psychometric functions. Specifically, averaging the two conditional functions results in an aggregated psychometric function,

$$G(c)=\frac{P(R_1|\langle cs\rangle)+P(R_2|\langle sc\rangle)}{2}. \tag{8}$$

At the *POE*, defined as $s = c$, this equation simplifies to

$$G(s)=\frac{P(R_1|\langle ss\rangle)+P(R_2|\langle ss\rangle)}{2}. \tag{9}$$

Since $R_1$ and $R_2$ are the only response alternatives, their associated response proportions must sum to one. Consequently,

$$G(s)=\frac{1}{2}, \tag{10}$$

that is, the average of the two order-conditional psychometric functions must pass through the point (*s*, 0.5). This restriction must be considered when fitting psychometric functions to the order-conditional data. Specifically, instead of estimating two independent psychometric functions, they must be fitted simultaneously and the number of the free parameters to be estimated for these two functions reduces to three (Ulrich 2010; Ulrich & Vorberg 2009). Matlab and R code for fitting logistic order-conditional psychometric functions under this restriction is provided by Bausenhart, Dyjas, Vorberg, and Ulrich (2012).

If a researcher chooses to let *s* and *c* vary along more than one dimension (e.g., in duration and stimulus size), then of course the constraint implied by Equation 10 does not hold, and the average function will pass through the point (*PSE*, 0.5) instead (García-Pérez & Alcalá-Quintana 2011). Then, the two order-conditional psychometric functions can be estimated independently from each other, just as outlined in the section on fixed order of standard and comparison stimuli above. The routines provided by Bausenhart et al. (2012) also provide the option to release the constraint at $s = c$ and therefore can also be employed for the analysis of order-conditional data coming from 2AFC tasks which vary along multiple stimulus dimensions.

## 3 Equality Judgments

Besides the comparative judgment task employed in the preceding methods, equality judgments as in the *temporal generalization method* are also often used in the domain of temporal cognition (e.g., Wearden 1992; Wearden, Edwards, Fakhri, & Percival, 1998). In the temporal generalization method, the standard *s* is usually presented for several times at the beginning of an experiment. After *s* has been initially presented, the participant receives in each trial a comparison duration $c_i$, as before spaced below and above the standard. After each presentation of a comparison duration, the participant has to judge whether this duration was the same as the standard or different, by responding with $R_{same}$ or $R_{different}$, respectively. Alternatively, the standard and the comparison may be presented in each trial, and the participant is also asked to judge whether the two stimulus durations are equal, $R_{same}$, or not equal, $R_{different}$ (see Birngruber et al. 2014; Dyjas & Ulrich 2014). Table 3.2 contains example data for such an equality judgment task. When the relative frequency of a same response is plotted against comparison duration, an approximately bell-shaped psychometric function emerges. As before, there are various methods available to summarize such data.

### 3.1 *Same-different Model with Constant Standard Deviation*

First, a parametric method has been suggested by Schneider and Komlos (2008). These authors have assumed that subjects base their judgment on the difference $\Delta = C - S$ between the internal representation of the comparison and the standard and respond with $R_{same}$ if $|\Delta + \varepsilon| < \gamma$ and otherwise with $R_{different}$. The parameter $\gamma$ denotes a constant threshold value and $\varepsilon$ the constant error.

TABLE 3.2 *Exemplary outcome of a psychophysical experiment with an equality judgment task. The experiment uses s = 500 ms and nine comparison levels $c_1,\ldots,c_9$ ranging from $c_1$ = 300 ms to $c_9$ = 700 ms. Row $n_i$ shows how often comparison level $c_i$ was repeated during the course of the experiment. The rows $n_{same,i}$ and $n_{different,i}$ give the number of $R_{same}$ and $R_{different}$ responses. The row $f_i$ contains the relative frequencies of the $R_{same}$ responses.*

| | $c_1$<br>300 | $c_2$<br>350 | $c_3$<br>400 | $c_4$<br>450 | $c_5$<br>500 | $c_6$<br>550 | $c_7$<br>600 | $c_8$<br>650 | $c_9$<br>700 |
|---|---|---|---|---|---|---|---|---|---|
| $n_i$ | 15 | 15 | 14 | 15 | 12 | 15 | 15 | 15 | 14 |
| $n_{same,i}$ | 0 | 5 | 6 | 11 | 8 | 4 | 1 | 0 | 0 |
| $n_{different,i}$ | 15 | 10 | 8 | 4 | 4 | 11 | 14 | 15 | 14 |
| $f_i$ | 0.00 | 0.33 | 0.43 | 0.73 | 0.67 | 0.27 | 0.07 | 0.00 | 0.00 |

If one assumes that $\Delta$ follows a normal distribution with mean $\mu_c = c - s + \varepsilon$ and standard deviation $\sigma$, then it can be shown that the probability of a $R_{same}$ response is given by

$$P\left(R_{same} \mid c,s\right) = \Phi\left(\frac{\gamma-(c-s)-\varepsilon}{\sigma}\right) - \Phi\left(\frac{-\gamma-(c-s)-\varepsilon}{\sigma}\right). \tag{11}$$

Again the maximum likelihood method can be used to obtain estimates of $\gamma$, $\varepsilon$, and $\sigma$ from the observed data. The supplementary Matlab script "MLESameDifferent.m" performs this analysis (see book's GitHub repository). Applying this procedure to the data of Table 3.2 yields $\gamma$ = 62.0 ms, *Se* = 7.1 ms, CI = [48.1, 75.9], $\varepsilon$ = −42.9 ms, *Se* = 9.1 ms, CI = [−60.6, −25.1], and $\sigma$ = 53.7 ms, *Se* = 7.6 ms, CI = [38.8, 68.7]. Figure 3.5 depicts the resulting psychometric

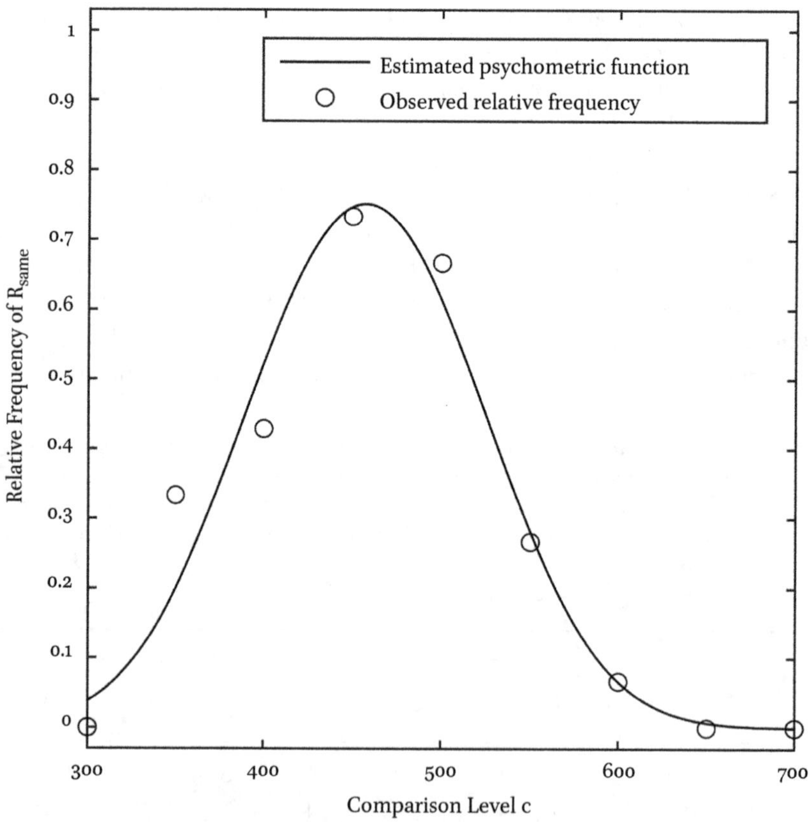


FIGURE 3.5 *Relative frequency of responding with $R_{same}$ (i.e., judging c and s as equally long). The solid line shows the best fitting psychometric function. This model assumes that the standard deviation $\sigma$ of the internal difference $\Delta$ is constant.*

function for these parameters. Note that the estimate of $\sigma$ can be regarded as a sensitivity measure and thus might be reexpressed as $DL = \sigma \cdot z_{0.75}$. Likewise, the $PSE$ can be obtained via $PSE = s + \varepsilon$. For the present example, this yields $PSE$ = 457.1 ms, $Se$ = 9.1 ms, $CI$ = [439.4, 474.9], and $DL$ = 36.3 ms, $Se$ = 5.1 ms, $CI$ = [26.2, 46.3].

### 3.2 *Same-different Model with Standard Deviation Dependent on Comparison Level*

The model underlying Equation 11 implies a symmetrical bell-shaped psychometric function. However, experiments employing the temporal generalization method or the standard procedure of presenting $s$ and $c$ in fixed order in each trial, typically generate asymmetrical psychometric functions with a positive skew (Birngruber et al. 2014; Wearden et al. 1998; Wearden 1992). In order to account for this asymmetrical shape, one may as before (i.e., Pseudo-Gaussian Model) assume that the standard deviation $\sigma$ in the preceding Equation 11 increases with the comparison level $c$, i.e., $\sigma_c = w \cdot c$ (see Birngruber et al. 2014),

$$P\left(R_{same} \mid c,s\right) = \Phi\left(\frac{\gamma-(c-s)-\varepsilon}{w \cdot c}\right) - \Phi\left(\frac{-\gamma-(c-s)-\varepsilon}{w \cdot c}\right). \tag{12}$$

Figure 3.6 displays the estimated function for this model variant when it is applied to the example data in Table 3.2. The parameter estimates, derived by the supplementary Matlab script "MLESameDifferent2.m" (see book's GitHub repository), are $\gamma$ = 61.5 ms, $Se$ = 7.1 ms, $CI$ = [47.7, 75.3], $\varepsilon$ = −52.3 ms, $Se$ = 9.0 ms, $CI$ = [−69.9, −34.6], and $w$ = 0.118, $Se$ = 0.016, $CI$ = [0.086, 0.150]. Because the predicted shape of this psychometric function is asymmetrical and influenced by the Weber fraction, it is difficult to properly define a measure of $PSE$. However, similar to the previous definition, one may again compute $PSE = s + \varepsilon$. Discrimination sensitivity is reflected in the parameter $w$, i.e., the Weber fraction. Due to the asymmetry of the underlying psychometric function, this measure should be used to index sensitivity.

### 3.3 *Waveform Moment Analysis*

The preceding two procedures involved a parametric approach to the analysis of data emerging from equality judgments. The *Waveform Moment Analysis* enables a non-parametric approach (Cacioppo & Dorfman 1987). Let $f_i$ be the observed relative frequency of a $R_{same}$ response associated with comparison level $c_i$. In a first step, these frequencies are converted to a probability distribution $p_i$, $i = 1,\ldots,k$, by the following transformation,

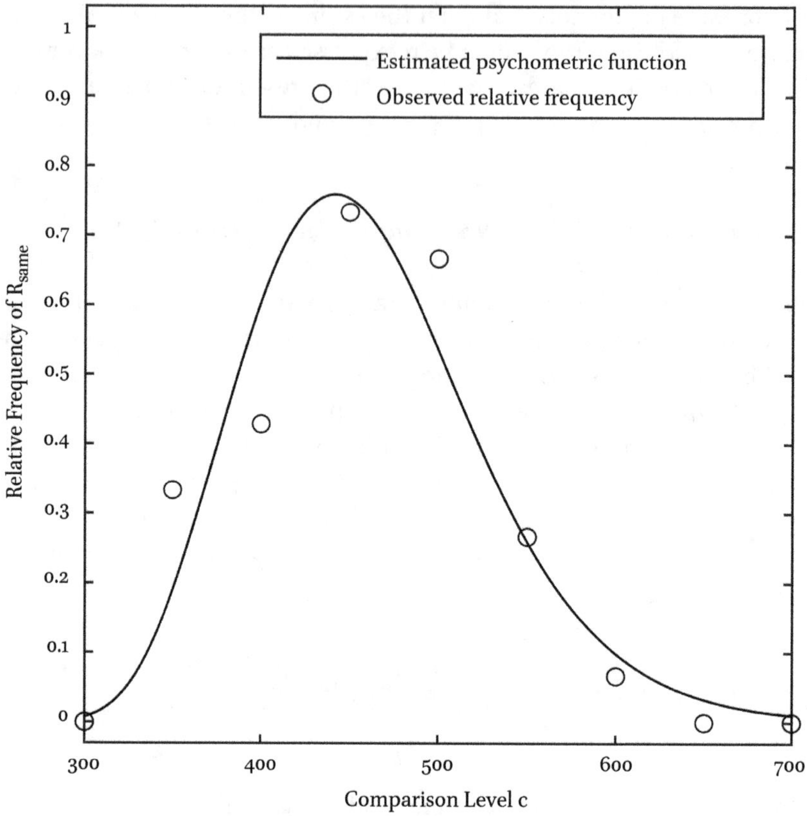


FIGURE 3.6 *Relative frequency of responding with $R_{same}$ (i.e., judging c and s as equally long). The solid line shows the best fitting psychometric function. This model assumes that the standard deviation $\sigma$ of the internal difference $\Delta$ increases with c.*

$$p_i = \frac{f_i}{\sum_{i=1}^{k} f_i}. \tag{13}$$

In a second step, the mean $\mu$ and the standard deviation $\sigma$ are computed for this "probability distribution", that is,

$$\mu = \sum_{i=1}^{k} p_i \cdot c_i \tag{14}$$

and

$$\sigma = \sqrt{\sum_{i=1}^{k} p_i \cdot (c_i - \mu)^2}. \tag{15}$$

The parameter $\mu$ assesses the location of the psychometric function on the abscissa and thus can be interpreted as *PSE*, whereas the parameter $\sigma$ captures the spread of this function and thus reflects discrimination performance with smaller values of $\sigma$ indicating a higher level of discrimination sensitivity. Applications of the waveform moment analysis in temporal discrimination have been reported by Birngruber et al. (2014) and by Dyjas and Ulrich (2014). A Matlab script for performing this analysis ("WaveformMoment.m") is available as supplementary material (see book's GitHub repository). For the data in Table 3.2, one obtains $\mu$ = 456.1 ms, $\sigma$ = 63.8 ms, and thus *CE* = −43.9 ms. This script also computes the standard error and confidence intervals for these parameters by the bootstrap method. For example, one obtains for $\mu$: *Se* = 8.7 ms, *CI* = [438.5, 472.5], for $\sigma$: *Se* = 5.2 ms, *CI* = [52.6, 72.8], and for *CE*: *Se* = 8.7 ms, *CI* = [−61.5, −27.5] (see also Figure 3.7).

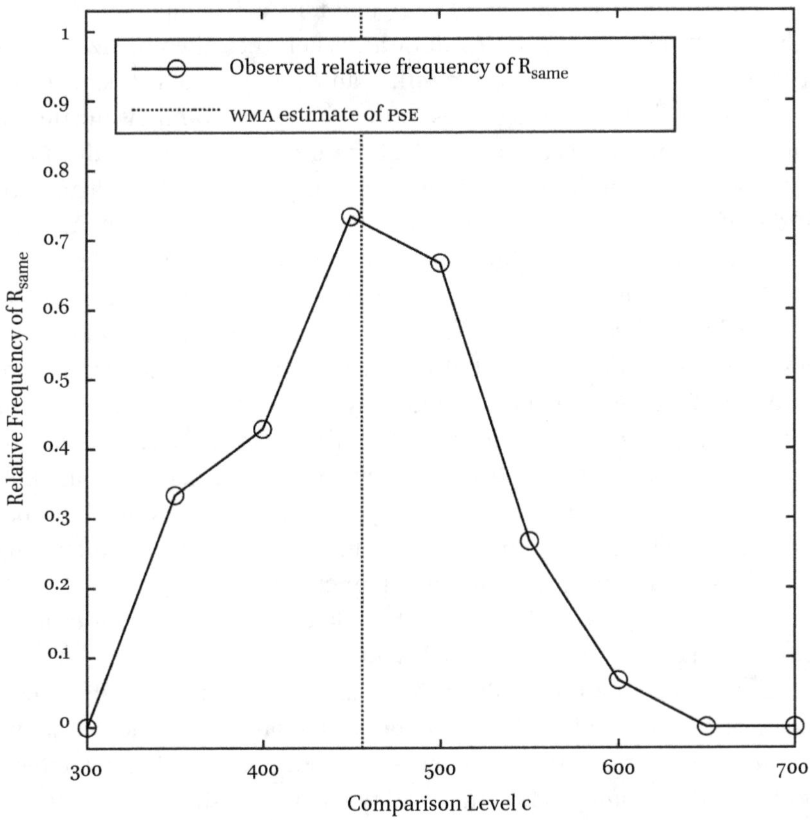


FIGURE 3.7 *Relative frequency of responding with $R_{same}$ (i.e., judging c and s as equally long). The dotted line indicates the PSE estimate derived by means of the Waveform Moment Analysis.*

## 4 Conclusions

In this chapter we have introduced and reviewed several psychophysical paradigms and analytical procedures for determining discrimination sensitivity and perceived duration. For each procedure, we provided Matlab example codes for estimating the respective parameters that beginners to the field of timing research might find helpful (see book's GitHub repository). We have discussed two major approaches differing in the type of judgment employed to assess timing performance, namely, comparative and equality judgments. The two major parameters of interest in this regard are *PSE* and *DL*. In experimental work, the absolute magnitude of these parameters is often of subordinate importance. Rather, the major interest lies in assessing differences in these parameters between experimental conditions. For example, an experimenter might be interested in whether or not the size of a visual stimulus affects perceived duration (e.g., Mo & Michalski 1972; Rammsayer & Verner 2014). In this case, the *PSE* should be estimated for large and for small comparison stimuli, using the same standard in both conditions. When the different size conditions are presented in random order within an experimental block, changes in *PSE* can be attributed to differences in the size of the comparison stimuli, since other influences on *PSE*, such as the time-order error, should affect *PSE* to an identical extent in both conditions. Therefore, a reminder design with fixed order of *s* and *c* is usually appropriate whenever an experimenter wants to investigate whether an experimental manipulation affects *PSE*.

It must be kept in mind, however, that perceived duration still can only be indirectly inferred from changes in the *PSE*, since the *PSE* reflects not only changes in perceived duration, but also decisional and response biases, and, therefore, this parameter should be cautiously interpreted in terms of *judged* duration rather than *perceived* duration. The use of a 2AFC task has the additional advantage that one can isolate the effects of secondary experimental manipulations from the time-order error by analyzing the order-conditional psychometric functions. Also, unbiased estimates of *DL* can be achieved by assessing the slope of the order-conditional functions.

Traditionally, comparative judgments have been used most often to measure both *DL* and *PSE*. However, equality judgments may of course also be employed, and might be especially useful to assess the robustness of experimental effects. For example, consider that one is interested in whether an experimental manipulation influences perceived duration. If similar PSE effects can be observed for comparative and equality judgments, this might strengthen the notion that the manipulation affects perceived duration rather than decisional processes (e.g. Birngruber et al. 2014; Dyjas & Ulrich 2014).

Supplementing this chapter, we provided basic Matlab scripts to illustrate the various psychophysical procedures for newcomers to the field of time perception (see book's GitHub repository). It must be mentioned, however, that elaborated psychophysical toolboxes are available for data analysis (see Table 3.3). We also refer the reader to comprehensive manuals on psychophysical methods for information about details of these toolboxes (Kingdom & Prins 2010; Lu & Dosher 2014). For example, the toolbox developed by Wichmann and Hill (2001) also allows the estimation of lapses in designs with comparative judgments. The toolbox *Palamedes* described in Kingdom and Prins (2010) also includes Matlab scripts for adaptive psychophysical procedures. Finally, the Matlab script by Bausenhart et al. (2012) is recommended for fitting psychometric functions conditional on stimulus order in 2AFC tasks.

In this chapter, we focused on psychophysical tools and procedures to obtain and analyze psychometric functions. This is sometimes considered as the classical psychophysical approach. An alternative approach for characterizing discrimination performance is offered by Signal Detection Theory (SDT; Green & Swets, 1966). Interestingly, in the domain of time perception, the psychophysical tools from SDT are much less often used than the classical tools described in this chapter. One major reason why time perception researchers usually prefer the classical tools is that SDT does not provide a parameter like the *PSE* that would allow to estimate judged duration. This is perhaps not surprising since SDT was mainly developed to identify near-threshold stimuli, an issue that does not apply to time perception. Furthermore, we did not address duration scaling methods as temporal reproduction, production, or verbal estimation, which are also often used to investigate duration perception (e.g., Allan 1979; Bindra & Waksberg 1956; see also Chapter 4 of this book). However,

TABLE 3.3 *Overview of advanced toolboxes and functions for psychometric function fitting.*

| Name | Website | References | Notes |
|---|---|---|---|
| Psychtoolbox | www.psychtoolbox.org | Brainard (1997), Pelli (1997) | See folder "psychometric". Function fitting (e.g., cumulative normal, Weibull, Naka-Rushton). For the cumulative normal and Weibull optimization toolbox is needed. |
| Palamedes | www.palamedestoolbox .org | Kingdom & Prins (2010) | Fits several types of psychometric function also to multiple conditions at the same time, performs model comparison, and contains adaptive procedures. |

TABLE 3.3 *Overview of advanced toolboxes and functions* (cont.)

| Name | Website | References | Notes |
|---|---|---|---|
| MLE2AFC | http://link.springer.com/article/10.3758%2Fs13428-012-0207-z | Bausenhart et al. (2012) | Fits psychometric functions conditional on stimulus order accounting for lapses and response errors. |
| Psignifit | psignifit.sourceforge.net | Fruend, Haenel, & Wichmann (2011) | Performs maximum-likelihood fits, tests the quality of the fit, and provides confidence intervals on the parameters of the fitted functions. |
| UML | http://hearlab.ss.uci.edu/UML/uml.html | Shen, Dai, & Richards (2015) | Increases efficiency of data collection by estimating the parameters of the psychometric and optimizing stimulus sampling. |
| model free | http://personalpages.manchester.ac.uk/staff/d.h.foster/software-modelfree/latest/index.html | Zychaluk & Foster (2009) | Non-parametric local linear fitting. |
| PMETRIC | www.psy.otago.ac.nz/miller/Software.htm | Miller & Ulrich (2004) | WIN EXE that performs probit analysis and Spearman-Kärber method and uses bootstrapping for standard errors of parameter estimates (a Matlab wrapper function is available from the authors). |
| Psychophysica | Available upon personal request to the authors | Watson & Solomon (1997) | Mathematica Notebooks, of which Psychometrica.nb fits and plots psychometric data. |
| quickpsy | dlinares.org/quickpsy.html | Linares & López-Moliner (2016) | R toolbox that fits and plots psychometric functions for multiple conditions. |
| GLMM | mixedpsychophysics.wordpress.com | Moscatelli, Mezzetti, & Lacquaniti (2012) | R toolbox to combine the analysis of the behavior at the level of single subject and population. |

data analyses for these approaches are performed with common statistical measures as mean and standard deviation of the observed data.

In sum, we hope that the present chapter will direct beginners with little or no background in psychophysics to the most important paradigms and psychophysical methods for assessing discrimination sensitivity and judged duration. The Matlab scripts provided as supplementary material should provide hands-on experience with these methods, although these scripts can not replace the elaborated toolboxes mentioned above.

## References


Allan, L.G. (1979). The perception of time. *Perception & Psychophysics*, 26, 340–354.

Allan, L.G., & K. Gerhardt (2001). Temporal bisection with trial referents. *Perception & Psychophysics*, *63*, 524–440.

Allan, L.G., & J. Gibbon (1991). Human bisection at the geometric mean. Special Issue: Animal timing. *Learning and Motivation*, *22*, 39–58.

Ayer, M., H.D. Brunk, G.M. Ewing, W.T. Reid, & E. Silverman (1955). An empirical distribution function for sampling with incomplete information. *Annals of Mathematical Statistics*, *26*, 641–647.

Bausenhart, K.M., O. Dyjas, D. Vorberg, & R. Ulrich (2012). Estimating discrimination performance in two-alternative forced choice tasks: Routines for matlab and R. *Behavior Research Methods*, *44*, 1157–1174.

Bausenhart, K.M., D. Bratzke, & R. Ulrich (2016). Formation and representation of temporal reference information. *Current Opinion in Behavioral Sciences, 8*, 46–52.

Bindra, D., & H. Waksberg (1956). Methods and terminology in studies of time estimation. *Psychological Bulletin*, *53*, 155–159.

Birngruber, T., H. Schröter, & R. Ulrich (2014). Duration perception of visual and auditory oddball stimuli: Does judgment task modulate the temporal oddball effect? *Attention, Perception, & Psychophysics*, *76*, 814–828.

Brainard, D. (1997). The psychophysics toolbox. *Spatial Vision*, *10*, 443–446.

Bruno, A., I. Ayhan, & A. Johnston (2012). Effects of temporal features and order on the apparent duration of a visual stimulus. *Frontiers in Psychology*, *3*, 1–7.

Cacioppo, J.T., & D.D. Dorfman (1987). Waveform moment analysis in psychophysiological research. *Psychological Bulletin*, *102*, 421–438.

Dyjas, O., & R. Ulrich (2014). Effects of stimulus order on discrimination processes in comparative and equality judgements: Data and models. *Quarterly Journal of Experimental Psychology*, *67*, 1121–1150.

Dyjas, O., K.M. Bausenhart, & R. Ulrich (2012). Trial-by-trial updating of an internal reference in discrimination tasks: Evidence from effects of stimulus order and trial sequence. *Attention, Perception, & Psychophysics*, *74*, 1819–1841.

Eisler, H., A.D. Eisler, & Å. Hellström (2008). Psychophysical issues in the study of time perception. In Grondin, S. (Ed.), *Psychology of time* (pp. 75–109). Bingley, UK: Emerald.

Fechner, G.T. (1889). *Elemente der Psychophysik I* (2nd ed.). Breitkopf & Härtel.

Finney, D.J. (1952). *Probit analysis: A statistical treatment of the sigmoid response curve.* Cambridge: Cambridge University Press.

Fründ, I., N.V. Haenel, & F.A. Wichmann (2011). Inference for psychometric functions in the presence of nonstationary behavior. *Journal of Vision, 11,* 16.

García-Pérez, M.A. (1998). Forced-choice staircases with fixed step sizes asymptotic and small-sample properties. *Vision Research, 38,* 1861–1881.

García-Pérez, M.A. (2014). Does time ever fly or slow down? The difficult interpretation of psychophysical data on time perception. *Frontiers in Human Neuroscience, 8,* 415.

García-Pérez, M.A., & R. Alcalá-Quintana (2011). Improving the estimation of psychometric functions in 2AFC discrimination tasks. *Frontiers in Psychology, 2,* 96.

García-Pérez, M.A., & R. Alcalá-Quintana (2013). Shifts of the psychometric function: Distinguishing bias from perceptual effects. *The Quarterly Journal of Experimental Psychology, 66,* 319–337.

Gescheider, G.A. (1997). *Psychophysics: The fundamentals* (3rd ed.). Hillsdale, NJ: Erlbaum.

Green, D., & J. Swets (1966). *Signal detection theory and psychophysics (rev. ed.).* Los Altos, CA: Peninsula Publishing, reprinted Edition 1998.

Grondin, S. (2001). Discriminating time intervals presented in sequences marked by visual signals. *Perception & Psychophysics, 63,* 1214–1228.

Grondin, S. (2010). Timing and time perception: A review of recent behavioral and neuroscience findings and theoretical directions. *Attention, Perception & Psychophysics, 72,* 561–582.

Guilford, J. (1954). *Psychometric methods* (2nd ed.). New York: McGraw-Hill Book Company, Inc.

Hegelmaier, F. (1852). Ueber das Gedächtniss für Linear-Anschauungen. *Archiv für physiologische Heilkunde, 11,* 844–853.

Hellström, Å. (1985). The time-order error and its relatives: Mirrors of cognitive processes in comparing. *Psychological Bulletin, 97,* 35–61.

Kaernbach, C. (1991). Simple adaptive testing with the weighted up-down method. *Perception & Psychophysics, 49,* 227–229.

Kärber, G. (1931). Beitrag zur kollektiven Behandlung pharmakologischer Reihenversuche. *Archiv für experimentelle Pathologie und Pharmakologie, 162,* 480–483.

Killeen, P.R., J.G. Fetterman, & L.A. Bizo (1997). Time's causes. In Bradshaw, C.M. & E. Szabadi (Eds.), *Time and behavior: Psychological and neurobiological analyses* (pp. 79–132). Amsterdam: Elsevier.

Kingdom, F.A.A., & N. Prins (2010). *Psychophysics.* Amsterdam: Elsevier.

Köhler, W. (1923). Zur Theorie des Sukzessivvergleichs und der Zeitfehler. *Psychologische Forschung, 4,* 115–175.

Leek, M.R. (2001). Adaptive procedures in psychophysical research. *Perception & Psychophysics, 63,* 1279–1292.

Linares, D., & J. López-Moliner (2016). quickpsy: An R Package to Fit Psychometric Functions for Multiple Groups. *The R Journal, 8,* 122–131.

Lord, F.M., M.R. Novick, & A. Birnbaum (1968). *Statistical theories of mental test scores.* Reading, Massachusetts: Addison-Wesley.

Lu, Z.-L., & B. Dosher (2014). *Visual psychophysics.* London, England: The mit Press.

Luce, R., & E. Galanter (1963). Discrimination. In Luce, R.D., R.R. Bush, & E. Galanter (Eds.), *Handbook of mathematical psychology* (Vol. 1, pp. 191–243). New York: John Wiley & Sons.

Matthews, W.J., & W.H. Meck (2016). Temporal cognition: Connecting subjective time to perception, attention, and memory. *Psychological Bulletin, 142,* 865–907.

Miller, J., & R. Ulrich (2001). On the analysis of psychometric functions: The Spearman-Kärber method. *Perception & Psychophysics, 63,* 1399–1420.

Miller, J., & R. Ulrich (2004). A computer program for Spearman-Kärber and probit analysis of psychometric function data. *Behavior Research Methods, Instruments, & Computers, 36,* 11–16.

Mo, S.S., & V.A. Michalski (1972). Judgment of temporal duration of area as a function of stimulus configuration. *Psychonomic Science, 27,* 97–98.

Moscatelli, A., M. Mezzetti, & F. Lacquaniti (2012). Modeling psychophysical data at the population-level: The generalized linear mixed model. *Journal of Vision, 12,* 26.

Nachmias, J. (2006). The role of virtual standards in visual discrimination. *Vision Research, 46,* 2456–2464.

Pelli, D. (1997). The VideoToolbox software for visual psychophysics: Transforming numbers into movies. *Spatial Vision, 10,* 437–442.

Rammsayer, T. (2010). Differences in duration discrimination of filled and empty auditory intervals as a function of base duration. *Attention, Perception, & Psychophysics, 72,* 1591–1600.

Rammsayer, T., & R. Ulrich (2001). Counting models of temporal discrimination. *Psychonomic Bulletin & Review, 8,* 270–277.

Rammsayer, T., & R. Ulrich (2012). The greater temporal acuity in the reminder task than in the 2AFC task is independent of standard duration and sensory modality. *Canadian Journal of Experimental Psychology, 66,* 26–31.

Rammsayer, T., & M. Verner (2014). The effect of nontemporal stimulus size on perceived duration as assessed by the method of reproduction. *Journal of Vision, 14,* 1–10.

Renz, T., & A. Wolf (1856). Versuche über die Unterscheidung differenter Schallstärken. *Archiv für physiologische Heilkunde, 15,* 185–193.

Schneider, K.A., & M. Komlos (2008). Attention biases decisions but does not alter appearance. *Journal of Vision, 8*, 1–10.

Shen, Y., W. Dai, & V.M. Richards (2015). A MATLAB toolbox for the efficient estimation of the psychometric function using the updated maximum-likelihood adaptive procedure. *Behavior Research Methods, 47*, 13–26.

Spearman, C. (1908). The method of "right and wrong cases" ("constant stimuli") without Gauss's formulæ. *Britisch Journal of Psychology*, 2, 227–242.

Sternberg, S., R.L. Knoll, & P. Zukofsky (1982). Timing by skilled musicians. In Deutsch, D. (Ed.), *The psychology of music* (pp. 181–239). New York: Academic Press.

Treutwein, B. (1995). Adaptive psychophysical procedures. *Vision Research, 35*, 2503–2522.

Treutwein, B., & H. Strasburger (1999). Fitting the psychometric function. *Perception & Psychophysics, 61*, 87–106.

Ulrich, R. (1987). Threshold models of temporal-order judgments evaluated by a ternary response task. *Perception & Psychophysics, 42*, 224–239.

Ulrich, R. (2010). DLs in reminder and 2AFC tasks: Data and models. *Attention, Perception, & Psychophysics, 72*, 1179–1198.

Ulrich, R., & J. Miller (2004). Threshold estimation in two-alternative forced-choice (2AFC) tasks: The Spearman-Kärber method. *Perception & Psychophysics, 66*, 517–533.

Ulrich, R., & D. Vorberg (2009). Estimating the difference limen in 2AFC tasks: Pitfalls and improved estimators. *Attention, Perception, & Psychophysics, 71*, 1219–1227.

Watson, A.B., & J.A. Solomon (1997). Psychophysica: Mathematica notebooks for psychophysical experiments. *Spatial Vision, 10*, 447–466.

Wearden, J.H. (1992). Temporal generalization in humans. *Journal of Experimental Psychology: Animal Behavior Processes, 18*, 134–144.

Wearden, J.H., P. Rogers, & R. Thomas (1997). Temporal bisection in humans with longer stimulus durations. *Quarterly Journal of Experimental Psychology, 50B*, 79–94.

Wearden, J.H., H. Edwards, M. Fakhri, & A. Percival (1998). Why "sounds are judged longer than lights": Application of a model of the internal clock in humans. *The Quarterly Journal of Experimental Psychology, 51B*, 97–120.

Wichmann, F.A., & N.J. Hill (2001). The psychometric function: I. Fitting, sampling, and goodness of fit. *Perception & Psychophysics, 63*, 1293–1313.

Woodworth, R.S., & H. Schlosberg (1954). *Experimental psychology* (*3rd ed.*). London: Methuen.

Żychaluk, K., & D.H. Foster (2009). Model-free estimation of the psychometric function. *Attention, Perception, & Psychophysics, 71*, 1414–1425.

CHAPTER 4

# Methodological Issues in the Study of Prospective Timing

*Giovanna Mioni*

## 1 Introduction

The ability to accurately estimate the passage of time plays an important role in daily activities, from the sleep-wake cycle to speaking, to the ability to play musical instruments. We process time across a wide range of intervals, which operates over the range of milliseconds to the 24-h (Buhusi & Meck, 2005; Fraisse, 1984; Grondin, 2010). As described by Block, Grondin, and Zakay (Chapter 2, this volume), the methods to investigate temporal processing can be distinguished in *prospective* and *retrospective paradigms*. In the *prospective paradigm*, participants know in advance that they will be asked to judge the duration of a time period. In the *retrospective paradigm*, participants do not know until the end of a time period that they will be asked to judge its duration. The way participants experience the passage of time and the various cognitive processes involved may nevertheless differ between the two paradigms. In the prospective paradigm, a person may intentionally encode temporal information as an integral part of the experience of the time period. In the retrospective paradigm, a person may incidentally encode temporal information, and whatever information is relevant may be later retrieved from memory.

Traditionally, processing duration in the milliseconds-seconds range has been explained using the pacemaker-accumulator model (Gibbon, Church, & Meck, 1984; Triesman, 1963). The model postulates that duration judgments go through three stages: clock, memory, and decision stage. The clock stage is composed of a pacemaker that emits pulses gated to an accumulator. A switch is placed before the accumulator that controls the flow of pulses from the pacemaker to the accumulator. The memory stage is conceptualized as the storing system that accumulates pulses in working memory for comparison with the content of reference memory. The reference memory contains a long-term memory representation of pulses accumulated across prior trials. The final stage is the decision stage, in which the current duration is compared with those stored in the reference memory. In Treisman's model (1963),

© GIOVANNA MIONI, 2018 | DOI 10.1163/9789004280205_005

the pacemaker directly receives inputs from an arousal center. As the arousal levels vary, the speed with which the pacemaker emits pulses change. An increased level of arousal leads to an increase in the speed of the pacemaker. For a given duration period, if the pacemaker runs faster, more pulses reach the accumulator, and this duration is judged to be longer (Droit-Volet & Meck, 2007). The Attentional-Gate Model (Block & Zakay, 1996; Zakay & Block, 1996) acknowledges the role of arousal in the pacemaker speed but it also adds an 'attentional gate' that is influenced by the amount of attention allocated to time. If less attention is devoted to time, the gate narrows and allows fewer pulses to transfer to the accumulator.

These models provide interesting frameworks to test research hypotheses. But how do we test them? Which method is more appropriate for a specific question? In this view, the purpose of this chapter is to facilitate a future researcher that approaches the study of timing to select the appropriate method with a specific focus on prospective timing. Here, the principal methods for studying time in the prospective paradigm will be described. Moreover, applications in clinical populations will be also described.

## 2 Methods for Studying Time in the Prospective Paradigm

Traditionally, researchers have utilized many different methods for investigating time (Block, 1990; Grondin, 2008; 2010; Zakay 1990; 1993), however, this chapter, will focus only on the most classical ones used in the prospective paradigm (i.e., time production, verbal estimation, and time reproduction).

*Time production* and *verbal estimation* tasks may be considered the two sides of the same coin and reflect similar underlying temporal processes and mechanisms (Block, 1990). In the production task, a participant has to produce an interval equal to an interval previously reported (i.e., "Produce 2 seconds"). In the verbal estimation task, after experiencing target duration, the participant has to translate this subjective duration into clock units. An adapted version of experiments utilizing these tasks can be downloaded from the GitHub repository of this book. These two methods involve the comparison of the experienced duration with internal information concerning conventional duration units, such as milliseconds and seconds (i.e., *reference memory*; Zakay, 1990). In both cases, a translation from an objectively labeled duration to a subjectively experienced duration (i.e., time production) and vice versa (i.e., verbal estimation) has to be completed. Time production and verbal estimation are appropriate ways for investigating individual differences related to the internal clock (its speed rate or variables influencing it), given that they best reflect the speed or rate of functioning of an internal clock (Baudouin et al., 2006; Glicksohn, & Hadad, 2011).

In the *time reproduction* tasks, participants are required to reproduce the duration of the temporal interval previously presented. The task is composed of two phases: the encoding and the reproduction phase. First, participants experience the target duration (i.e., encoding phase), and then they are asked to delimit a time interval (by pressing a designed key) equivalent to the target duration previously presented (i.e., reproduction phase; Mioni, Stablum, McClintock, & Grondin, 2014c). An adapted version of such experiments can be downloaded from the GitHub repository of this book. Compared to the production or verbal estimation tasks, a reproduction task is used less in the investigation of individual differences at the internal clock level. In fact, the speed rate of the internal clock is the same when experiencing the target duration and when reproducing it. This method, which relies on a comparison of previously experienced temporal intervals, has the potential disadvantage that it might be mainly an index of the consistency of the subjective time experience, providing no information about variation at the pacemaker rate. Even if the rate of physiological and cognitive processes varies, the same rate will subserve a person's experience of the target duration (encoding phase) and its reproduction (reproduction phase). Thus, the reproduction method may detect individual differences only if it is used in the framework of psychophysical studies, in which duration is varied. In addition, judgments obtained by using the reproduction method (as well as the production method) may be confounded by extraneous variables such as the desire to terminate the experiment sooner, impatience, or the inability to delay a response.

Researchers often use these three methods interchangeably without providing an explanation for the selection of a given method. It is important to remember that each method activates different timing-related processes and presents some specific perceptual errors. For example, participants tested with the verbal estimation method are prone to respond to the estimated duration using rounded numbers and data are often characterized by a great amount of variability compared to the other methods (Grondin, 2010; Zakay, 1990). Time reproduction is considered to be more accurate and reliable than time production and verbal estimation; however, it is less useful for investigating variations in the pacemaker rate. Time production and verbal estimation show more inter-participant variability than time reproduction, but can be successfully used in studies where the rate of the internal pacemaker is manipulated.

Moreover, it is important to consider that each method requires cognitive resources to be processed. In fact, a reproduction task recruits attention and working memory to keep active the reference duration in order to subsequently reproduce it. Conversely, the production and verbal estimation tasks require other cognitive processes such as language processing and access to long-term memory (Block, Zakay, & Hancock, 1998). Time production and reproduction tasks require a motor action[1] to produce and reproduce, respectively, the temporal intervals after the presentation of a given target duration.

## 3 Scoring Methods

Data collected from the reproduction, production, and verbal estimation tasks may be scored in term of the: (1) $S_d/O_d$ ratio, (2) absolute discrepancy ($|S_d-O_d|$), (3) absolute error ($|S_d-O_d|/O_d$), and (4) coefficient of variation ($SD^2/S_d$; Glicksohn & Hadad, 2012; Mioni et al., 2014). $S_d$ represents the subjective duration expressed by the participants and $O_d$ represents the objective target duration presented.

In computing the $S_d/O_d$ ratio, the time estimation is expressed in terms of proportion of physical duration and then the results are compared across different temporal intervals. The absolute discrepancy ($|S_d-O_d|$) reflects the magnitude of temporal error without giving any information regarding the duration (i.e., over- or under-estimation). The absolute error ($|S_d-O_d|/O_d$) may be the most sensitive of all indices and reflects a more generalized disruption of timing (Brown & Boltz, 2002). Finally, the coefficient of variance (CV) is an index of timing variability over a series of trials.

Interestingly, Glicksohn and Hadad (2012) tried to compare and contrast these different indices (with exclusion of CV) while studying sex difference in time perception. The authors only considered the time production task, but the results are a good starting point for extending the conclusions to the time reproduction and verbal estimation methods. Briefly, the results showed that when $S_d/O_d$ ratio is considered, female participants had a lower mean ratio in comparison to the male participants. In contrast, results reported using $|S_d-O_d|$ and $|S_d-O_d|/O_d$ indices were seriously compromised by wide individual differences. Regarding the CV, it is an interesting index when comparing clinical and healthy groups and when group and individual variability is under investigation. With clinical population, higher temporal variability is a robust finding (Cester, Mioni, & Cornoldi, 2017; Jones & Jahanshahi, 2014; Mioni,

1 Verbal speech can also be considered a motor action but here we refer as "motor action" only the context of physical motor action. It includes the preparation and the execution of a physical movement.

2 SD = standard deviation.

Grondin, & Stablum, 2014a; Mioni, Stablum, Prunetti, & Grondin, 2016b) and it has been explained as a manifestation of difficulties in maintaining a stable representation of duration. This difficulty is accentuated in patients due to attentional, working memory, and executive dysfunction, and could be viewed as a manifestation of failure to fully attend to temporal information (Cester et al., 2017; Jones & Jahanshahi, 2014; Mioni, Grondin, & Stablum, 2014a; Mioni, Stablum, Prunetti, & Grondin, 2016b).

What are the benefits of one measure over the other? If multiple temporal intervals are included within the same experimental design $S_d/O_d$ ratio should be used. This index also provides information regarding the direction of duration (over- or under-estimation); if an index of amount of error is needed, the absolute discrepancy ($|S_d\text{-}O_d|$) is more appropriate. The absolute error ($|S_d\text{-}O_d|/O_d$) includes a mix of the other two indices reflecting a more generalized disruption of timing. The CV can be included if different groups (i.e., clinical vs. healthy or children vs. adults) are compared and it is an interesting measure of temporal variability.

## 4 Critical Considerations

Taking into consideration that each method activates different time-related processes (attentional or working memory resources) and activate different representation of time (subjective or objective), one way to select the appropriate method is to take into account other variables involved such as the temporal interval, the motor component, cognitive load, and cognitive strategies as well as the group of participants under investigation. A general consideration is also that, given the tendency we have to round off time estimates with chronometric units, verbal estimations produce more variability and are less accurate (greater discrepancy with respect to the standard duration) than time production and reproduction methods.

### 4.1 *Temporal Range*

Regarding the temporal range, a general tendency in timing literature, mainly in neuroscience researchers, is to emphasize a distinction between intervals above and below 1s, which is based on differential pharmacological effects (Rammsayer, 2008) and on patient studies with various cases of brain damage (see Allman & Meck, 2012; Meck, 2005; Mioni, Grondin, & Stablum, 2014a; Piras et al., 2014). Moreover, researchers claim that the processing of smaller intervals is more sensory-based or benefits from some automatic processing, whereas the processing of longer intervals requires the support of cognitive resources (also see Lewis & Miall, 2003; Mioni, Stablum, & Grondin, 2014b). Even if this '1s' transition period remains somewhat arbitrary, there is certainly some turning point on the time continuum. This is evident considering the benefit observed from adopting explicit counting strategy for processing long temporal intervals (around seconds range) with respect to brief temporal interval (around milliseconds range) (Grondin, Meilleur-Wells, & Lachance, 1999; Grondin, Ouellet, & Roussel, 2004; Mioni, Stablum, & Grondin, 2014b). Indeed, there are empirical reasons to believe that this transition occurs around 1200 ms. Using a time discrimination task,[3] Mioni and colleagues (2014) tested participants with duration ranging from 400 to 1600 ms to investigate if temporal discrimination is influenced by the temporal range and/or by the context.[4] Results suggest that context influences time discrimination performance only when the temporal range under investigation is below 1300 ms and the temporal intervals varied within blocks. In the case of temporal intervals longer than 1300 ms, participants presented a tendency to respond "long" independently of the procedure used. Researchers claimed that processing of smaller intervals is more sensory based, or benefits from some automatic processing, whereas the processing of longer intervals requires the support of cognitive resources (Hellström & Rammsayer, 2004; Lewis & Miall, 2003). This is also confirmed by results observed from patients' studies with brain lesions (see later sections in this chapter and also Allman & Meck, 2012; Jones & Jahanshahi, 2014; Mioni, Grondin, & Stablum, 2014a; Piras et al., 2014).

Therefore, if the theoretical question involves exploring "pure" temporal abilities, without (or with reduced) influence of cognitive processes, brief temporal intervals (below 1200–1300 ms) should be privileged.

### 4.2 *Concurrent Secondary Task*

In most cases, when performing temporal tasks, participants are also engaged in an additional exercise, called the "secondary task". The aim of adding a secondary task is to: (1) avoid counting strategies that are often used when experiencing long temporal intervals (see Section 4.3) and (2) understand the effects of cognitive load on duration judgments. As mentioned before, attention plays a major role in prospective duration experience. The Attentional-Gate Model (Block & Zakay, 1996; Zakay & Block, 1996) proposes that the

---

3 In a time discrimination task, participants are required to judge the relative duration of two time intervals presented successively and indicate whether the second stimulus is presented longer or shorter with respect to the first one.

4 The context effect here refers to the presentation of temporal intervals randomized within or across blocks (see also Jones and McAuley 2005).

experienced duration of a time period depends on the amount of information encoded during the temporal information processor and by a non-temporal information processor. Task demands (variation in the cognitive load) determine the way in which a person divides attention between the two processors. In fact, employing a concurrent secondary task gives the opportunity of testing the effects of variation of cognitive resources on temporal processing (Block, Hancock, & Zakay, 2010). It is thought that when a person is working on a difficult or attention-demanding task, time seems to pass quickly, but if a person is working on an easy or less attentional-demanding task, time seems to pass slowly. Therefore, time processing is adversely affected by the attentional or workload demands of any non-temporal (secondary) task (Brown, 1997, 2008; Brown & Boltz, 2002; Zakay & Block, 2004). It is possible to test the effects of the secondary task by: (1) manipulating the cognitive load of the secondary task by employing easy or more complex secondary tasks (Brown, 1985, 1997), or (2) varying the instruction by asking participants to focus attention mainly on the temporal task, or on the secondary task or both.

Generally, dual-task conditions (temporal task + secondary task) typically cause time judgments to become less accurate than single-task conditions in which participants judge time alone, but the effects also vary depending on the temporal task used. Verbal estimates and reproductions usually decrease (shorter temporal intervals estimated or reproduced) with greater processing demands during a time period, the reverse usually holds for time productions. During time reproduction tasks, temporal reproductions are longer if the timing task is defined as the primary task (attention fully focused on performing the timing task) than if it is defined as the secondary task (attention is focused on the secondary task).

## 4.3 *Counting during Timing Tasks*

Adding a secondary task during the temporal task, not only gives the opportunity to test variations of attention and cognitive resources on time processing, but it also prevents the participants from adopting strategies, such as counting, during time processing (Grondin et al., 1999, 2004). It has been demonstrated that explicit counting improves temporal sensitivity in children as well as in adults (Clément & Rattat, 2006) and also in clinical groups (Perbal, Couillet, Azouvi, & Pouthas, 2003; Perbal, Couillet, Azouvi, & Pouthas, 2005). An explanation of this counting advantage, rely on the knowledge that segmenting information into smaller parts helps the processing of this information. Moreover, counting reduces the contribution of memory to the overall variance in the timing process. This reduction is made possible by the remembering of the number counted rather than by remembering an interval representation.

However, letting participants use strategies during the timing task interferes with the meaning of testing "time". Therefore, it is preferred to not allow participants in adopting counting or any other strategy. One option is to employ brief temporal intervals in which counting is not advantageous (Grondin et al., 1999, 2004). Grondin (2004) showed, in a time discrimination task, benefits from explicit counting at 1.6 s, but not at .8 s, when temporal intervals were marked by auditory as well as visual stimuli (see also Mioni, Stablum, & Grondin, 2014b; Mioni et al., 2016a). Rattat and Droit-Volet (2012) compared the effects of the three classic methods to avoid.counting: (1) instructions not to count, (2) articulatory suppression, and (3) administration of an interference task in temporal generalization, time bisection, and reproduction tasks with two duration ranges (i.e., 1–4 s and 2–8 s). Results showed that all three no-counting conditions prevented participants from counting, and interestingly, the instructions not to count actually constituted the simplest and more efficient method of preventing counting in timing tasks. An alternative way to prevent participants from counting is to employ a motor task (i.e., continuous finger tapping). This method has the advantage to be employed independently of the modality used to present the temporal intervals. In some cases using a counting or articulatory suppression strategies with auditory stimuli might create and additional modality interference (Mioni et al., 2016a).

### 4.4 *Methods for Producing and Reproducing Time*

In the case of time production and reproduction, an additional issue concerns the motor action required to produce or reproduce temporal intervals. Participants need to integrate their motor action in order to produce a precise button press to process the temporal interval (Droit-Volet, 2010; Mioni et al., 2014). Moreover, preparing and executing a motor action requires cognitive resources that might result in additional variance (Caldara et al., 2004). For instance, we can assume that people suffering from deficits in planning, preparation, and execution of motor movements could have poor temporal performance in a temporal reproduction task due to their motor-related deficits rather than to their inability to estimate time (Bloxham, Dick, & Mooret, 1987; Stuss et al., 1989). Mioni and colleagues (2014) compared the effects of three classical methods used to reproduce time with temporal intervals ranging from 1 to 18 s: (1) pressing a designed key at the end of the reproduction, (2) pressing a designed key to start and stop the reproduction, and (3) continuous pressing a designed key to reproduce the duration. Results showed that temporal performance depends on the method employed and on the interval range under investigation. In fact, the second method (i.e., pressing to start and end the reproduction) led to better accuracy ($|S_d\text{-}O_d|$) than did

Methods 1 and 3 with longer durations. However, when short durations were employed (i.e., 1 s), method 2 generated a higher absolute discrepancy ($|S_d$-$O_d|$), indicating lower performance. Participants using method 1 showed good performance only when reproducing 1-s intervals. Of interest are the results obtained with CVs. Although participants using method 3 showed the lowest level of accuracy ($|S_d$-$O_d|$) and under-reproduced temporal durations ($|S_d/O_d|$) more than participants using method 1 or 2, method 3 led to the lowest CVs. Interestingly, pressing continuously to reproduce the duration, generated less variability than the other methods, in particular when participants reproduce longer durations.

### 4.5 *Modality and Stimulus Type*

Finally, when building a timing task the modality and the type of temporal interval should be taken into account. The modality refers to the sensory modality used to present the temporal intervals (visual, auditory, or tactile). It is known that time perception is influenced by the sensory modality used for marking the time intervals (Grondin, 2003). Timing is more precise when stimuli are presented in the auditory rather than the visual modality (Grondin, 2003, 2010; Ulrich, Nitschkle, & Rammsayer, 2006), and this auditory superiority might be due to the automaticity of temporal processing in audition.

Also, the way the temporal intervals are presented influence temporal performance. Temporal intervals can be defined as "filled" or "empty". An interval is defined "filled" when there is one given continuous signal between onset and offset; the signals' onset and offset marks the interval's beginning and end. An "empty" interval is a silent duration, that is, without stimulation, included within two sensory signals that mark the beginning and end of the interval. Generally, duration estimates lengthen if duration is filled as opposed to unfilled. However, the differences between performance with filled and empty intervals are used depends on the type of markers, on the range of duration investigated, and the method employed (Grondin, 2008, 2010).

## 5 Prospective Timing in Clinical Populations

It is useful to bear in mind that there is no human clinical condition that can be defined solely as a disorder of timing and time perception per se. However, distortions in ones' timing ability are present, to varying degrees, in many patient populations, and may or may not accompany differences in other aspects of sensory processing, as well as developmental, cognitive, and behavioural profiles. From a clinical perspective, examinations of the timing ability in patients with certain psychiatric or behavioural disorders may help to ameliorate understanding of the psychological experience of these disorders and their potential remediation. In this regard, temporal distortion in Parkinson's patients (PD) and traumatic brain injury patients (TBI) will be discussed. The selection of these two clinical groups concerns the presentation of two different patterns of temporal dysfunction and will help in the understanding of the source of their temporal misperception. Readers who are more interested in psychopathologies will also find a large number of time-related articles in the neuropsychological or psychiatric literature (Allman & Meck, 2012; Meck, 2005): Developmental findings in healthy (Droit-Volet, 2013) and clinical children (Autism: Allman & Falter, 2015; Attention-deficit hyperactivity disorder: Hart, Radua, Mataix-Cols, & Rubia, 2012; Toplak, Dockstader, & Tannock, 2006), depressed (Thönes, & Oberfeld,2015) and anxious (Mioni et al., 2016a) patients; patients with schizophrenia (Thoenes & Oberfeld, 2017).

### 5.1 *Timing in Parkinson's Patients*

PD is a movement disorder characterized by bradykinesia, tremor, rigidity, and postural instability. Disease symptoms occur as a result of selective, progressive, and chronic degeneration of the nigrostriatal and mesocorticolimbic dopamine systems (Allman & Meck, 2012; Jankovic & Tolosa, 2007; Meck, 2005). Dysfunctions in dopamine transmission have also been identified as the main cause of temporal impairment in PD patients. Evidence suggests that PD patients show dysfunctions in time perception in various temporal tasks (Jones & Jahanshahi, 2014).

PD patients have been tested due to their basal ganglia dysfunction and dopaminergic alteration. Dopaminergic antagonists produce a deceleration of the subjective clock speed (Buhusi & Meck, 2005). These results traditionally elected the dopaminergic system and the basal ganglia as a "house" of the pacemaker-clock system. However, the pathophysiology of PD is complex and evolves in the course of the illness. The basal ganglia seem to be particularly involved in timekeeping functions, and a strict connection is hypothesized between the basal ganglia and cortical structures supporting conscious representation and memory for time. The frontal cortex, mainly the dorsolateral prefrontal cortex, has an established role in a range of cognitive processes, such as working memory and decision-making, which the clock model proposes as necessary for efficient interval timing. Therefore, it can be difficult in PD patients to tease apart whether the timing difficulties are driven by a core timing dysfunction, or by the disruption of general cognitive processes.

When investigating perceptual timing in patients with movement disorders the most effective tasks are those not involving reaction time and responses based on motor components. Time discrimination and time bisection tasks fit these criteria; however, studies have been conducted also with time production, verbal estimation, and time reproduction tasks. Despite the motor limitation (that can be controlled including a simple motor task), results from studies conducted with time reproduction tasks showed both increased and reduced variability and over- and under-reproduction in PD patients compared to controls. It is possible that some of the contradictory findings reported in the literature on timing and time perception with PD patients may be explained by differences in temporal range and stimuli modality.

An interesting phenomenon that is often observed is the "migration effect" (Malapani, Deweer, & Gibbon, 2002; Malapani et al., 1998). When temporal intervals in different temporal ranges are used within the same block, short intervals are over-reproduced and long intervals are under-reproduced (see also the Vierordt's law; Lejeune, & Wearden, 2009). What is the source of this effect in PD patients? Malapani et al. (2002) suggested that the memory for learned durations is the source of the temporal impairment rather than variation at the clock level.

Some other critical factors that have to be taken into account when testing PD patients is the effect of medication and the role of cognitive factors. Regarding the first point, Jones and Jahanshahi (2014) reported that half of the studies that reported a direct comparison between "ON" and "OFF" medication indicated a beneficial effect of medication on perceptual timing. However, some studies have also reported the opposite findings, reflecting a negative effect of dopaminergic remediation of relatively preserved basal ganglia circuits. Moreover, patients can vary in severity and duration of the illness, which are factors that can mediate the medication impact (Merchant et al., 2008).

Finally, regarding the involvement of cognitive function on temporal processing, this leads to an important area of debate. Does the temporal dysfunction observed in PD patients reflect a dysfunction in critical timing regions or it is mediated by global cognitive impairment? Traditionally, the way for testing the extent to which cognitive impairment is correlated with timing performance is to compare performance in neuropsychological and temporal tasks. Even though most studies report a complete neuropsychological evaluation (Jones & Jahanshahi, 2014), very few studies run correlational analyses. Jones et al. (2008) conducted a factor analysis and reported a common factor between measures of attention and time production that was distinct from time reproduction. However, Merchant et al. (2008) found that performance on a range of cognitive tasks did not discriminate those with PD who did well or poorly on a range or motor and perceptual timing tasks. Importantly, in all these studies, the group of PD patients was treated as a whole sample, however, different levels of impairment exist at the individual level. A different approach was followed by Mioni et al. (2015, 2017). That is, by using Litvan's criteria (Litvan et al., 2012), patients with a diagnosis of PD who failed at least two tasks (below two standard deviations) in the same domain or in different domains were defined as PD with mild cognitive impairment (MCI) and tested separately from PD patients without MCI. Results showed that PD-MCI patients were less accurate and more variable with respect to PD-no-MCI and controls that demonstrated similar temporal abilities. It is, therefore, evident, that the level of cognitive function can influence temporal performance in PD patients, but this does not preclude that "pure" clock dysfunction is also present.

### 5.2 *Time in Traumatic Brain Injury Patients*

Neuropathological evidence suggest a marked heterogeneity of injuries across TBI patients. However, it is apparent that diffuse axonal injury is common, and that damage occurs most frequently in the frontal and temporal lobes. Temporal impairments in patients with TBI are expected considering the disruption of cognitive function involved in temporal processing (Mioni, Grondin, & Stablum, 2014a; Piras et al., 2014).

Reviewing the studies conducted to investigate temporal dysfunction in TBI patients (Mioni, Grondin, & Stablum, 2014a), four used a time reproduction task (Meyers & Levin, 1992; Mioni, Stablum, McClintock, & Cantagallo, 2012; Mioni, Mattalia, & Stablum, 2013a; Perbal et al., 2003), three a verbal estimation task (Anderson & Schmitter-Edgecombe, 2011; Meyers & Levin, 1992; Schmitter-Edgecombe & Rueda, 2008), two a time production task (Mioni, Mattalia, & Stablum, 2013a; Perbal et al., 2003), and two a time discrimination task[5] (Mioni, Mattalia, & Stablum, 2013a; Mioni, Stablum, & Cantagallo, 2013b).

The studies conducted with the time reproduction task showed that TBI patients were as accurate as controls ($S_d/O_d$ ratio) and showed higher variability (CV) in their timing judgments, indicating dysfunction in maintaining a stable representation of temporal intervals. In the study conducted by Perbal et al. (2003), participants were also asked to perform a secondary (non-temporal) task together with the time reproduction task. Similar under-reproduction was observed in TBI patients and controls in both simple (time reproduction only) and concurrent (time reproduction + non-temporal task) conditions, in particular under the latter condition.

5 In time discrimination tasks, participants are required to compare the relative duration of two intervals that are sequentially presented (standard-comparison) and then judge, which one was longer or shorter. Since this chapter does not cover this method, please refer to the original papers for a detailed presentation of the results (Mioni et al. 2013a, 2013b).

The studies conducted with a time production task confirmed and extended the results obtained with the time reproduction task. Again, TBI patients were as accurate as controls ($S_d/O_d$ ratio) and showed higher temporal variability (CV; Mioni et al., 2013b; Perbal et al., 2003). Regarding the impact of a concurrent non-temporal task, no effect was found (time production only vs. time production + non-temporal task) and this finding applies to both groups (i.e., TBI and controls). TBIs and controls showed the same performances ($S_d/O_d$ ratio and CV) in both simple and concurrent conditions (Perbal et al., 2003).

Three studies were conducted with a verbal estimation task but performance was only analysed in two of them. Indeed, in Meyers and Levin's (1992) study, performance at a verbal estimation task was not analysed due to the extreme variability noted in the TBI sample. Schmitter-Edgecombe and Rueda (2008), as well as Anderson and Schmitter-Edgecombe (2011), reported lower accuracy ($|S_d\text{-}O_d|$), higher under-estimation ($S_d/O_d$ ratio), and more variability (i.e., CV) in TBI patients than controls.

In brief, TBI patients and controls showed similar performances ($|S_d\text{-}O_d|$ or $S_d/O_d$ ratio) when time reproduction and time production tasks are employed. However, TBI patients performed less accurately than controls when verbal estimation was used. Moreover, in all studies, variability is higher with TBI patients than with controls.

In sum, the revision of the existing literature investigating time perception in TBI patients showed that temporal dysfunctions in TBI patients were related to deficits in cognitive function involved in temporal processing such as working memory, attention, and executive function rather than an impairment in time estimation per se. In fact, temporal dysfunctions were observed when the tasks employed required high cognitive functions to be performed (Mioni, Mattalia, & Stablum, 2013a; Mioni, Stablum, & Cantagallo, 2013b). The consistent higher temporal variability observed is a sign of impaired frontally mediated cognitive functions that affect temporal representation.

## 6 Conclusions and Practical Considerations

This chapter provided a general overview of the typical perceptual timing tasks used under the prospective paradigm. The tasks that have been described were the time production, verbal estimation, and time reproduction tasks. Each of these tasks highlights different temporal characteristics and requires different cognitive processes. Briefly, time production and verbal estimation tasks are suitable methods for investigating individual differences in the speed rate of the internal clock, while time reproduction is useful if it is used in the framework of psychophysical studies where duration is varied.

Different factors should be taken into account when selecting a timing task. First of all, the temporal range is critical. Brief temporal ranges should be preferred to reduce additional cognitive factors that are required when performing long temporal intervals (i.e., attention and working memory). Moreover, a motor component can alter temporal performance in time production and time reproduction, when brief temporal intervals are employed. If longer temporal intervals are used (around 1.2 s), participants have the natural tendency to use strategies (i.e., counting) during the temporal task. Segmenting a long temporal interval using counting improves temporal performance, however in this case the results can be less interpreted as an index of temporal ability and more as an index of counting. To prevent participants from counting, various methodologies can be used, and simply asking participants not to count has been demonstrated as a good method to eliminate counting. Finally, this chapter briefly presented a review of the main findings observed in timing literature when PD and TBI patients are tested with some specific clinical and methodological suggestions.

## References

Allman, M., & C. Falter (2015). Abnormal timing and time perception in autism spectrum disorder? A review of the evidence. In Vatakis, A., & M.J. Allman (Eds.). *Time distortions in mind-temporal processing in clinical populations* (pp. 37–56). Leiden, The Netherlands: Brill Academic Publishers

Allman, M.J., & W.H. Meck (2012). Pathophysiological distortions in time perception and timed performance. *Brain, 135,* 656–677.

Anderson, J.W., & M. Schmitter-Edgecombe (2011). Recovery of time estimtion following moderate to severe traumatic brain injury. *Neuropsychology, 25,* 36–44.

Baudouin, A., S. Vanneste, M. Isingrini, & V. Pouthas (2006). Differential involvement of internal clock and working memory in the production and reproduction of duration: A study on older adults. *Acta Psychologica, 121,* 285–296.

Block, R.A. (1990). Models of psychological time. In Block, R.A. (Ed.), *Cognitive models of psychological time* (pp. 1–35). Hillsdale, NJ: Lawrence Erlbaum Associates Inc.

Block, R.A., & D. Zakay (1996). Models of psychological time revised. In Helfrich, H. (Ed.), *Time and mind* (pp. 171–195). Kirkland, WA: Hogrefe and Huber.

Block, R.A., D. Zakay, & P.A. Hancock (1998). Human aging and duration judgments: A meta-analytic review. *Psychology & Aging, 13,* 584–596.

Block, R.A., P.A. Hancock, & D. Zakay (2010). How cognitive load affects duration judgments: A meta-analytic review. *Acta Psychologica, 134*, 330–343.

Bloxham, C.A., D.J. Dick, & J.M. Mooret (1987). Reaction times and attention in Parkinson's disease. *Journal of Neurology, Neurosurgery, and Psychiatry, 50*, 1178–1183.

Brown, S.W. (1985). Time perception and attention: The effects of prospective versus retrospective paradigms and task demands on perceived duration. *Perception & Psychophysics, 38*(2), 115–124.

Brown, S.W. (1997). Attentional resources in timing: Interference effects in concurrent temporal and nontemporal working memory tasks. *Perception & Psychophysics, 59*, 1118–1140.

Brown, S.W. (2008). Time and attention: Review of the literature. In Grondin, S. (Ed.), *Psychology of Time* (pp. 111–138). Bingley, England: Emerald.

Brown, S.W., & M.G. Boltz (2002). Attentional processes in time perception: Effects of mental workload and event structure. *Journal of Experimental Psychology: Human Perception and Performance, 28*, 600–615.

Buhusi, C.V., & W.H. Meck (2005). What makes us stick? Functional and neural mechanisms of interval timing. *Nature Review Neuroscience, 6*, 755–765.

Caldara, R., M.-P. Deiber, C. Andrey, G.M. Michel, G. Thut, & C.-A. Hauert (2004). Actual and mental motor preparation and execution: A spatiotemporal ERP study. *Experimental Brain Research, 159*, 389–399.

Cester, I., G. Mioni, & C. Cornoldi (2017). Time processing in children with mathematical difficulties. *Learning and Individual Differences, 58*, 22–30.

Clément, A., & S. Droit-Volet (2006). Counting in a time discrimination task in children and adults. *Behavioural Processes, 71*, 164–171.

Droit-Volet, S. (2010). Stop using time reproduction tasks in a comparative perspective without further analyses of the role of the motor response: The example of children. *European Journal of Cognitive Psychology, 22*(1), 130–148.

Droit-Volet, S. (2013). Time perception in children: A neurodevelopmental approach. *Neuropsychologia, 51*(2), 220–234.

Droit-Volet, S., & W.H. Meck (2007). How emotions colour our perception of time. *Trends in Cognitive Sciences, 11*(12), 504–513.

Fraisse, P. (1984). Perception and estimation of time. *Annual Review of Psychology, 35*, 1–36.

Gibbon, J. (1977). Scalar expectancy theory and Weber's law in animal timing. *Psychological Review, 84*, 279–385.

Gibbon, J., R.M. Church, & W.H. Meck (1984). Scalar timing in memory. In Gibbon, J. & L.G. Allan (Eds.), *Timing and time perception* (pp. 52–77). New York: The New York Academy of Science.

Glicksohn, J., & Y. Hadad (2011). Sex differences in time production revisited. *Journal of Individual Differences, 33*(1), 35–42.

Grondin, S. (2003). Sensory modalities and temporal processing. In Helfrich, H. (Ed.), *Time and mind II* (pp. 61–77). Göttingen: Hogrefe & Huber.

Grondin, S. (2008). Methods for studying psychological time. In Grondin, S. (Ed.), *Psychology of time* (pp. 51–74). Bingley, UK: Emerald Group Publishing.

Grondin, S. (2010). Timing and time perception: A review of recent behavioral and neuroscience findings and theoretical directions. *Attention, Perception & Psychophysics, 72*(3), 561–582.

Grondin, S., G. Meilleur-Wells, & R. Lachance (1999). When to start explicit counting in a time-intervals discrimination task: A critical point in the timing process of humans. *Journal of Experimental Psychology: Human, Perception and Performance, 25*, 993–1004.

Grondin, S., B. Ouellet, & M.-E. Roussel (2004). Benefits and limits of explicit counting for discriminating temporal intervals. *Canadian Journal Experimental Psychology, 58*, 1–12.

Hart, H., J. Radua, D. Mataix-Cols, & K. Rubia (2012). Meta-analysis of fMRI studies of timing in attention-deficit hyperactivity disorder (ADHD). *Neuroscience & Biobehavioral Reviews, 36*(10), 2248–2256.

Hellström, Å., & T.H. Rammsayer (2004). Effects of time-order, interstimulus interval, and feedback in duration discrimination of noise bursts in the 50- and 1000-ms ranges. *Acta Psychologica, 116*, 1–20.

Jankovic, J., & E. Tolosa (Eds.) (2007). *Parkinson's disease and movement disorders.* Philadelphia: Lippincott Williams

Jones, C.R.G., & M. Jahanshahi (2014). Motor and perceptual timing in Parkinson's disease. In Merchant, H. & V. de Lafuente (Eds.), *Neurobiology of interval timing. Advances in experimental medicine and biology* (pp. 265–290). New York: Springer Verlag.

Jones, M.R., & J.D. McAuley (2005). Time judgments in global temporal contexts. *Perception & Psychophysics, 67*, 398–417.

Jones, C.R.G., T.J. Malone, G. Dirnberger, M. Edwards, & M. Jahanshahi (2008). Basal ganglia, dopamine and temporal processing: Performance on three timing tasks on and off medication in Parkinson's disease. *Brain & Cognition, 68*(1), 30–41.

Karaminis, T., G.M. Cicchini, L. Neil, G. Cappagli, D. Aagten-Murphy, D. Burr, & E. Pellicano (2016). Central tendency effects in time interval reproduction in autism. *Scientific Reports, 28*(6), 28570.

Lejeune, H., & J.H. Wearden (2009). Vierordt's the experimental study of the time sense (1868) and its legacy. *European Journal of Cognitive Psychology, 21*, 941–960.

Lewis, P.A., & R.C. Miall (2003). Distinct systems for automatic and cognitively controlled time measurement: Evidence from neuroimaging. *Current Opinion in Neurobiology, 13*, 250–255.

Litvan, I., J.G. Goldman, A.I. Tröster, B.A. Schmand, D. Weintraub, R.C. Petersen, B. Mollenhauer, C.H. Adler, K. Marder, C.H. Williams-Gray, D. Aarsland, J. Kulisevsky, M.C. Rodriguez-Oroz, D.J. Burn, R.A. Barker, & M. Emre (2012). Diagnostic criteria for mild cognitive impairment in Parkinson's disease: Movement disorder society task force guidelines. *Movement Disorders, 27*(3), 349–356.

Malapani, C., B. Rakitin, R. Levy, W.H. Meck, B. Deweer, B. Dubois, & J. Gibbon (1998). Coupled temporal memories in Parkinson's disease: A dopamine-related dysfunction. *Journal of Cognitive Neuroscience, 10*(3), 316–331

Malapani, C., B. Deweer, & J. Gibbon (2002). Separating storage from retrieval dysfunction of temporal memory in Parkinson's disease. *Journal of Cognitive Neuroscience, 14*(2), 311–322.

Meck, W.H. (2005). Neuropsychology of timing and time perception. *Brain & Cognition, 58*, 1–8.

Merchant, H., M. Luciana, C. Hooper, S. Majestic, & P. Tuite (2008). Interval timing and Parkinson's disease: Heterogeneity in temporal performance. *Experimental Brain Research, 184*(2), 233–248.

Meyers, C.A., & H.S. Levin (1992). Temporal perception following closed head injury: Relationship of orientation and attention span. *Neuropsychiatry, Neuropsychology, and Behavioral Neurology, 1*, 28–32.

Mioni, G., F. Stablum, M.S. McClintock, & A. Cantagallo (2012). Time-based prospective memory in severe traumatic brain in jury patients: The involvement of executive functions and time perception. *Journal of International Neuropsychology Society, 18*, 697–705.

Mioni, G., G. Mattalia, & F. Stablum (2013a). Time perception in severe traumatic brain injury patients: A study comparing different methodologies. *Brain & Cognition, 81*, 305–312.

Mioni, G., F. Stablum, & A. Cantagallo (2013b). Time discrimination in traumatic brain injury patients. *Journal of Clinical Experimental Neuropsychology, 35*, 90–102.

Mioni, G., S. Grondin, & F. Stablum (2014a). Temporal dysfunction in traumatic brain injury patients: Primary or secondary impairment? *Frontiers in Human Neuroscience, 8*(269).

Mioni, G., F. Stablum, & S. Grondin (2014b). Interval discrimination across different duration ranges with a look at spatial compatibility and context effects. *Frontiers in Psychology, 5*(717).

Mioni, G., F. Stablum, S.M. McClintock, & S. Grondin (2014c). Different methods for reproducing time, different results. *Attention, Perception & Psychophysics, 76*, 675–681.

Mioni, G., L. Meligrana, S. Grondin, F. Perini, L. Bartolomei, & F. Stablum (2015). Effects of emotional facial expression on time perception in patients with Parkinson's disease. *Journal of the International Neuropsychological Society, 22*(9), 890–899.

Mioni, G., M. Grassi, V. Tarantino, F. Stablum, S. Grondin, & P.S. Bisiacchi (2016a). The impact of a concurrent motor task on auditory and visual temporal discrimination tasks. *Attention, Perception & Psychophysics, 78*, 742–748.

Mioni, G., F. Stablum, E. Prunetti, & S. Grondin (2016b). Time perception in anxious and depressed patients: A comparison between time reproduction and time production tasks. *Journal of Affective Disorders, 196*, 154–163.

Mioni, G., S. Grondin, L. Meligrana, F. Perini, L. Bartolomei, & F. Stablum (2017). Effects of happy and sad facial expressions on the perception of time in Parkinson's disease patients with mild cognitive impairment. *Journal of Clinical and Experimental Neuropsychology, 22*, 1–16.

Perbal, S., J. Couillet, P. Azouvi, & V. Pouthas (2003). Relationship between time estimation, memory, attention, and processing speed in patients with severe traumatic brain injury. *Neuropsychologia 41*, 1599–1610.

Perbal, S., B. Deweer, B. Pillon, M. Vidailhet, B. Dubois, & V. Pouthas (2005). Effects of internal clock and memory disorders on duration reproductions and duration productions in patients with Parkinson's disease. *Brain & Cognition, 58*(1), 35–48.

Piras, F., F. Piras, V. Ciullo, E. Danese, C. Caltagirone, & G. Spalletta (2014). Time dysperception perspective for acquired brain injury. *Frontiers in Neurology, 4*, 217.

Rammsayer, T.H. (2008). Neuropharmacological approaches to human timing. In Grondin, S. (Ed.), *Psychology of Time* (pp. 295–320). Bingley: Emerald Group.

Rattat, A.C. & S. Droit-Volet (2012). What is the best and easiest method of preventing counting in different temporal tasks? *Behavioural Research Methods, 44*(1), 67–80.

Schmitter-Edgecombe, M., & A.D. Rueda (2008). Time estimation and episodic memory following traumatic brain injury. *Journal of Clinical Experimental Neuropsychology, 30*, 212–223.

Stuss, D.T., L. Stethem, H. Hugenholtz, T. Picton, J. Pivik, & M.T. Richard (1989). Reaction time after head injury: Fatigue, divided and focused attention, and consistency of performance. *Journal of Neurology, Neurosurgery, and Psychiatry, 52*, 742–748.

Thoenes, S., & D. Oberfeld (2017). Meta-analysis of time perception and temporal processing in schizophrenia: Differential effects on precision and accuracy. *Clinical Psychology Review, 54*, 44–64.

Thönes, S., & D. Oberfeld (2015). Time perception in depression: A meta-analysis. *Journal of Affective Disorders, 175*, 359–372.

Toplak, M.E., C. Dockstader, & R. Tannock (2006). Temporal information processing in ADHD: Findings to-date and new methods. *Journal of Neuropsychological Methods, 15*, 15–29.

Triesman, M. (1963). Temporal discrimination and the indifference interval. Implication for model of the "internal clock". *Psychological Monographs, 77*, 1–31.

Ulrich, R., J. Nitschke, & T. Rammsayer (2006). Crossmodal temporal discrimination: Assessing the predictions of a general pacemaker-counter model. *Perception & Psychophysics, 68*, 1140–1152.

Zakay, D. (1990). The evasive art of subjective time measurement: Some methodological dilemmas. In Block, R.A. (Ed.), *Cognitive models of psychological time* (pp. 59–84). Lawrence Erlbaum Associates.

Zakay, D. (1993). Time estimation methods: Do they influence prospective duration estimates? *Perception, 22*, 91–101.

Zakay, D., & R.A. Block (1996). The role of attention in time estimation processes. In Pastor, M.A. & J. Artieda (Eds.), *Time, internal clocks and movement* (pp. 143–163). Amsterdam: Elsevier.

Zakay, D., & R.A. Block (2004). Prospective and retrospective duration judgments: An executive-control perspective. *Acta Neurobiologiae Experimentalis, 64*, 319–328.

CHAPTER 5

# Duration Bisection: A User's Guide

*Trevor B. Penney and Xiaoqin Cheng*

## 1 Introduction

Duration bisection is a prospective, perceptual timing task. Prospective because the participant is aware in advance that duration judgments will be required (Chapters 2 and 4, this volume) and perceptual because the participant is instructed to classify the presentation durations of probe stimuli relative to a standard or standards rather than to produce or reproduce a given duration. The major advantage of prospective timing tasks is that multiple trials can be conducted during a test session, thereby allowing robust estimates of timing behavior and, equally important, psychophysical analysis. A major advantage of duration perception tasks is that the motor response does not contribute to the estimate of timing accuracy or variability.

Duration bisection has seen extensive use since its introduction in modern form by Church and Deluty (1977). Although originally applied to non-human animals (i.e., rats), following early work by Allan and Gibbon (1991) and Wearden (1991), the task has proven popular for use in human studies. Indeed, since 1991 more than 90 articles reporting data from the duration bisection task have been published (see Appendix – Table 5.1). These data come from participants ranging in age from the very young (e.g., Droit-Volet & Wearden, 2001) to the very old (e.g., Lustig & Meck, 2011), and although the preponderance of work has been in samples of healthy individuals, typically college students, numerous studies have applied the bisection task to clinical samples (e.g., Melgire et al., 2005; Nichelli et al., 1996). In some cases, the task has been used to address fundamental questions about the perceptual, cognitive, and neural mechanisms that underlie interval timing (e.g., How does subjective time scale with objective time?; Church & Deluty, 1977), whereas other studies have used the task to elucidate whether and how interval timing interacts with other perceptual and cognitive processes (e.g., Droit-Volet, Fayolle, & Gil, 2016).

In this chapter, we provide a brief overview of the duration bisection task, describe the analysis approaches used for duration bisection data, and conclude the chapter with some guidelines for use. Basic stimulus presentation programs and analysis code, written in MATLAB and R, are available at the book's GitHub repository.

 | DOI 10.1163/9789004280205_006

## 2 Duration Bisection – Origins

In an early study of temporal discrimination in non-human animals (Cowles & Finan, 1941), rats received reinforcement for running down one alleyway of a Y shaped discrimination box after a 10 s restraint period and the other alleyway after 30 s of restraint. After 600 trials of training, six of nine animals showed evidence of having learned the temporal discrimination. Although this experiment demonstrated that (some) rats could learn to discriminate the durations, it did not speak to the limits of temporal discrimination, nor to the processes underlying it. Later the same decade, Heron (1949) used a similar temporal restraint design, but a different maze layout, in an "*attempt to determine the differential limen for temporal discrimination*". Eleven rats were initially trained to discriminate 5 and 45 s and then progressed to more difficult discriminations. Eight rats learned to discriminate 5 vs. 35 s and 5 vs. 25 s, but only one learned to discriminate 5 vs. 10 s.

Approximately twenty years later, Stubbs (1968) reported a procedure that examined timing of a range of durations during a single test session. Specifically, he reinforced pigeons for responding on one keylight for durations from 1 to 5 s and a second keylight for durations from 6 to 10 s (Experiment 1). Subsequently, Stubbs (1976) developed a task that measured the bird's subjective experience of time by not reinforcing intermediate durations. Pigeons controlled the illumination color (green or red) of a reinforcement key by pecking on a separate changeover key. Whether pecking on the reinforcement key was rewarded depended on the key's color and how long it had been illuminated. For example, in one condition, reinforcement followed pecks on the green key after 2 to 4 s of illumination and pecks on the red key after 60 to 64 s of illumination. However, pecking during the period from 4 to 60 s was not reinforced. The critical measure of interest was when pigeons switched the color of the reinforcement key from green to red by pecking on the changeover key, thereby changing the response from the short interval key (green) to the long interval key (red).

## 3 Duration Bisection – Modern Form

Church and Deluty (1977) reported the first use of the duration bisection task in what is now its most common form. In an initial training phase, rats received reinforcement for pressing one lever following presentation of a *short* duration stimulus (short *anchor*) and a second lever following presentation of a *long* duration stimulus (long *anchor*). In the test phase, these short and long anchor durations were presented and correct responses reinforced, but unreinforced durations intermediate to the short and long anchors were also presented. Hence, classification of the intermediate durations provided information about the animal's subjective judgment of time.

The human version of duration bisection is similar. In the test phase, the anchor durations and intermediate durations are presented, but feedback is provided for the anchor durations only, which ensures that the task measures subjective perception of duration. However, the training/anchor learning phase is not as extensive for human participants. It usually involves a relatively limited number of presentations of each of the anchor durations only (i.e., 4 to 10), with each presentation followed by feedback indicating whether it was the long or short anchor, and a single training session prior to testing rather than multiple sessions.

It is worth noting that there is a variant of duration bisection in which participants are not explicitly trained on the anchor durations. Rather, in this *partition* version of the duration bisection task for humans (e.g., Droit-Volet & Rattat, 2007; Wearden & Ferrara, 1995; Wiener et al., 2014) participants are merely asked to classify each presented stimulus as short or long. Importantly, after the participants have had some experience with the range of probe durations, the response functions calculated from subsequent trials in the test session are sigmoidal. This indicates that participants are able to categorize the probe durations appropriately in the absence of explicit training with the anchor durations.

In animal studies, the anchor durations typically comprise 50% of the trials (25% each for short and long), whereas in human studies all durations, whether anchor or intermediate, are usually presented an equal number of times. Notably, the presentation probability of the anchor durations does affect stimulus classification (Akdoğan & Balcı, 2016). For example, when fewer short than long anchor stimuli are presented in the test phase, mice are more likely to classify a given probe duration as long as compared to when the short and long anchors are presented with equal frequency during the test session. Moreover, in animal work, correct classification of the anchor durations in the test phase is normally reinforced on less than 100% of the correct trials. This ensures that animals continue to respond on the intermediate probe trials, which are never reinforced, rather than learning to discriminate reinforced anchor durations from unreinforced probe durations (i.e., learning not to respond on the intermediate probes as they do not lead to reward). Indeed, with extensive training and 100% reinforcement on the anchor durations, rats learn not to respond following intermediate probe durations (Brown et al., 2011).

In contrast, human participants are usually given feedback on all anchor duration trials, unless the researcher has a specific reason for omitting feedback (e.g., Wiener et al., 2014). Human participants are also frequently told to consider the classification task as indicating whether the probe duration is "closer to" short or long judgment rather than as an endorsement that the probe duration is exactly the same as the short or long exemplar.

Neither human nor non-human animal participants are normally reinforced for classifications of the intermediate probe durations (but see Kim et al., 2013, and Ward et al., 2009, for exceptions). Consequently, the classification is a measure of the participant's subjective judgment of time rather than a test of whether the participant remembers being reinforced for making a particular response for a particular duration.

## 4 Data Presentation

Data from the duration bisection task are most often plotted as the probability with which a given test duration is classified as 'long', although in some early papers the 'short' classification probability was plotted (e.g., Gibbon, 1981). If the participant has learned the temporal discrimination, the probability of a long response should be zero or close to zero for the short anchor duration and 100% or close to 100% for the long anchor duration. Due to variance in timing, classification of the intermediate probes as 'long' increases relatively smoothly with increasing duration (see Figure 5.1). That is to say, the psychometric response function usually has the form of a sigmoidal function rather than a step function. Absent timing variation, a step function should manifest because all durations below some magnitude would be consistently classified as short on all trials and all durations above that magnitude would be classified as long on all trials.

The parameters of interest that can be derived from the psychometric response function include the bisection point (BP; also known as the Point of Subjective Equality or PSE), the difference limen (DL), and the Weber fraction (WF; also known as the Weber ratio or WR). The BP is the duration value at which the participant is equally likely to classify the stimulus (i.e., test duration) as short or long. The BP can be determined in a variety of ways from the response function. For example, Church & Deluty (1977) fitted a straight line to the three most central durations in the range presented to participants to determine the duration corresponding to 50% long classifications. Fitting a mathematical function (e.g., sigmoidal, pseudo-logistic, Weibull) and determining the BP from that fitted function is also a common

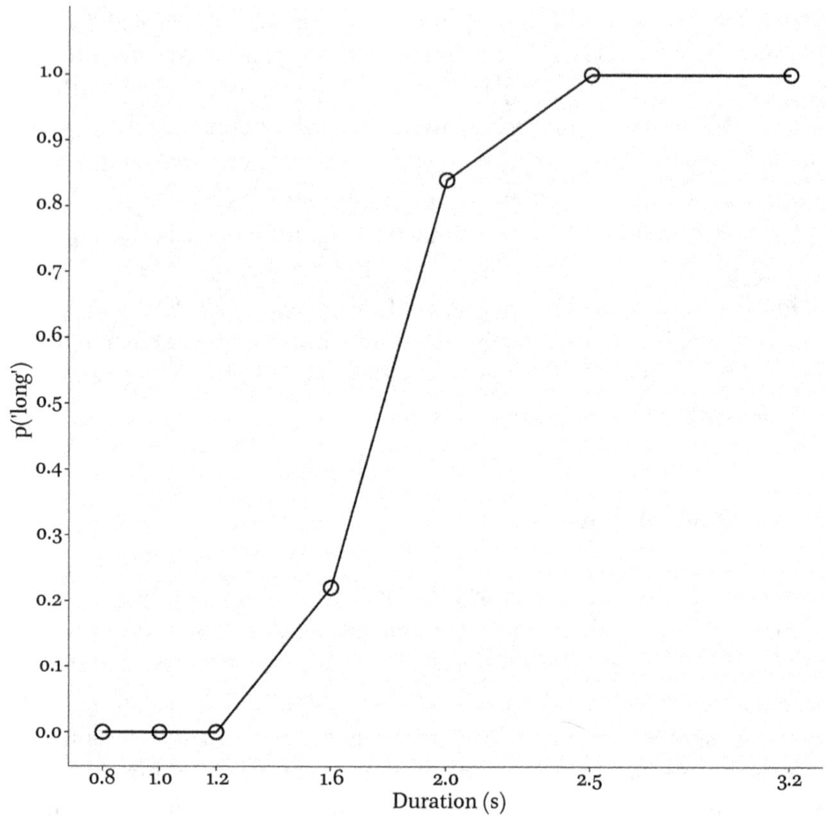


FIGURE 5.1 *Mean proportion of 'long' responses by a single human participant trained on anchor durations of .8 s and 3.2 s. The psychometric response function shows the typical sigmoidal form obtained with the duration bisection task.*

approach. When the psychometric response functions are smooth, the method of determination of the BP is unlikely to greatly influence the values obtained (Figure 5.2). The maximum smoothness or resolution of the response function generated for individual participants is a consequence of the number of trials presented at each probe duration (i.e., 10 trials for each probe duration equals p('long') steps of .1). Hence, ensuring presentation of an adequate number of probe trials is important, but it is also important to ensure that the experimental session is not so long that performance declines due to participant fatigue and boredom.

The steepness of the psychometric response function reflects temporal sensitivity and can be characterized by the difference limen (DL), with smaller DLs (steeper slopes) indicating greater temporal sensitivity. The DL is usually defined as one half of the difference between the duration corresponding to a p('long') of 75% and the duration corresponding to a p('long') of 25%. The

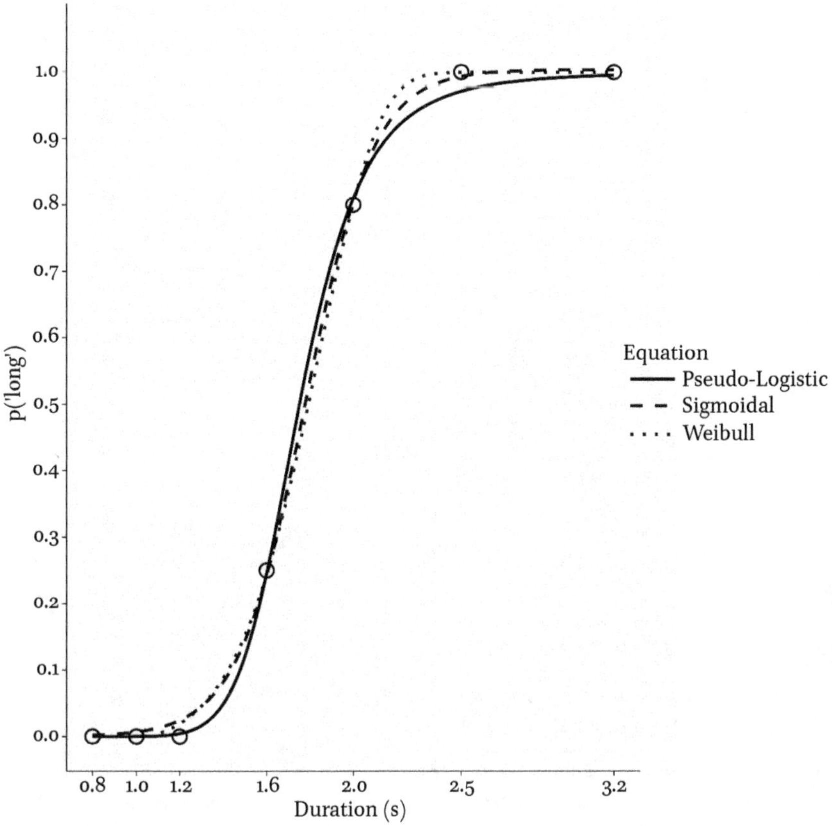


FIGURE 5.2 *Fits of pseudo-logistic, sigmoidal, and Weibull functions to duration bisection data of a single subject.*

75% and 25% p('long') durations also can be determined in several different ways. Church and Deluty (1977) used the same fitted line they used to determine the BP (see above), but other fitted functions are equally appropriate. The WF, which is obtained by dividing the DL by the BP, provides a measure of temporal sensitivity corrected for the magnitude of the durations used in that experimental condition. If Weber's Law holds for time, then the WF should be constant for different anchor duration pairs (Gibbon, 1981) and the psychometric response functions should superimpose (see Figure 5.3).

## 5 Location of the Bisection Point (BP)

The location of the BP is important theoretically because it constrains the models that can account for bisection performance. For example, whether

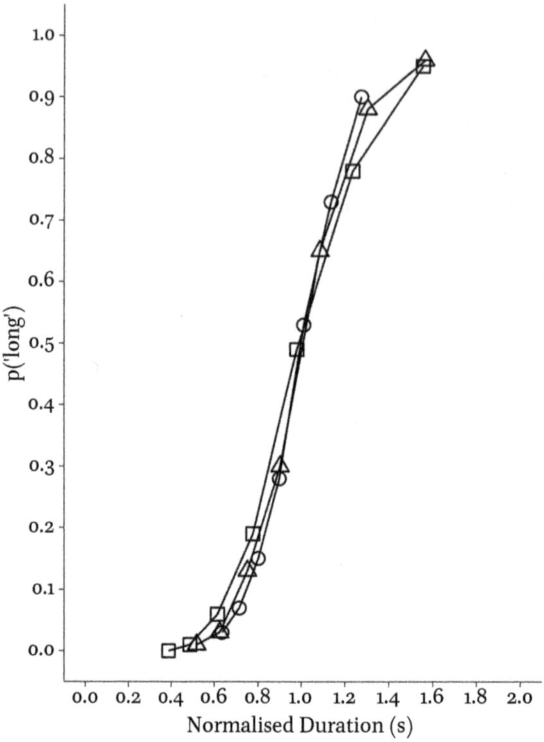


FIGURE 5.3 *Group mean psychometric response functions from three long (L)/short (S) anchor duration ratios following normalization by the group function bisection point. Data replotted from the 'No Distracter' condition in Experiment 2 of Penney, Yim, and Ng (2014).*

bisection occurs at the geometric mean (GM) or the arithmetic mean (AM) has been interpreted as an indicator of whether the temporal scale is linear or logarithmic and whether the response decision rule operates on ratios or differences between the currently elapsing time on a trial and memory representations of previously reinforced durations. Church and Deluty (1977) presented probe durations at the GM, the harmonic mean, and the AM of the anchor durations with the goal of determining the location of classification indifference in rats. They found that across four pairs of short (S)/ long (L) durations (i.e., 1s:4s, 2s:8s, 3s:12s, 4s:16s), the BP was closer to the GM than either the AM or the harmonic mean. The authors interpreted this as indicating that the underlying temporal scale was logarithmic, rather than linear. The basic idea being that if animals select a response lever based on the arithmetic difference between the current duration and the memory representations of the S and L durations, then the midpoint will be at the GM when the time scale is logarithmic.

However, they also noted that a linear timescale with a different decision rule could result in bisection at the GM rather than the AM.

Gibbon (1981) provided a detailed analysis of the combination of timescale and decision processes that can account for the form and location of the duration bisection psychometric function. He noted that Weber's Law for time requires that "*the discriminability of two durations remains constant at constant ratios of these durations, regardless of their absolute durations*" and that "*the subjective "middle" between two time values appears to lie at the geometric mean of these values*" (p. 59). Accordingly, different probe durations, $T_1$ and $T_2$, from two different test conditions will be classified 'short' with the same probability when the T/S ratio is equivalent in both conditions and the S/L ratio is constant. For example, if the 'short' report probability is .75 after 3 s has elapsed when S is 2 s and L is 8 s, then we would expect the short report probability to be .75 at 6 s when the S and L durations are 4 and 16 s, respectively. Indeed, Church and Deluty (1977) found that the Weber Fraction was constant across anchor duration conditions that had GMs of 2, 4, 6, and 8 s, indicating that timing variability scaled with the duration being timed. Although Gibbon's analyses ruled out certain combinations of timescale and decision process (e.g., Poisson timing with a likelihood decision rule and Scalar timing with a likelihood decision rule), they did not allow him to discriminate among log timing with a likelihood decision rule, log timing with a similarity decision rule, and Scalar timing with a similarity decision rule. That said, subsequent work by Gibbon and colleagues generally modeled duration bisection as a Scalar timing process with a similarity decision rule.

Although early experiments in rats and pigeons (e.g., Church & Deluty, 1977; Platt & Davis, 1983), as well as some work in humans (Allan & Gibbon, 1991), indicated that the BP was closer to the GM than the AM, this result has not been found consistently for human participants. Indeed, bisection at or close to the AM of the short and long anchor durations (e.g., Wearden, 1991; Wearden & Ferrara, 1995) is found as frequently, if not more frequently, than bisection at or close to the GM. It is clear from the literature that various experimental design factors, such as stimulus spacing (Wearden & Ferrara, 1995) and stimulus range (Wearden & Ferrara, 1996), have a substantial effect on the location of the BP. Indeed, studies reporting GM bisection in humans have typically used logarithmic spacing of probe stimuli whereas those reporting AM bisection have typically used linear spacing. Indeed, Allan (2002) advised that because the BP was susceptible to bias due to paradigm features such as stimulus spacing, L/S ratio, and duration range, it should not be interpreted as indicating the time value that is equally confused with the S and L anchor durations. We briefly describe the effects of stimulus spacing and stimulus range below.

### 5.1 *Stimulus Spacing*

In rats, Church and Deluty (1977) failed to obtain evidence for an effect of stimulus spacing on the location of the BP with L/S pairs of 4:1. However, Raslear (1983, 1985), also with rats, found a significantly larger BP for linear as compared to logarithmic spacing with an L/S ratio of 100:1, but no spacing effects for less extreme L/S ratios.

Although Allan and Gibbon (1991) found the BP was closer to the GM than the AM for both arithmetic (Exp. 1) and logarithmic spacing (Exp. 2) conditions in humans, the same year Wearden (1991) reported duration bisection at the AM for conditions using 200 vs. 800 ms and 100 vs. 900 ms anchor duration pairs. Subsequently, Wearden and Ferrara (1995) explicitly compared logarithmic and linear spacing. They found leftward shifts in the response functions for logarithmic as compared to linear spacing at both duration ranges tested (Exp. 1: 200 vs. 800 ms and 100 vs. 900 ms). In a second experiment, they used unequal arithmetic spacing, such that in some conditions there was a larger number of shorter than longer durations. Response functions in these conditions were shifted to the left compared to those conditions with a larger number of longer than shorter durations.

Wearden, Rogers, and Thomas (1997) used longer durations (1 vs. 4 s and 2 vs. 8 s), but included a concurrent task to prevent counting (verbally shadowing visually presented digits). Response functions appeared leftward shifted in the logarithmic spacing condition as compared to linear spacing, but there was no statistical difference between the BPs. They also examined L/S ratios of 2:1 and 5:1, but found little effect on the location of the BP, although the BP was closer to the AM than the GM in all conditions. Interestingly, participants showed greater timing sensitivity for the more difficult L/S ratios (see Kopec & Brody, 2010, for a brief review).

Wearden and Ferrara (1995) proposed that stimulus spacing affects the location of the BP because participants calculate the midpoint of the distribution of probe durations presented in the test phase and decide whether to respond short or long based on the magnitude of the current duration relative to the midpoint. The midpoint of a logarithmically spaced S/L range will be smaller than the midpoint of a linearly spaced S/L range. For example, the midpoint of a log spaced 2:8 s range (i.e., 2, 2.52, 3.17, 4.00, 5.04, 6.35, and 8.00) is 4.44, whereas the midpoint of a linearly spaced 2:8 s range (i.e., 2, 3, 4, 5, 6, 7, 8) is 5.00. Hence, the psychometric response functions will differ between these two spacing conditions. Although this explanation accounts well for some spacing effects, it doesn't provide an explanation for all findings. As noted above, the L/S ratio impacts the presence or absence of stimulus spacing effects (Raslear, 1983, 1985). Indeed, task difficulty, indicated by L/S ratio, was greater in Allan and Gibbon (1991) than in Wearden (1991).

## 5.2 *Stimulus Range*

The presence of spacing effects for some L/S ratios, but not others, suggests that the stimulus range (i.e., ratio of the short and long anchor durations) the participant experiences also impacts the location of the BP. Wearden and Ferrara (1996) examined stimulus range effects in humans using the partition method of bisection (see above). For three groups of participants, they examined L/S ratios of 5:1 and 2:1 in situations where the difference between the S and L values was kept constant at 400 ms. For another three groups the L/S ratio was kept constant at 4:1, but the difference between short and long values ranged from 300 to 600 ms. They obtained bisection at the AM for all groups except the 2:1 group, which showed GM bisection. Experiment 2 of the same study explored linear and logarithmic spacing, L/S ratios of 2:1 (450 vs. 900 ms) and 19:1 (50 vs. 900 ms), and explicitly trained the participants on the S and L anchor durations instead of using the partition method. They obtained an effect of stimulus spacing for the large 19:1 ratio, but not the 2:1 ratio, and a BP close to the GM for the 2:1 ratio. Hence, they concluded that GM bisection manifests for small L/S ratios and that linear versus logarithmic spacing effects manifest only when L/S ratios are large. They also showed that Wearden's modified difference model (Wearden, 1991) accounted for the obtained data patterns reasonably well. Unfortunately from the perspective of finding generally applicable models, it does not account for the stimulus spacing effects described in the preceding section.

With a view toward developing a single model that could account for the idiosyncrasies found in the duration bisection literature, including effects of stimulus spacing and stimulus range such as those described above, Kopec and Brody (2010) analyzed data from 148 experiments reported in 18 different studies of human duration bisection. They developed a two-step decision model in which the participant first determines whether the probe duration is one of the anchor durations and, if not, conducts the second step in which the relative distance of the probe duration from the anchor durations is compared.

## 5.3 *Timing Precision*

As noted above, the DL and WF provide measures of temporal sensitivity in the duration bisection task. The steeper the participant's response function, the more precise (i.e., less variable) the participant's timing on a trial-to-trial basis. Several studies in human and non-human animals have shown more precise timing performance with more difficult L/S ratios (e.g., Church & Deluty, 1977). However, when the L/S ratios are extremely difficult (e.g., 3 vs. 3.6 s), timing anomalies may arise. For example, Penney et al. (2008) found reversals in the psychometric response functions, meaning that participants classified a duration close to the GM as shorter more often than a probe duration that was more distant from the GM and closer to the S anchor value. This pattern held in humans, mice, and pigeons, although the difficulty level eliciting the anomaly differed among species.

## 6 Analysis of Duration Bisection Data

### 6.1 *Atheoretical*

Data from the duration bisection task are often analyzed in what could be termed an atheoretical manner. The simplest approach is to compare the probability of a long classification, p('long'), at each test phase duration between conditions. In other words, probe duration and experimental condition can be treated as factors in an ANOVA. This analytic approach can reveal whether duration classification differs between conditions, but it has several shortcomings.

First, even when there are substantial condition effects at intermediate probe durations the stimulus classifications at or very close to the anchor durations may not be different. Both human and non-human subjects often classify the shortest and longest stimuli as 'short' and 'long' with perfect, or near perfect, accuracy (e.g., Droit-Volet & Wearden, 2001; Meck, 1983). Given a main effect of Condition collapses across the test durations, a condition difference among intermediate probe durations could be concealed. Although a Condition difference may manifest as a Duration x Condition interaction, detecting a significant interaction often requires greater statistical power as compared to detecting a significant main effect.

Second, in the event a Duration x Condition interaction manifests, one would typically then test the effect of Condition at each level of test duration. This may result in significant differences at some intermediate test durations, but not others, particularly when one corrects for multiple comparisons. However, inconsistent effects at the intermediate probe durations may be difficult to interpret in a meaningful manner.

Consequently, rather than analyzing p('long') values, researchers often analyze the BP, DL, and WF values derived from the response function. Differences in the BP indicate whether the manipulation of interest shifted the response function to the left or right and changes in temporal sensitivity are revealed by differences in the DL or WF.

Whether a manipulation shifts the psychometric response function in an additive or multiplicative manner provides critical information for understanding the putative psychological mechanisms underlying effects on interval timing. For example, within the Scalar Timing Theory (STT; Gibbon, Church,

& Meck, 1984) framework, if a manipulation affects the pacemaker rate, then the shift in BP will be multiplicative across duration ranges. More concretely, if the BP changes from 4 to 4.4 s for a 2 vs. 8 s anchor duration pair, then it should shift from 4.24 to 4.67 s for a 3 vs. 6 s anchor pair and from 6.93 to 7.63 s for a 4 vs. 12 s anchor pair. In contrast, if the manipulation affects detection of the onset of the timing stimulus, then the shift in BP will be additive (i.e., a constant absolute amount) across anchor duration pairs because the magnitude of the timing signal will not materially impact the gain (loss) of timing signal due to improved (worsened) stimulus detection. Hence, if a manipulation allows the onset of the timing stimulus to be detected 200 ms faster, then the corresponding perceived duration will be lengthened by 200 ms whether the timing stimulus is objectively 4 s long or 8 s long.

Ideally, to determine whether a manipulation shifts the BP in an additive or multiplicative manner, the experiment should comprise at least three anchor duration pairs. This allows the between condition BP difference to be calculated for each anchor duration pair and whether the shift is a constant absolute amount or a constant proportional amount (i.e., a linear increase), or some other functional form, to be determined. In practice, very few duration bisection studies have used more than one or two pairs of anchor durations. Consequently, most discussions of additive versus multiplicative shifts in the bisection function have centered on whether there is evidence for superimposition of the response functions after normalization by their respective bisection points. We return to this issue below.

The DL, which reflects the steepness of the psychometric response function, provides a measure of the participant's temporal sensitivity or acuity. The more consistent a participant is in categorizing the same stimulus duration in the test phase, then the steeper the participant's response function and the smaller the corresponding DL. For example, continuing with the 2 vs. 8 s anchor pair example described above, a participant who has a 25% long value of 3.17 s and a 75% long value of 5.04 s has a much sharper response function and smaller DL than a participant with a 25% long value of 2.52 and a 75% long value of 6.35 (.93 vs. 1.92).

As noted above, the WF provides a measure of temporal sensitivity that is corrected for the magnitude of the timed duration, which in the case of duration bisection means dividing the DL by the BP. If Weber's Law holds, then the WF is constant across conditions. However, to determine the constancy of the WF, researchers usually statistically test whether there is a between condition difference in the WF and, if not, conclude that Weber's Law holds. Obviously, the failure to find a statistical difference among WF values does not comprise particularly strong evidence that the WFs are equivalent.

It is worth noting that the Scalar Property for time imposes a stricter requirement on response functions than mere equivalence of WF values. Rather, when normalized by T/BP, where T is a probe duration, the entire psychometric response function should superimpose (Gibbon, 1981; Penney et al., 2000). However, response function superimposition is usually tested "by eye" in combination with the WF analysis approach just described. One could subject the normalized p('long') values for all test durations to a statistical analysis to determine whether values differ between conditions, but this approach also entails the "accepting the null hypothesis" problem.

To determine whether superimposition held in the presence of a BP difference between auditory and visual timing stimuli, Penney et al. (2000) compared the superimposition of the response functions when normalized in different ways. These were multiplicative normalization in which p('long') was plotted against T/BP (i.e., the typical approach) and what the authors termed "lateral" normalization in which they rescaled the objective time axis by adding one half of the difference between the auditory and visual BPs to each T value for the auditory modality and subtracted half of the BP difference for the visual modality. Sigmoidal equations were fit to the resulting response functions and the quality of the fit used as a measure of the degree of superimposition. A statistical test revealed that fit quality was better for the multiplicative normalization than the lateral normalization, a result that was taken as support for a clock speed interpretation of the shift rather than a timing onset interpretation.

Balcı and colleagues introduced a more principled method for examining shifts in psychometric response functions obtained from the duration bisection task (Balcı & Gallistel, 2006; Çoşkun et al., 2015). Although not explicitly used for this purpose in their papers, the approach can easily be applied to test superimposition of two, or more, response functions. First, the best fitting cumulative normal distribution function (or another sigmoidal function that best describes the data) is determined for the p('long') data in each condition and the log likelihood of each data point for each distribution fit is calculated. These log likelihoods are summed to obtain the likelihood of the data under different cumulative distributions with different mean and variance parameters (i.e., the non-superimposition model). The superimposition model can be obtained, for instance, by applying one of the best fitting distributions obtained in the first step to the data from all conditions. As above, log likelihoods of each data point are obtained and summed to obtain the likelihood of the data under the single cumulative distribution model. The superimposition and non-superimposition models can then be compared. The likelihood of the data will always be higher for the latter than the former model, but this difference might not be large enough to legislate choosing the non-superimposition model over the superimposition model after the difference in number of free-parameters is considered. Thus, the difference between the sum of log likelihoods should be penalized by some function of the extra number of parameters that the non-superimposition model has. This constitutes the essence of many model comparison statistics widely used in the literature (e.g., Bayesian information criterion, Akaike information criterion).

### 6.2 *Theory-based Analysis*

Many different theoretical models have been applied to data from the duration bisection task. Perhaps the most popular applications have been of models that fall within the framework provided by Scalar Expectancy Theory (SET; Gibbon, 1977) and its information processing companion model, Scalar Timing Theory (STT; Gibbon, Church, & Meck, 1984). There are several different duration bisection models within the SET/STT framework, with the differences among models centering on what parameters are allowed to vary and the decision rules assumed to apply. For example, Meck (1983) applied Gibbon's Referents Known Exactly (RKE) model to account for pharmacological and electric shock induced effects on timing. As the name suggests, this model assumes that perceptual variance is greater than the variance in the values stored in reference memory. However, the Sample Known Exactly (SKE) model has been more commonly used. It was originally proposed by Gibbon (1981) and, subsequently, modified by various researchers to account for a broader range of experimental results (e.g., Meck, 1984; Penney, Gibbon, & Meck, 2000, 2008). The SKE model posits that the participant maintains a veridical representation of the probe duration presented on the test trial (i.e., the current sample is known exactly), whereas there is variability in the memory representations for S and L. In its simplest form, the model has two free parameters: g, which reflects variation in the S and L memory representations, and b, which represents bias to respond long. The decision rule to respond "long" in the SKE model is $[m(T)^2] > (w_S w_L)/b$, where m(T) represents mean subjective time, and $w_S$ and $w_L$ are samples from the memory distributions for S and L. In the absence of bias, this decision rule results in bisection at the geometric mean of S and L. Of course, many other duration bisection models exist in the literature (e.g., Allan, 2002; Balcı & Simen, 2014; Kopec & Brody, 2010; Machado & Pata, 2005; Rodríguez-Gironés & Kacelnik, 2001; Wearden, 1991).

The major benefit of a theoretical analysis is that it provides an interpretive framework for understanding the pattern of results obtained in an experiment. For example, determining whether a between condition difference is due to a clock speed effect or a memory storage effect. Meck (1983) used the duration bisection procedure in the context of examining pharmacological (methamphetamine, haloperidol, physostigmine, atropine) and stress (foot-shock) effects on interval timing. He found that when animals were trained in a non-drug state (i.e., following saline administration), dopamine (DA) agonists shifted the bisection response function to the left, whereas DA antagonists shifted the response function to the right. For the cholinergic drugs, the agonist (i.e., physostigmine) decreased the remembered duration of the reinforced interval, such that the psychophysical response function was left shifted, whereas an ACh antagonist (i.e., atropine) increased the remembered duration of the reinforced interval, such that the response function was right shifted. Interpreting this result within the STT framework, Meck concluded that DA agonists selectively increased clock speed and antagonists selectively decreased it, whereas ACh agonists increased memory storage speed while antagonists decreased it.

## 7 Implementation Recommendations

What constitutes best practice in the bisection task depends on the question being asked. For example, if one wishes to unambiguously demonstrate that a shift in the psychometric response function is multiplicative rather than additive, then three sets of anchor duration pairs should be used in a within-subjects design. However, interference between sets of learned anchor durations is a concern when more than one duration pair is presented within the same test session. Moreover, particularly when using seconds range durations, the test session may be rather lengthy. Consequently, very few duration bisection papers published within the past decade have used more than two pairs of anchors durations and most have used a single anchor duration pair.

The selection of L/S range is also important. The task should not be too difficult as this may result in poor performance. However, it also should not be too easy as this may result in participants not being particularly attentive. Perhaps the most commonly used L/S range has been 4:1, although most of these studies came from the same research group or closely related groups. Choice of stimulus spacing is less critical, but if one intends to compare the results with previous findings in the literature, then it should be considered carefully because, as described above, it can influence the location of the BP.

We recommend presenting at least 10 trials at each probe duration during the test phase of the task. The goal is to have enough resolution in the timing measure to provide relatively smooth response functions, while not inducing task fatigue in the participants. Of course, how many trials can be run without participant fatigue depends on the participants themselves. We have conducted duration bisection EEG experiments in which 56 trials were presented at each probe duration (Ng et al., 2011). This was necessary to ensure adequate trials for the EEG analysis, in spite of our concerns about participant fatigue. However, as shown in Figure 5.4, even at the individual participant level data quality from the fourth quarter of the test session was quite good. Notably, the participants were university students and the test durations were relatively short (anchor durations of .8 and 3.2 s). It seems unlikely that children, or individuals with neurological problems, would be as attentive from start to finish of such a lengthy trial sequence.

For normal, adult participants, we recommend using seven durations in the test phase (i.e., two anchor durations and five intermediate probes). This number of durations is more likely to reveal the true form of the response

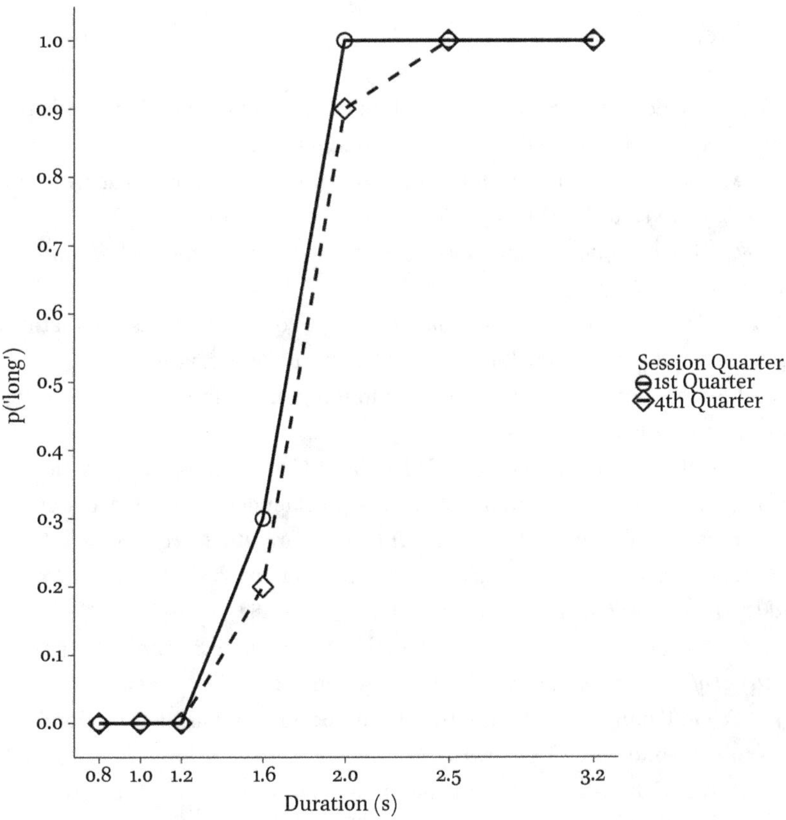


FIGURE 5.4 *Data from a single subject illustrating performance early and late in a duration bisection test session that comprised 392 trials. Each response function was generated by randomly selecting 10 trials for each probe duration from the first and the last quarter of the test session.*

function and provide a more accurate estimate of the BP and DL/WF. However, in situations where the amount of time available for testing is limited or one is concerned about participant fatigue (e.g., with certain patient populations or young children), then it may be necessary to reduce the number of intermediate probe durations to three for a total of five test phase durations.

In sum, the duration bisection task is relatively easy to implement across a wide range of participant populations and, therefore, has seen substantial use, particularly in situations where one wants to eliminate the influence of motor responding on the duration estimate. However, the details of task implementation can have a profound impact on the data obtained. Consequently, task parameters must be carefully considered in light of the particular psychological question being asked.

## References

Akdoğan, B., & F. Balcı (2016). Stimulus probability effects on temporal bisection performance of mice (Mus musculus). *Animal Cognition, 19*, 15–30.

Allan, L.G. (2002). The location and interpretation of bisection point. *The Quarterly Journal of Experimental Psychology, 55B*, 43–60.

Allan, L.G., & J. Gibbon (1991). Human bisection at the geometric mean. *Learning and Motivation, 22*, 39–58.

Balcı, F., & C.R. Gallistel (2006). Cross-domain transfer of quantitative discriminations: Is it all a matter of proportion? *Psychonomic Bulletin & Review*, 13, 636–642.

Balcı, F., & P. Simen (2014). Decision processes in temporal discrimination. *Acta Psychologica, 149*,157–168.

Brown, B.L., S. Höhn, A. Faure, S. Hörsten, P. Le Blanc, N. Desvignes, N. El Massioui, & V. Doyère (2011). Temporal sensitivity changes with extended training in a bisection task in a transgenic rat model. *Frontiers of Integrative Neuroscience, 5*, 44.

Church, R.M., & M.Z. Deluty (1977). Bisection of temporal intervals. *Journal of Experimental Psychology: Animal Behaviour Processes, 3*, 216–228.

Çoşkun, F., Z.C. Sayalı, E. Gürbüz, & F. Balcı (2015). Optimal time discrimination. *Quarterly Journal of Experimental Psychology, 68*, 381–401.

Cowles, J.T., & J.L. Finan (1941). An improved method for establishing temporal discrimination in white rats. *The Journal of Psychology, 11*, 335–342.

Droit-Volet, S., & A.C. Rattat (2007). A further analysis of time bisection behaviour in children with and without reference memory: The similarity and the partition task. *Acta Psychologica, 125*, 240–256.

Droit-Volet, S., & J.H. Wearden (2001). Temporal bisection in children. *Journal of Experimental Child Psychology, 80*, 142–159.

Droit-Volet, S., S. Fayolle, & S. Gil (2016). Emotion and time perception in children and adults: The effect of task difficulty. *Timing and Time Perception, 4*, 7–29

Gibbon, J. (1977). Scalar expectancy theory and Weber's law in animal timing. *Psychological Review, 84*, 279–325.

Gibbon, J. (1981). On the form and location of the psychometric bisection function for time. *Journal of Mathematical Psychology, 24*, 58–87.

Gibbon, J., R.M. Church, & W.H. Meck (1984). Scalar timing in memory. In Gibbon, J. & L. Allan (Eds.), *Timing and time perception* (pp. 58–87). New York: New York Academy of Sciences.

Heron, W.T. (1949). Time discrimination in the rat. *Journal of Comparative and Physiological Psychology, 42*, 27–31.

Kim, J., J.W. Ghim, J.H. Lee, & M.W. Jung (2013). Neural correlates of interval timing in rodent prefrontal cortex. *The Journal of Neuroscience, 33*, 13834–13847.

Kopec, C.D., & C.D. Brody (2010). Human performance on the temporal bisection task. *Brain and Cognition, 74*, 262–272.

Lustig, C., & W.H. Meck (2011). Modality differences in timing and temporal memory throughout the lifespan. *Brain and Cognition, 77*, 298–303.

Machado, A., & P. Pata (2005). Testing the scalar expectancy theory (SET) and the learning-to-time model (LeT) in a double bisection task. *Animal Learning and Behaviour, 33*, 111–122.

Meck, W.H. (1983). Selective adjustment of the speed of internal clock and memory processes. *Journal of Experimental Psychology: Animal Behaviour Processes, 9*, 171–201.

Meck, W.H. (1984). Attentional bias between modalities: Effects on the internal clock, memory, and decision stages used in animal time discrimination. In Gibbon, J. & L. Allan (Eds.), *Timing and time perception* (pp. 528–541). New York: New York Academy of Sciences.

Melgire, M., R. Ragot, S. Samson, T.B. Penney, W.H. Meck, & V. Pouthas (2005). Auditory/visual duration bisection in patients with left or right medial-temporal lobe resection. *Brain and Cognition, 58*, 119–124.

Ng, K.K., S. Tobin, & T.B. Penney (2011). Temporal accumulation and decision processes in the duration bisection task revealed by contingent negative variation. *Frontiers of Integrative Neuroscience, 5*, 10.

Nichelli, P., D. Alway, & J. Grafman (1996). Perceptual timing in cerebellar degeneration. *Neuropsychologia, 34*, 857–873.

Penney, T.B., J. Gibbon, & W.H. Meck (2000). Differential effects of auditory and visual signals on clock speed and temporal memory. *Journal of Experimental Psychology: Human Perception and Performance, 26*, 1770–1787.

Penney, T.B., J. Gibbon, & W.H. Meck (2008). Categorical scaling of duration bisection in pigeons (Columba livia), mice (Mus musculus), and humans (Homo sapiens). *Psychological Science, 19*, 1103–1109.

Penney, T.B., E.N.K. Yim, & K.K. Ng (2014). Distractor expectancy effects on interval timing. *Timing & Time Perception*, 2, 1–19.

Platt, J.R. & E.R. Davis (1983). Bisection of temporal intervals by pigeons. *Journal of Experimental Psychology: Animal Behavior Processes*, 9, 160–170.

Raslear, T.G. (1983). A test of the Pfanzagl bisection model in rats. *Journal of Experimental Psychology: Animal Behaviour Processes, 9,* 49–62.

Raslear, T.G. (1985). Perceptual bias and response bias in temporal bisection. *Perception and Psychophysics, 38,* 261–268.

Rodríguez-Gironés, M.A., & A. Kacelnik (2001). Relative importance of perceptual and mnemonic variance in human temporal bisection. *The Quarterly Journal of Experimental Psychology: Section A, 54,* 527–546.

Stubbs, A. (1968). The discrimination of stimulus duration by pigeons. *Journal of the Experimental Analysis of Behavior, 11,* 223–238.

Stubbs, D.A. (1976). Scaling of stimulus duration by pigeons. *Journal of the Experimental Analysis of Behavior, 26,* 15–25.

Ward, R.D., S.T. Barrett, R.N. Johnson, & A.L. Odum (2009). Nicotine does not enhance discrimination performance in a temporal bisection procedure. *Behavioural Pharmacology, 20,* 99–108.

Wearden, J.H. (1991). Human performance on an analogue of an interval bisection task. *The Quarterly Journal of Experimental Psychology, 43,* 59–81.

Wearden, J.H., & A. Ferrara (1995). Stimulus spacing effects in temporal bisection by humans. *The Quarterly Journal of Experimental Psychology, 48,* 289–310.

Wearden, J.H., & A. Ferrara (1996). Stimulus range effects in temporal bisection by humans. *The Quarterly Journal of Experimental Psychology: Section B, 49,* 24–44.

Wearden, J.H., P. Rogers, & R. Thomas (1997). Temporal bisection in humans with longer stimulus durations. *The Quarterly Journal of Experimental Psychology: Section B, 50,* 79–94.

Wiener, M., J.C. Thompson, & H.B. Coslett (2014). Continuous carryover of temporal context dissociates response bias from perceptual influence for duration. *PLoS ONE, 9,* e100803.

## Appendix

TABLE 5.1 *These publications present data from variants of the duration bisection task. Duration range (milliseconds – ms or seconds – s) and subject species are indicated for each publication.*

| Duration | Subjects | Reference |
|---|---|---|
| s | Pigeons | Stubbs, A. (1968). The discrimination of stimulus duration by pigeons. *Journal of the Experimental Analysis of Behavior, 11,* 223–238. |
| s | Pigeons | Stubbs, D.A. (1976). Scaling of stimulus duration by pigeons. *Journal of the Experimental Analysis of Behavior, 26,* 15–25. |
| s | Rats | Church, R.M., & Deluty, M.Z. (1977). Bisection of temporal intervals. *Journal of Experimental Psychology: Animal Behavior Processes, 3,* 216–228 |
| s | Rats | Meck, W.H. (1983). Selective adjustment of the speed of internal clock and memory processes. *Journal of Experimental Psychology: Animal Behavior Processes, 9,* 171–201. |
| s | Pigeons | Platt, J.R., & Davis, E.R. (1983). Bisection of temporal intervals by pigeons. *Journal of Experimental Psychology: Animal Behavior Processes, 9,* 160–170. |
| ms & s | Rats | Raslear, T.G. (1983). A test of the Pfanzagl bisection model in rats. *Journal of Experimental Psychology: Animal Behavior Processes, 9,* 49–62. |
| s | Rats | Meck, W.H. (1984). Attentional bias between modalities: Effects on the internal clock, memory, and decision stages used in animal time discrimination. In J. Gibbon & L. Allan (Eds.), *Timing and time perception* (pp. 528–541). New York: New York Academy of Sciences. |
| s | Rats | Siegel, S.F., & Church, R.M. (1984). The decision rule in temporal bisection. In J. Gibbon & L. Allan (Eds.), *Timing and time perception* (pp. 643–645). New York: New York Academy of Sciences. |
| ms & s | Rats | Raslear, T.G. (1985). Perceptual bias and response bias in temporal bisection. *Perception & Psychophysics, 38,* 261–268. |
| s | Humans – adults | Allan, L.G., & Gibbon, J. (1991). Human bisection at the geometric mean. *Learning and Motivation, 22,* 39–58. |

TABLE 5.1 *These publications present data from variants of the duration bisection task.* (cont.)

| Duration | Subjects | Reference |
|---|---|---|
| ms | Humans – adults | Wearden, J.H. (1991). Human performance on an analogue of an interval bisection task. *The Quarterly Journal of Experimental Psychology, 43*, 59–81. |
| ms | Humans – adults | Wearden, J.H., & Ferrara, A. (1995). Stimulus spacing effects in temporal bisection by humans. *The Quarterly Journal of Experimental Psychology, 48*, 289–310. |
| ms & s | Humans – adults | Nichelli, P., Alway, D., & Grafman, J. (1996). Perceptual timing in cerebellar degeneration. *Neuropsychologia, 34*, 857–873. |
| ms | Humans – adults | Wearden, J.H. (1996). Stimulus range effects in temporal bisection by humans. *The Quarterly Journal of Experimental Psychology: Section B, 49*, 24–44. |
| s | Rats | Al-Zahrani, S.S.A., Ho, M.Y., Al-Ruwaitea, A.S.A., Bradshaw, C.M., & Szabadi, E. (1997). Effect of destruction of the 5-hydroxytryptaminergic pathways on temporal memory: quantitative analysis with a delayed interval bisection task. *Psychopharmacology, 129*, 48–55. |
| s | Humans – adults | Wearden, J.H., Rogers, P., & Thomas, R. (1997). Temporal bisection in humans with longer stimulus durations. *The Quarterly Journal of Experimental Psychology: Section B, 50*, 79–94. |
| s | Humans – adults | Penney, T.B., Gibbon, J., & Meck, W.H. (2000). Differential effects of auditory and visual signals on clock speed and temporal memory. *Journal of Experimental Psychology: Human Perception and Performance, 26*, 1770–1787. |
| ms | Humans – adults | Allan, L.G. & Gerhardt, K. (2001). Temporal bisection with trial referents. *Perception and Psychophysics, 63*, 524–540. |
| s | Humans – children | Droit-Volet, S., & Wearden, J.H. (2001). Temporal bisection in children. *Journal of Experimental Child Psychology, 80*, 142–159. |
| s | Humans – children | Rattat, A.C., & Droit-Volet, S. (2001). Variability in 5-and 8-year-olds' memory for duration: An interfering task in temporal bisection. *Behavioural Processes, 55*, 81–91. |

| Duration | Subjects | Reference |
|---|---|---|
| ms & s | Humans – adults | Rodríguez-Gironés, M.A., & Kacelnik, A. (2001). Relative importance of perceptual and mnemonic variance in human temporal bisection. *The Quarterly Journal of Experimental Psychology: Section A, 54*, 527–546. |
| s | Humans – adults | Allan, L.G. (2002). The location and interpretation of bisection point. *The Quarterly Journal of Experimental Psychology, 55B*, 43–60. |
| s | Humans – adults | Droit-Volet, S., Clément, A., & Fayol, M. (2003). Time and number discrimination in a bisection task with a sequence of stimuli: A developmental approach. *Journal of Experimental Child Psychology, 84*, 63–76. |
| ms | Humans – adults & children | Droit-Volet, S., Tourret, S., & Wearden, J. (2004). Perception of the duration of auditory and visual stimuli in children and adults. *Quarterly Journal of Experimental Psychology Section A, 57*, 797–818. |
| ms | Humans – adults | Brown, G.D., McCormack, T., Smith, M., & Stewart, N. (2005). Identification and bisection of temporal durations and tone frequencies: common models for temporal and nontemporal stimuli. *Journal of Experimental Psychology: Human Perception and Performance, 3*, 919–938. |
| s | Pigeons | Machado, A., & Pata, P. (2005). Testing the scalar expectancy theory (SET) and the learning-to-time model (LeT) in a double bisection task. *Animal Learning & Behavior, 33*, 111–122. |
| ms & s | Humans – adults | Melgire, M., Ragot, R., Samson, S., Penney, T.B., Meck, W.H., & Pouthas, V. (2005). Auditory/visual duration bisection in patients with left or right medial-temporal lobe resection. *Brain and Cognition, 58*, 119–124. |
| s | Humans – adults | Penney, T.B., Meck, W.H., Roberts, S.A., Gibbon, J., & Erlenmeyer-Kimling, L. (2005). Interval-timing deficits in individuals at high risk for schizophrenia. *Brain and Cognition, 58*, 109–118. |
| s | Humans – adults | Balcı, F., & Gallistel, C.R. (2006). Cross-domain transfer of quantitative discriminations: Is it all a matter of proportion? *Psychonomic Bulletin & Review, 13*, 636–642. |

TABLE 5.1 *These publications present data from variants of the duration bisection task.* (cont.)

| Duration | Subjects | Reference |
|---|---|---|
| ms & s | Humans – adults | Effron, D.A., Niedenthal, P.M., Gil, S., & Droit-Volet, S. (2006). Embodied temporal perception of emotion. *Emotion, 6*, 1–9. |
| ms & s | Humans – adults & children | Droit-Volet, S., & Rattat, A.C. (2007). A further analysis of time bisection behavior in children with and without reference memory: The similarity and the partition task. *Acta Psychologica, 125*, 240–256. |
| s | Humans adults & children, | Droit-Volet, S., Meck, W.H., & Penney, T.B. (2007). Sensory modality and time perception in children and adults. *Behavioural Processes, 74*, 244–250. |
| ms & s | Humans – adults | Lee, K.H., Egleston, P.N., Brown, W.H., Gregory, A.N., Barker, A.T., & Woodruff, P.W. (2007). The role of the cerebellum in subsecond time perception: Evidence from repetitive transcranial magnetic stimulation. *Journal of Cognitive Neuroscience, 19*, 147–157. |
| ms & s | Humans – adults | Ortega, L., & Lopez, F. (2008). Effects of visual flicker on subjective time in a temporal bisection task. *Behavioural Processes, 78*, 380–386. |
| s | Pigeons, Mice, Humans – adults | Penney, T.B., Gibbon, J., & Meck, W.H. (2008). Categorical scaling of duration bisection in pigeons (Columba livia), mice (Mus musculus), and humans (Homo sapiens). *Psychological Science, 19*, 1103–1109. |
| ms & s | Humans – adults | Tipples, J. (2008). Negative emotionality influences the effects of emotion on time perception. *Emotion, 8*, 127–131. |
| ms & s | Humans – adults | Gil, S., & Droit-Volet, S. (2009). Time perception, depression and sadness. *Behavioural Processes, 80*, 169–176. |
| s | Pigeons | Ward, R.D., Barrett, S.T., Johnson, R.N., & Odum, A.L. (2009). Nicotine does not enhance discrimination performance in a temporal bisection procedure. *Behavioural Pharmacology, 20*, 99–108. |
| s | Pigeons | Johnson, R.N., Ward, R.D., & Odum, A.L. (2010). Baseline training history and effects of methamphetamine on performance of pigeons on an interval-bisection task. *Behavioural Processes, 84*, 484–489. |
| ms & s | Humans – adults | Tipples, J. (2010). Time flies when we read taboo words. *Psychonomic Bulletin & Review, 17*, 563–568. |

| Duration | Subjects | Reference |
|---|---|---|
| s | Rats | Brown, B.L., Höhn, S., Faure, A., von Hörsten, S., Le Blanc, P., Desvignes, N., El Massioui, N., & Doyère, V. (2011). Temporal sensitivity changes with extended training in a bisection task in a transgenic rat model. *Frontiers of Integrative Neuroscience, 5*, 10. |
| ms & s | Humans – adults & children | Gil, S., & Droit-Volet, S. (2011). Time perception in response to ashamed faces in children and adults. *Scandinavian Journal of Psychology, 52*, 138–145. |
| s | Humans – adults & children | Lustig, C., & Meck, W.H. (2011). Modality differences in timing and temporal memory throughout the lifespan. *Brain and Cognition, 77*, 298–303. |
| ms & s | Humans – adults | Ng, K.K., Tobin, S., & Penney, T.B. (2011). Temporal accumulation and decision processes in the duration bisection task revealed by contingent negative variation. *Frontiers of Integrative Neuroscience, 5*, 10. |
| ms & s | Humans – infants | Provasi, J., Rattat, A.C., & Droit-Volet, S. (2011). Temporal bisection in 4-month-old infants. *Journal of Experimental Psychology: Animal Behavior Processes, 37*, 108–113. |
| ms & s | Humans – adults | Tipples, J. (2011). When time stands still: Fear-specific modulation of temporal bias due to threat. *Emotion, 11*, 74–80. |
| ms & s | Humans – adults & children | Zélanti, P.S., & Droit-Volet, S. (2011). Cognitive abilities explaining age-related changes in time perception of short and long durations. *Journal of Experimental Child Psychology, 109*, 143–157. |
| s | Pigeons | Carvalho, M.P.D., & Machado, A. (2012). Relative versus absolute stimulus control in the temporal bisection task. *Journal of the Experimental Analysis of Behavior, 98*, 23–44. |
| ms & s | Humans – children | Gil, S., Chambres, P., Hyvert, C., Fanget, M., & Droit-Volet, S. (2012). Children with autism spectrum disorders have "the working raw material" for time perception. *PLoS ONE, 7*, e49116. |
| ms & s | Humans – adults | Shi, Z., Jia, L., & Müller, H.J. (2012). Modulation of tactile duration judgments by emotional pictures. *Frontiers in Integrative Neuroscience, 6*. 24. |
| ms & s | Humans – adults & children | Droit-Volet, S., & Zélanti, P. (2013). Development of time sensitivity: duration ratios in time bisection. *The Quarterly Journal of Experimental Psychology, 66*, 671–686. |

TABLE 5.1 *These publications present data from variants of the duration bisection task.* (cont.)

| Duration | Subjects | Reference |
|---|---|---|
| ms | Humans – adults | Fallow, K.M., & Voyer, D. (2013). Degree of handedness, emotion, and the perceived duration of auditory stimuli. *Laterality: Asymmetries of Body, Brain and Cognition, 18,* 671–692. |
| ms | Humans – adults | Iordanescu, L., Grabowecky, M., & Suzuki, S. (2013). Action enhances auditory but not visual temporal sensitivity. *Psychonomic Bulletin & Review, 20,* 108–114. |
| s | Rats | Kim, J., Ghim, J.W., Lee, J.H., & Jung, M.W. (2013). Neural correlates of interval timing in rodent prefrontal cortex. *The Journal of Neuroscience, 33,* 13834–13847. |
| ms | Humans – adults | Kroger-Costa, A., Machado, A., & Santos, J.A. (2013). Effects of motion on time perception. *Behavioural Processes, 95,* 50–59. |
| s | Humans – adults | Lindbergh, C.A., & Kieffaber, P.D. (2013). The neural correlates of temporal judgments in the duration bisection task. *Neuropsychologia, 51,* 191–196. |
| s | Humans – adults | Martínez-Cascales, I., Fuente, J.D.L., & Santiago de Torres, J. (2013). Space and time bisection in schizophrenia. *Frontiers in Psychology, 4,* 823. |
| ms & s | Humans – adults | Nicol, J.R., Tanner, J., & Clarke, K. (2013). Perceived duration of emotional events: Evidence for a positivity effect in older adults. *Experimental Aging Research, 39,* 565–578. |
| s | Humans – adults | Spínola, I., Machado, A., de Carvalho, M.P., & Tonneau, F. (2013). What do humans learn in a double, temporal bisection task: Absolute or relative stimulus durations? *Behavioural Processes, 95,* 40–49. |
| ms & s | Humans – adults | Balcı F. & Simen, P. (2014) Decision processes in temporal discrimination. *Acta Psychologica, 149,* 157–168. |
| ms & s | Humans – adults | Fayolle, S.L., & Droit-Volet, S. (2014). Time perception and dynamics of facial expressions of emotions. *PLoS ONE, 9,* e97944. |
| s | Humans – adults | Jozefowiez, J., Polack, C.W., Machado, A., & Miller, R.R. (2014). Trial frequency effects in human temporal bisection: Implications for theories of timing. *Behavioural Processes, 101,* 81–88. |

| Duration | Subjects | Reference |
|---|---|---|
| s | Humans – adults | Penney, T.B., Yim, E.N., & Ng, K.K. (2014). Distractor expectancy effects on interval timing. *Timing & Time Perception*, *2*, 1–19. |
| ms | Humans – adults | Wiener, M., Thompson, J.C., & Coslett, H.B. (2014). Continuous carryover of temporal context dissociates response bias from perceptual influence for duration. *PLoS ONE*, *9*, e100803. |
| s | Humans – adults | Çoşkun, F., Sayalı, Z.C., Gürbüz, E., Balcı, F. (2015). Optimal time discrimination. *Quarterly Journal of Experimental Psychology*, 68, 381–401. |
| ms & s | Humans – adults | Droit-Volet, S., Lamotte, M., & Izaute, M. (2015). The conscious awareness of time distortions regulates the effect of emotion on the perception of time. *Consciousness and Cognition*, *38*, 155–164. |
| s | Humans – adults | Herbst, S.K., Chaumon, M., Penney, T.B., & Busch, N.A. (2015). Flicker-induced time dilation does not modulate EEG correlates of temporal encoding. *Brain Topography*, *28*, 559–569. |
| ms | Humans – adults | Jia, L., Shi, Z., Zang, X., & Müller, H.J. (2015). Watching a real moving object expands tactile duration: The role of task-irrelevant action context for subjective time. *Attention, Perception, & Psychophysics*, *77*, 2768–2780. |
| ms & s | Humans – adults | Jusyte, A., Schneidt, A., & Schönenberg, M. (2015). Temporal estimation of threatening stimuli in social anxiety disorder: Investigation of the effects of state anxiety and fearfulness. *Journal of Behavior Therapy and Experimental Psychiatry*, *47*, 25–33. |
| ms & s | Humans – adults | Kliegl, K.M., Watrin, L., & Huckauf, A. (2015). Duration perception of emotional stimuli: Using evaluative conditioning to avoid sensory confounds. *Cognition and Emotion*, *29*, 1350–1367. |
| s | Humans – adults | Levy, J.M., Namboodiri, V.M., & Shuler, M.G.H. (2015). Memory bias in the temporal bisection point. *Frontiers in Integrative Neuroscience*, *9*, 44. |
| ms | Humans – adults | Voyer, D., & Reuangrith, E. (2015). Perceptual asymmetries in a time estimation task with emotional sounds. *Laterality: Asymmetries of Body, Brain and Cognition*, *20*, 211–231. |

TABLE 5.1 *These publications present data from variants of the duration bisection task.* (cont.)

| Duration | Subjects | Reference |
|---|---|---|
| ms & s | Humans – adults | Zhang, Z., Jia, L., & Ren, W. (2015). Time changes with feeling of speed: an embodied perspective. *Frontiers in Neurorobotics, 8.* |
| s | Mice | Akdoğan, B., & Balcı, F. (2016). Stimulus probability effects on temporal bisection performance of mice (Mus musculus). *Animal Cognition, 19*, 15–30. |
| s | Humans – adults | Akdoğan, B., & Balcı, F. (2016). The effects of payoff manipulations on temporal bisection performance. *Acta Psychologica, 170*, 74–83. |
| ms | Humans – adults | Colonnello, V., Domes, G., & Heinrichs, M. (2016). As time goes by: Oxytocin influences the subjective perception of time in a social context. *Psychoneuroendocrinology, 68*, 69–73. |
| ms & s | Humans – adults | Droit-Volet, S., Fayolle, S., & Gil, S. (2016). Emotion and time perception in children and adults: The effect of task difficulty. *Timing & Time Perception, 4*,7–29. |
| ms & s | Humans – adults | Eberhardt, L.V., Huckauf, A., & Kliegl, K.M. (2016). Effects of neutral and fearful mood on duration estimation of neutral and fearful face stimuli. *Timing & Time Perception, 4*, 30–47. |
| s | Rats | Es-seddiqi, M., El Massioui, N., Samson, N., Brown, B.L., & Doyère, V. (2016). The amygdalo-nigrostriatal network is critical for an optimal temporal performance. *Learning & Memory, 23*, 104–107. |
| s | Pigeons | Fox, A.E., Prue, K.E., & Kyonka, E.G. (2016). What is timed in a fixed-interval temporal bisection procedure? *Learning & Behaviour, 44*, 366–377. |
| ms | Humans – adults | Ishikawa, K., & Okubo, M. (2016). Overestimation of the subjective experience of time in social anxiety: Effects of facial expression, gaze direction, and time course. *Frontiers in Psychology, 7*, 711. |
| s | Pigeons | Laude, J.R., Daniels, C.W., Wade, J.C., & Zentall, T.R. (2016). I can time with a little help from my friends: Effect of social enrichment on timing processes in Pigeons (Columba livia). *Animal Cognition, 19*, 1205–1213. |

| Duration | Subjects | Reference |
|---|---|---|
| ms & s | Humans – adults | Millot, J.L., Laurent, L., & Casini, L. (2016). The influence of odors on time perception. *Frontiers in Psychology, 7*, 181. |
| ms | Humans – adults | Mioni, G., Grondin, S., Forgione, M., Fracasso, V., Mapelli, D., & Stablum, F. (2016). The role of primary auditory and visual cortices in temporal processing: A tDCS approach. *Behavioural Brain Research, 313*, 151–157. |
| s | Rats | Peterson, J.R., & Kirkpatrick, K. (2016). The effects of a time-based intervention on experienced middle-aged rats. *Behavioural Processes, 133*, 44–51. |
| ms & s | Humans – adults | Pillai, J.S., & McLoughlin, A. (2016). Exercise and time perception: An exploration of the impact of high intensity cardio exercise (Zumba) on human timing. *Timing & Time Perception, 4*, 343–353. |
| s | Pigeons | Pinheiro de Carvalho, M., Machado, A., & Tonneau, F. (2016). Learning in the temporal bisection task: Relative or absolute? *Journal of Experimental Psychology: Animal Learning and Cognition, 42*, 67–81. |
| ms & s | Humans – adults | Righi, S., Galli, L., Paganini, M., Bertini, E., Viggiano, M.P., & Piacentini, S. (2016). Time perception impairment in early-to-moderate stages of Huntington's disease is related to memory deficits. *Neurological Sciences, 37*, 97–104. |
| ms & s | Humans – adults | Schirmer, A., Ng, T., Escoffier, N., & Penney, T.B. (2016). Emotional voices distort time: Behavioral and neural correlates. *Timing & Time Perception, 4*, 79–98. |
| ms & s | Humans – adults | Terhune, D.B., Sullivan, J.G., & Simola, J.M. (2016). Time dilates after spontaneous blinking. *Current Biology, 26*, R459–R460. |
| s | Humans – adults | Yamamoto, K., & Miura, K. (2016). Effect of motion coherence on time perception relates to perceived speed. *Vision Research, 123*, 56–62. |
| ms | Humans – adults | Zhang, J., Nombela, C., Wolpe, N., Barker, R.A., & Rowe, J.B. (2016). Time on timing: Dissociating premature responding from interval sensitivity in Parkinson's disease. *Movement Disorders, 31*, 1163–1172. |
| s | Pigeons | Araiba, S., & Brown, B.L. (2017). The effect of the long anchor duration on performance in the temporal bisection procedure. *Behavioural Processes, 135*, 76–86. |

TABLE 5.1 *These publications present data from variants of the duration bisection task.* (cont.)

| Duration | Subjects | Reference |
|---|---|---|
| ms | Humans – children | Casini, L., Pech-Georgel, C., & Ziegler, J.C. (2017). It's about time: Revisiting temporal processing deficits in dyslexia. *Developmental Science, e12530.* |
| ms | Humans – adults | Charras, P., Droit-Volet, S., Brechet, C., & Coull, J.T. (2017). The spatial representation of time can be flexibly oriented in the frontal or lateral planes from an early age. *Journal of Experimental Psychology: Human Perception and Performance, 43*, 832–845. |
| s | Dogs | Domeniconi, C., & Machado, A. (2017). Temporal bisection task with dogs: An exploratory study. *Psychology & Neuroscience, 10*, 101–108. |
| ms & s | Humans – adults & children | Droit-Volet, S. (2017). Time dilation in children and adults: The idea of a slower internal clock in young children tested with different click frequencies. *Behavioural Processes, 138*, 152–159. |
| ms & s | Humans – adults | Gable, P.A., Neal, L.B., & Poole, B.D. (2017). Sadness speeds and disgust drags: Influence of motivational direction on time perception in negative affect. *Motivation Science, 2*, 238–255. |
| ms & s | Humans – adults | Gonidis, L., & Sharma, D. (2017). Internet and Facebook related images affect the perception of time. *Journal of Applied Social Psychology, 47*, 224–231. |
| ms | Humans – adults | Jones, C.R., Lambrechts, A., & Gaigg, S.B. (2017). Using time perception to explore implicit sensitivity to emotional stimuli in autism spectrum disorder. *Journal of Autism and Developmental Disorders*, 47, 2054–2066. |
| ms | Humans – adults | Makwana, M., & Srinivasan, N. (2017). Intended outcome expands in time. *Scientific Reports, 7*, 6305. |
| ms & s | Humans – adults | Lamotte, M., & Droit-Volet, S. (2017). Aging and time perception for short and long durations: A question of attention? *Timing & Time Perception, 5*, 149–167. |
| ms & s | Humans – adults | Sarigiannidis, I., Ernst, M., Grillon, C., Roiser, J., & Robinson, O. (2017). Induced anxiety leads to underestimating time. *Biological Psychiatry, 81*, S352. |

| Duration | Subjects | Reference |
|---|---|---|
| ms & s | Humans – adults | Zhang, M., Zhang, L., Yu, Y., Liu, T., & Luo, W. (2017). Women overestimate temporal duration: Evidence from Chinese emotional words. *Frontiers in Psychology, 8*, 4. |
| ms | Humans – adults | Wiener, M., Parikh, A., Krakow, A., & Coslett, H.B. (2017). Causal role of beta oscillations in time estimation. *bioRxiv*, 165233. |

CHAPTER 6

# Temporal Decision-making: Common Procedures and Contemporary Approaches

*David Freestone and Fuat Balcı*

## 1 Introduction

Various procedures have been developed to investigate interval timing and time-based choice behavior. These methods are typically first developed and validated in non-human animals and then adapted for human testing. Several researchers in particular have contributed to the adaptation of prominent procedures to humans and, thus, to our cross-species understanding of interval timing behavior (e.g., Allan & Gibbon 1991; Rakitin et al. 1998; Wearden 2002). Each of these procedures has advantages and disadvantages in terms of their sensitivity to measuring psychological variables. Thus, the choice of the most suitable procedure for a given research question entails a deep understanding of the procedure and its analysis.

In this chapter, we describe three different interval timing procedures: the peak interval task, the switch task, and the differential reinforcement of low rates of responding (DRL) task. Our aim is to provide a wide range of data acquisition and analysis tools to our readers. The peak interval task is a temporal differentiation task, the switch task is a real-time temporal discrimination task, and the DRL task is a response inhibition task. The peak interval procedure is used most widely, whereas the switch and DRL tasks are used relatively less often. However, together, these three tasks add complementary results about human timing ability. Furthermore, performance on the switch and DRL tasks (and, maybe, the peak interval task) can be evaluated with respect to objective optimal functions to investigate the temporal risk assessment (for review see Balcı et al. 2011).

We will describe each procedure separately. We will first provide background information regarding the use of each procedure, followed by the description of the analysis. Programming code written in Python to collect and analyze the data accompany this chapter. These applications include line by line descriptions and tutorials for readers who are interested in following each step of the experimental procedure through data processing. The Python code can be found

DOI 10.1163/9789004280205_007

at our github page for this chapter (https://github.com/freestone-lab/timing_tasks.git) and at the book's GitHub repository.

## 2 Peak Interval Procedure

### 2.1 *Background*

Experimental psychologists have been using fixed interval (FI, operant conditioning) and fixed time (FT, classical conditioning) schedules for decades. The response patterns across species in both tasks is similar, and have led to general and important results regarding animal timing ability (Ferster & Skinner 1957; Pavlov, 1927; Schneider, 1969). Freestone, MacInnis, and Church (2013) have argued that the timing mechanisms exert their effect on both operant and classically conditioned responses in the same way. Pavlov showed that the delay to the onset of conditioned response is correlated with the interval between conditioned stimulus and unconditioned stimulus. This observation was termed as the "inhibition of delay" by Pavlov (Pavlov 1927; Drew et al. 2005). In the original version of the FI schedule of reinforcement, the participant is reinforced for the first response after the fixed interval. This cycle continues over the course of a session without any interruption by inter-trial intervals (ITIs). At steady state, the rate of responding abruptly increases a little more than half way through the trial (e.g., Schneider, 1969). When averaged over trials, the response rate gradually increases throughout the interval. The discrete version of the task, where trials are signaled by a conditioned stimulus (and separated by an ITI), shows the same pattern of results.

The peak interval procedure extends the discrete FI procedure, and is arguably one of the most widely used tasks in interval timing research. This procedure was originally developed by Catania (1970) and Roberts (1981) for animal testing and it has been used to characterize the temporal characteristics of anticipatory responses for reinforcers that are available after a fixed delay after the onset of a conditioned stimulus. Unlike discrete FI procedures, the peak interval procedure also contains test trials in which the conditioned stimulus lasts much longer than the FI and reinforcement is not given. These trials allow us to observe anticipatory responses as a function of trial time based on reinforcement expectancy without any contamination by the reinforcement delivery. The average rate of responding forms a curve with a peak (hence the name). This curve is roughly normally distributed, centered around the typical reinforcement time, which presumably reflects the animal's expectation about when food is typically delivered. The data analysis can be conducted at the level of average response curves or individual trials. This task has been adapted to human testing in various ways (e.g., Balcı et al., 2013; Rakitin et al. 1998). We use the human version of the task below because it is straightforward for readers to follow along using the code we provide with this chapter.

### 2.2 *Procedural Details*

#### 2.2.1 Fixed Interval Training

In the FI procedure, a timing stimulus is presented and a response option is available (this could be a key on a keyboard or response box or a keyboard). This starts a trial. The first response following the fixed interval results in a reinforcer (e.g., monetary reward), responses prior to this time do not payoff, but are not penalized. The stimulus turns off when the reinforcer is delivered. The FI trials are separated by an intertrial interval (ITI), which should not be predictable. In the limited hold version of the task, the trial ends without reinforcement if too much time has elapsed, that is, the reinforcement is only held for a limited time. This ensures that the experienced delay to the outcome is not much longer than the scheduled delay (e.g., Stoddard et al., 1988).

#### 2.2.2 Peak Interval Testing

Testing is composed of both discrete FI and peak interval (PI) trials. In PI trials, the stimulus is presented much longer than the fixed interval to reinforcement (typically 3×FI) and responses are not reinforced. Peak trials allow the experimenters to observe the timed behavior not only prior to the typical reinforcement time (as in FI trials) but how expectation declines after the typical reinforcement time has elapsed without reinforcement. The proportion of PI trials to total number of trials often ranges from .33 to .5. Experimenters might choose first give 100% FI trials (training) before adding peak trials, or they may skip training and mix the trial types from the start. The training phase allows participants to learn the reinforcement time on every trial before being tested on peak trials, but this limits the total number of PI trials that can be collected in a single session. There is no restriction on the number of intervals participants may time. For example, two PI procedures can each be associated with a different response, running either concurrently or in separate trials in the same session. The timing stimulus can be a visual cue (e.g., change in the color of a square or the screen) or an auditory signal. And these stimuli can be the same or different for different intervals.

Participants can be either asked to emit multiple responses (Rakitin et al. 1998) or press and hold the key when they think the reinforcement time approaches, then release it when they think the reinforcement was omitted (e.g., Balcı et al., 2013). In order to capture the biomechanical cost of responding in animal studies, an explicit cost can be introduced per response (in the multiple response version) and duration of response (in the press-release version). The tutorial that accompanies this chapter (see Github repository) presents the version of the task that requires multiple responses during the stimulus. Each response is time-stamped and recorded.

#### 2.2.3 Procedure Code

The task was written using PsychoPy (version 1.83.03), and is available on our github page. When the program loads, it reads a file "peak_session_information .csv" that specifies the task parameters like the FI and peak intervals, the reward amount, and the cost per response, along with parameters like the session duration and break duration (it is often useful to give participants a short break to reduce the effects of boredom and fatigue).

In Psychopy, procedures are written as a series of routines. Each routine performs some function, and either waits for a response or some amount of time before moving to the next routine. Typically, there will be a routine for the initial instructions, for closing the experiment, and one or more routines for running the actual trial. PsychoPy provides a graphical user interface for creating these routines. The critical routine for the PI procedure is shown in Figure 6. 1. The top image shows the procedure schematic, and the bottom image shows the primary PsychoPy routine that runs a trial. Stimuli and responses are added as components to each routine. A peak interval trial consists of a keyboard component (for responses), a text component (for the fixation point), and a code component. The code component records the time of each response, and keeps track of the score per response. Not shown in the schematic is the ITI component, but it is important. It not only waits a random time before starting a new trial (randomly drawn between 0.5 and 1.5 s), it also randomizes which trial type to show. Conditional branching in Psychopy is done by creating Routine "Loops" around each trial type that executes either zero or one times per trial. For example, on PI trials, the "Fixed Interval" loop will execute zero times, the "Peak Interval" loop will execute once, the "feedback" loop will execute once, and the "break" loop will execute only if it is time for the participant to take a break (the time between breaks is a variable specified in the "peak_session_information.csv" file).

### 2.3 *Data Processing and Analysis*

#### 2.3.1 Average Response Curves

Data gathered from the PI trials are typically expressed in the form of average response curves. In order to build the response curves, the trial duration is divided into fixed-width bins (typically 1 s and the number of responses that fall

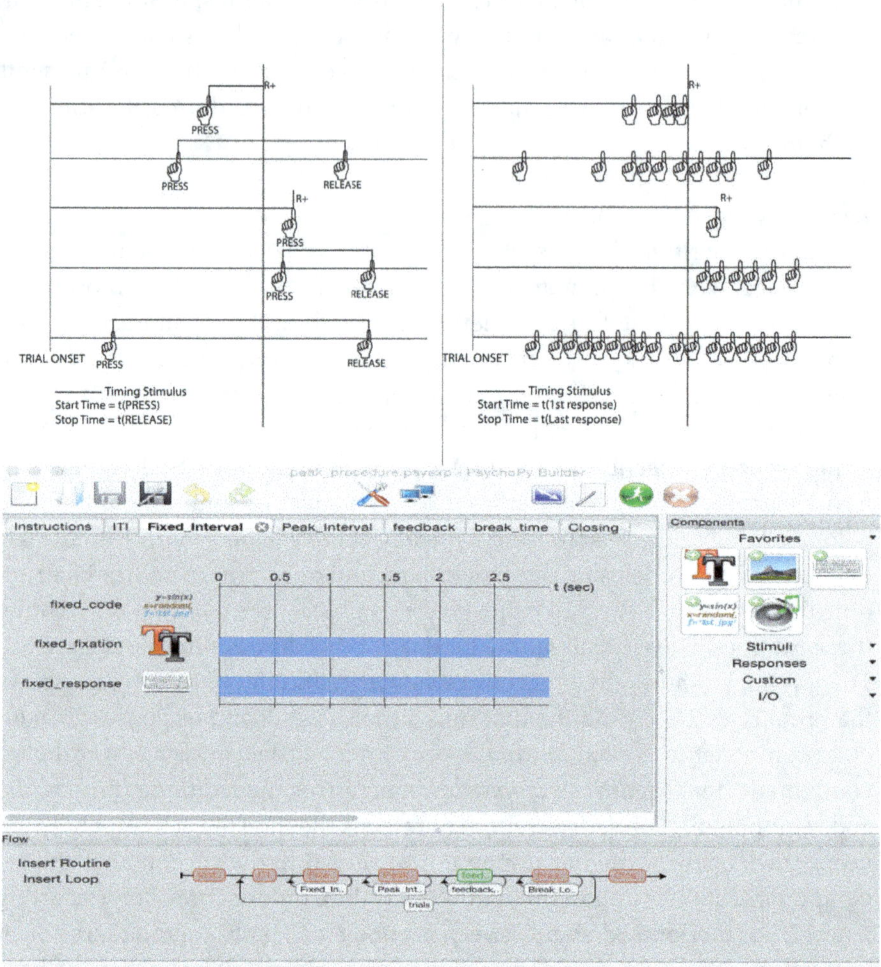


FIGURE 6.1 *The peak interval task. Top: A schematic of the task. Bottom: The PsychoPy builder screenshot of a fixed interval trial.*

within each bin is recorded for each trial. The average over trials is computed. The result is a scaled histogram that gives the average number of responses per second, the response rate in each one-second bin. The average response curve is roughly normally distributed with its peak around the typical reinforcement time. The reward expectancy increases smoothly as the participant approaches the reinforcement time, and decreases smoothly after The more data collected, the smoother the curve (we show data from a single participant

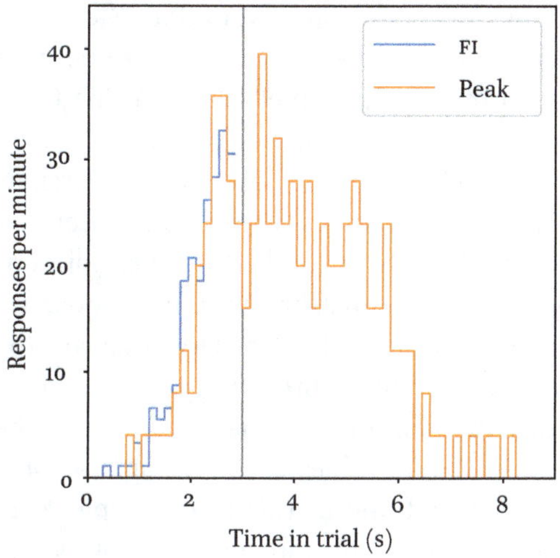


FIGURE 6.2
*Example average response curves for Fixed Interval (blue) and Peak Interval (red) trials for a short 30-minute session for a single participant. The target duration was 3s in this example.*

in a short session in figure 6.2). Many studies also report a slight right skew in the peak response curves, which may theoretically important (e.g., Balcı & Simen 2016; Simen et al. 2011; 2013).

A number of key parameters can be estimated from the average response curve using parametric and nonparametric statistics. The primary parameters are: (i) the location of the peak (peak time), which indicates timing accuracy. The closer the peak time is to the typical food time, the more accurate the participant. And (ii) the width of the response curve, which indicates timing precision. The wider the distribution, the worse a timer a participant is. The first parameter can be estimated by simply locating the trial time of the maximum of the response curve, although this might prove difficult if there is substantial amount of noise in the data. In these cases, the data can be smoothened using methods that would minimize the shift in the actual curve. Alternatively, researchers can take the average of the trial times at which the response curve first exceeds and then first fell below the 90% of the curve's peak Similarly, the second parameter can be estimated by finding the distance in time between the distributions 25th and 75th percentiles. These constitute nonparametric approaches to the estimation of the timing accuracy and precision.

The parametric approach finds the best fitting distribution function to the response curve. For instance, the best-fit mean $\mu$ parameter of the normal distribution function is a measure of the peak time (accuracy), and the best-fit standard deviation $\sigma$ is a measure of spread (timing precision). The ratio of the standard deviation to the mean is the coefficient of variation ($\sigma / \mu$), a critically important measure of precision in the timing literature because it is scale-free, it does not depend on the interval timed. Before using these methods, researchers should study the shape of the peak response curves to choose the best function possible to characterize their data. Researchers should default to using all of the available data unless there is a strict outlier criterion specified in advance. A reasonable metric for the fitted distributions is how well the model captures the data (e.g., a measure like omega squared). In animal data a second gradual increase is sometimes observed toward the end of a PI trial. This increase is thought to reflect anticipatory responding to the next trial. Increasing the stimulus interval on peak trials and making the ITI less predictable can minimize this. When it does occur, fitting a second, increasing function to the unimodal distribution function (for example, a cumulative distribution function) might be necessary. Finally, the amplitude of the peak response curve (peak rate) is another meaningful parameter; it correlates with the subjective value of the anticipated reward in animal studies (Roberts, 1981).

### 2.3.2 Single Trial Responding

Although the average response curve suggests that participants smoothly ramp up their response rate, the responses in individual trials suggest that participants start responding abruptly at some variable time in each trial. Participants seldom respond early in the trial, then abruptly start rapidly responding. When food is not delivered, participants abruptly lower their response rate again (called the "break-run-break" pattern). This pattern is shown in Figure 6.3. In animal studies, where subjects emit many responses in a trial, experimenters need to apply change detection algorithms to estimate the trial time in which there is an abrupt increase (start time) and an abrupt decrease in response rates (stop time; see Church, Meck, & Gibbon 1994). In human studies, the first response time can often be treated as the start time and the last response time can be treated as the stop time. In the press-release version of this task, the press time is the start time, and the release is the stop time. The time between the start and stop called the spread, is a trial-by-trial index of timing precision and their midpoint can be treated as the trial-based timing accuracy. The distribution of both the start and stop times can give additional measures of timing accuracy and precision (Freestone, MacInnis, & Church 2013). The data analysis tutorial that accompanies this chapter implements these analyses.

There are an enormous number of change detection algorithms that can be used to estimate the start and stop times. Many of these are computationally

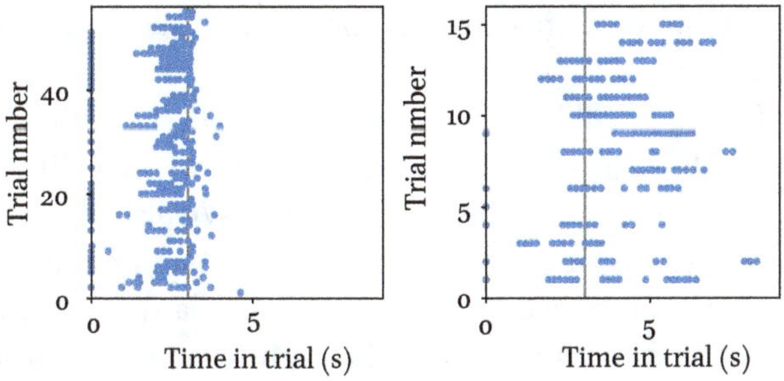


FIGURE 6.3 *Example single trial raster plots for Fixed Interval (left) and Peak Interval (right) trials for a short 30-minute session for a single participant.*

intensive and were not designed to detect changes in hundreds or thousands of trials quickly or efficiently. A few change detection algorithms are uniquely designed to find the start (and sometimes stop) times in FI schedules. The Church, Meck, and Gibbon (1994) method finds the set of transition times that maximize the total distance between each segment's response rate and the overall rate. In other words, it finds the response that maximally segments the data into low and high response rates. The usual implementation of this method is to run through each combination of possible start and stop times on every trial and compute the distance metric just described. The computational cost of this approach is high and depends heavily on the number of trials and number of responses per trial. Gallistel, Fairhurts, and Balsam's (2004) method finds the point that maximizes the distance between the cumulative response data and the null hypothesis line connecting the first point to the last. This method is computationally fast for detecting the start times, but when detecting multiple transitions, the usual implementation is to iterate over the responses until a significant change-point is found, then segment the data and start over. This reduces the speed of the algorithm, and requires a researcher-defined significance value for deciding when a change has been found. For finding start times, these two methods are conceptually different but mathematically identical. They both find the location, where the cumulative residuals against the mean peaks. Framed this way leads to an algorithm that is computationally very fast; the start times for thousands of trials can be estimated in less than a second. A more general method is to use the cumulative residuals against a regression line, called the OLS-CUSUM algorithm (Ploberger & Kramer 1992). These three methods differ for the PI task when both start and stop times need to be estimated. More formal methods for jointly estimating the start and stop times exist, but they are computationally intensive, even for a relatively small number of a trials.

Because the Church, Meck, and Gibbon (1994) implementation is still the most heavily used, we include it in the accompanying Python code (Just-In-Time compiled to be faster; "see the Github repository"). We also include an algorithm that combines aspects of the above discussion. We fit a regression line to the cumulative inter-response times (not the cumulative record), and compute the residuals. The minimum residual is the stop time. The start time is the maximum residual in the data record up to the stop time. That is, start times are conditioned on stop times. This method is computationally efficient, and seems to do a robust job on our data. It should be noted that any algorithm has pros, cons, and edge-cases in which the algorithm fails to find reasonable change-points. The experimenter should examine their particular results to assess the degree to which the algorithm provides a good description of the data.

### 2.3.3 Analysis Code

The analysis code uses the Python programming language with the scientific stack, most notably Pandas. Python is used because it is free, open-source, cross-platform, and includes libraries that allow for quick prototyping of the task (PsychoPy). The Pandas package contains a rich set of functions that operate on the split-apply-combine framework (Wickham 2011). This allows one to write code for a single instance and then apply it to groups of data in a single line of code, both maximizing readability and minimizing errors. The Jupyter Notebook allows for figures, code, and markdown/latex text in the same document, effectively creating a human readable analysis script. All of the necessary software was downloaded using the Python Anaconda distribution from Continuum Analytics (https://www.continuum.io/). The analysis uses Python (version 3.5), Jupyter Notebook (v4.2), Pandas (v0.18), Numpy (v1.10.4), and Matplotlib (v1.5.1). The R programming language with the "tidyverse" ecosystem also implements this analysis philosophy, and RStudio includes the ability to integrate figures, code, and text into a notebook.

The code that analyzes the PI task is called "Peak_Analysis.ipynb." It renders as HTML in the browser, but can be used as an analysis script when downloaded. It shows the stages of analysis from start to finish: from loading in the data, to data wrangling (cleaning) to analysis. The outputs are figures and tables. It shows both the averaged and single trial analysis methods discussed in this chapter.

## 3 Switch Procedure

### 3.1 *Background*

The switch procedure, rooted in the free-operant temporal discrimination task Platt and Davis (1983), is closely related to the PI procedure described above and the temporal bisection task (Chapter 5, this volume). The temporal bisection task is arguably one of the most commonly used procedures in the study of temporal judgments of humans and non-human animals (Church & Deluty 1977; Allan & Gibbon 1991; Wearden, 1991). In this temporal bisection task, participants are initially trained to discriminate two reference intervals as short and long (e.g., 200 and 800 ms). Once the participants learn to discriminate accurately (e.g., 85% correct), intermediate durations are presented intermixed with the reference durations. Participants are asked to classify these durations as short or long depending on their subjective similarity to the reference durations. No feedback is provided for the categorization of intermediate intervals to avoid explicitly training them on intermediate intervals.

Experimenters can estimate both timing precision and the point of subjective equality (the interval that participants judge as equally distant from the referents) by fitting a logistic function to the participant's trial-by-trial choices to predict the proportion of long responses as a function of test durations. (see Chapter 5, this volume). Although the responses are emitted after the termination of the timing stimulus in the temporal bisection task, recent work showed that both human and non-human decision processes actually evolve dynamically during the timing stimulus (e.g., Balcı & Simen, 2014; Machado & Keen 2003).

The switch task was developed specifically to capture this dynamically evolving belief state (Balcı et al. 2008, 2009; Kheifets & Gallistel 2012; Kheifets, Freestone, & Gallistel, 2017). This Switch task is the prospective analogue of the temporal bisection task in which participants behaviorally invest in the "short" and "long" latency options freely over the course of the trial (during the stimulus). In this task, only the referent durations are presented. The participants are reinforced for "catching" the reinforcement at the right location at the right time. For example, a mouse will be reinforced for poking its nose into the left port at or after short trials, and reinforced for poking its nose into the right port at or after long trials. The mice do not know in advance whether it is a short or long trial, they learn to switch from the short to the long port between the intervals in order to catch the reinforcer no matter the trial type. Humans may be asked to hold down the left key and then switch to the right when they believe the short duration has elapsed without reinforcement. This allows the experimenter to observe, on an individual trial, when the belief state of the participant switches from short to long. In other words, instead of estimating the PSE from binary response data, the criterion is directly measured trial-by-trial via the switch times. Below, we present the procedural implementation and analysis for this task.

### 3.2 *Procedural Details*

Participants are trained to anticipate the reinforcement at two different locations associated with two different intervals (short and long). These are often two different feeding hoppers located at two different sides of the operant chamber in animals and visual targets presented at two different sides of the computer screen in humans. In a given trial, only one of these options is active. The active option is not signaled to the participants; they can only rely on the elapsed time to guide their responding. After experiencing the task parameters in the first trials/sessions, participants often begin at the short location and switch to the long location once they believe the short duration has elapsed without reinforcement. This switch time is the main unit of analysis. In human experiments, participants indicate their choices by pressing one key to indicate their preference for the short option and a second key to indicate their preference for the long option.

### 3.3 *Procedure Code*

As before, the task was written using PsychoPy (version 1.83.03), and is available on our Github page. When the program loads, it reads a file "Switch_session_information.csv" that specifies the task parameters like the short and long durations, payoffs, and probabilities, along with parameters like the session duration and break duration.

The critical PsychoPy routine is shown in the bottom panel of Figure 6.2, and is described by the procedure schematic in the top panel. A trial consists of a keyboard component (for responses), two shape components (to give visual feedback about which response is currently being recorded), two text components (one that controls the fixation cue that starts the trial and one that keeps track of the participant's total score), interstimulus interval (ISI; duration without any stimuli drawn from a uniform interval between 0.5 and 1.5 seconds), and a code component. The code component draws a trial type (short or long), and records whether or not the final response was correct.

### 3.4 *Data Processing and Analysis*

The trial time at which the participant leaves the short latency option for the long latency option is calculated for each trial. The switch latencies aggregated

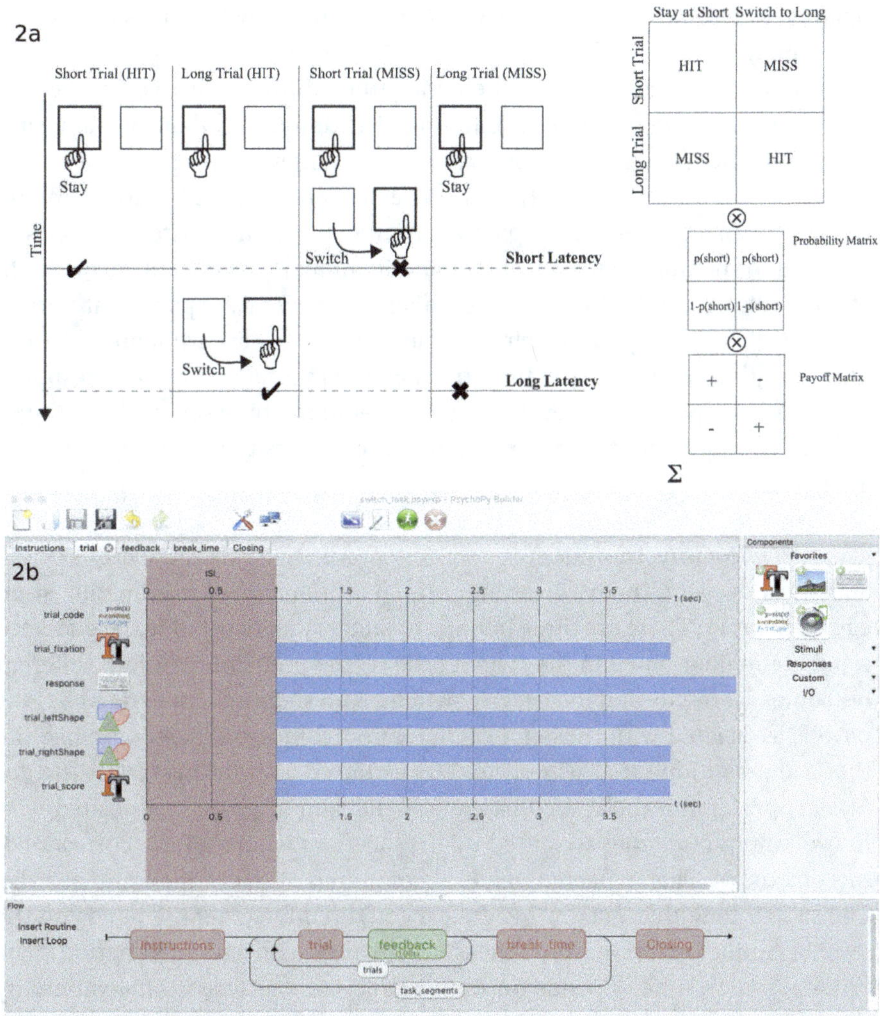


FIGURE 6.4 *The switch task. Top: A schematic of the task. Bottom: The PsychoPy builder screenshot of the trial.*

across multiple long-latency trials are nearly normally distributed. The mean (or median) switch time can be treated as the PSE, whereas the coefficient of variation (or interquartile range) of the switch times can be treated as an index of precision in temporal judgments. Note that only the data from the long-latency trials are used in the analysis since participants often do not switch on short trials (they shouldn't). The primary advantage of the switch task over the temporal bisection task is that the belief state of the participant can be evaluated in real time rather than being evaluated at arbitrarily chosen decision points (i.e., test durations). Figure 6.5 shows about 40 switch times for a single participant.

Although it applies more to the data gathered from animals, another advantage of the switch task over temporal bisection task is that the data gathered from the switch task allows the experimenter to treat responses that originate from timed vs. non-timing processes separately. It is fairly common to observe an exponential (impulsive) component in the switch latencies in addition to the normally distributed data in animals (Balcı et al. 2009; Khefiets & Gallistel, 2012). Fitting mixture distributions (e.g., exponential normal mixture distribution) to switch latencies allows the experimenter to work only with the data that come from trials with temporal control over responding. This is simply not achievable with the temporal bisection task. The tutorial that accompanies this chapter allows the readers to easily conduct these analyses.

### 3.5 *Optimality Analysis*

Statistical decision theory gives an optimal reinforcement-maximizing strategy on this task. There are three important factors that determine the expected reinforcement attained in the switch task: (1) the probability of four different outcomes (correctly and incorrectly switching on short and long trials), (2) the payoffs associated with those four different outcomes and (3) the probability that a short or long trial will occur. The expected gain in this task is the dot (scalar) product of these vectors; namely the sum of payoffs associated with four different outcomes weighted by the joint probability of the corresponding outcomes. This expected value is computed at every moment in time. The switch time that maximizes the expected reinforcement depends on the level of timing precision (often measured by the coefficient of variation). The optimal switch time depends on timing precision because timing variability determines the probability of switching between the short and the long intervals. Worse timers should switch earlier. The tutorial that accompanies this

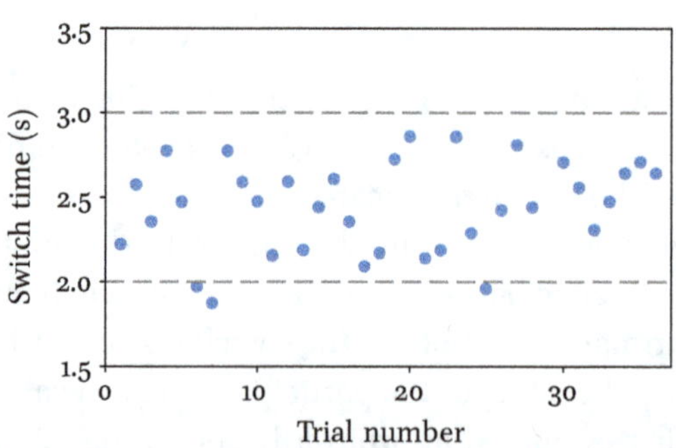


FIGURE 6.5
*Example switch times from a participant on the switch procedure. The short duration was 2 s and the long duration was 3 s.*

chapter allows the readers to conduct the optimality analysis of the data gathered from the switch task.

## 3.6 *Analysis Code*

The analysis code is contained in "Switch_Analysis.ipynb" on our Github repository. It walks the user through the stages of analysis, from reading in the data file to assessing optimality in the participants. A detailed mathematical treatment of the optimal solution accompanies its implementation. Briefly, it is possible to construct performance curves that specify the expected reinforcement for any given response mean switch time. Three example curves are given in Figure 6.6. Each curve shows a different level of timing precision. A coefficient of variation of 0.20 (blue curve), for example, means that most of the switches on long trials will be within about 20% of the mean switch time. The optimal time to switch from the short to the long location is when this curve peaks. Notice that as timing precision grows (from the blue to the red to the orange curve), the peak of the curve shifts earlier. The black line shows the optimal performance curve –the curve that specifies what the optimal switch time should be for any given level of timing precision.

From here, experimenters can assess the degree to which their participants are optimal: find the optimal switch time for each participant (because their timing precision varies), and then compare it to their actual mean switch time. Figure 6.7 shows two such comparisons. First, how closely the participants match the optimal switch time as the ratio of the actual to the optimal switch time. (left), and second, how much reinforcement they earn compared to what they would earn if they were optimal, again as a ratio (Balcı, Freestone, & Gallistel 2009 see also Freestone et al. 2015).

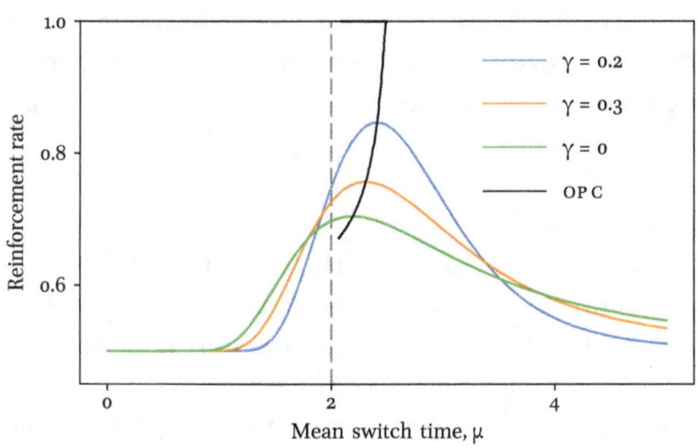


FIGURE 6.6 *Example performance curves for the switch task. The black optimal performance curve shows that the optimal switch time decreases as the coefficient of variation increases.*

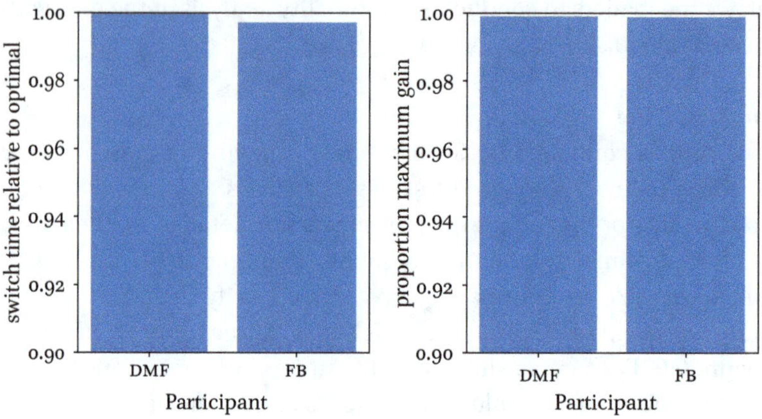


FIGURE 6.7 *Example ways of assessing optimality in participants. Notice the y-limits start at .90, humans are often very close to optimal on this task.*

## 4 The Differential Reinforcement of Low Rates Procedure

### 4.1 *Background*

The DRL procedure has been widely used in psychopharmacological animal studies because it is sensitive to anti-depressant pharmacological agents (e.g., Paterson et al. 2011). This task requires participants to wait some amount of time before a response will be reinforced, early responses reset the clock. That is, the DRL task requires participants to withhold responding for at least a fixed interval. Responses after this fixed interval are reinforced, responses earlier than this fixed interval are not reinforced. Both types of responses reset clock. For instance, in a DRL-20s schedule, the minimum wait time since the previous response is set to 20 s. Typically, the average inter-response time in this task is longer than the DRL schedule, and roughly positioned to maximize reinforcement rate (Balcı et al. 2011; Çavdaroğlu et al. 2014; Freestone, Balcı, Simen, & Church 2015; Wearden 1990).

### 4.2 *Procedural Details*

The implementation of this task in humans is fairly straightforward. Participants are presented with a stimulus in the middle of the computer screen. They can be instructed to wait for a minimum interval since their previous response before responding again. The minimum interval can be presented at the beginning of the test session for brief training. In other versions of the task, instructions are not provided. If the participant does not wait long enough before responding, the stimulus can turn red briefly, whereas if they have waited for the minimum interval, the stimulus can turn green briefly, both to indicate the outcome of their response.

### 4.3 *Procedure Code*

When the experiment starts, it reads the file "DRL_session_information.csv," which contains the task parameters like the DRL interval, the magnitude of the reward, the cost of a response, as well as information about the experiment, like the session and break durations. The top panel of Figure 6.8 depicts a schematic of the task, the bottom panel shows a screenshot of the PsychoPy builder for the experiment, focused on the "DRL_task" routine. It contains fixation and feedback text components, along with a keyboard component for recording responses. A code component keeps track of each inter-response time and controls the reinforcement.

### 4.4 *Data Processing and Analysis*

The primary unit of analysis in the DRL task is the inter-response time (IRT). Similar to the Peak procedure, the distributional of IRTs provide information accuracy and precision of timing. In animal data, the IRTs typically originate from two different generative processes leading to a mixture distribution (e.g., exponential and an inverse Gaussian). In these cases, the exponential portion of the IRTs is assumed to originate from those responses for which there was no temporal control over behavior, and the inverse Gaussian portion of the IRTs is assumed to originate from trials in which the responses were under temporal control (the inverse Gaussian distribution). The central tendency relative to the DRL schedule gives timing accuracy, and the spread gives precision. The proportion of inverse Gaussian IRTs can also be used as an index of the degree of temporal control. These parameters are sensitive to both motivational and pharmacological manipulations (Paterson et al., 2011; Doughty & Richards 2002). In the human data, the proportion of exponentially distributed IRTs is virtually zero after training, showing that humans have stronger temporal control over their waiting behavior. Consequently, often fitting a single distribution to human IRTs is sufficient to estimate the parameters of performance. Readers can use the tutorial that accompanies this chapter to fit the DRL data.

### 4.5 *Optimality Analysis*

The mean IRTs are typically longer than the DRL schedule. This is an adaptive strategy to maximize reinforcement rate given timing imprecision. Like the switch task, it is possible to mathematically describe optimal IRTs depending

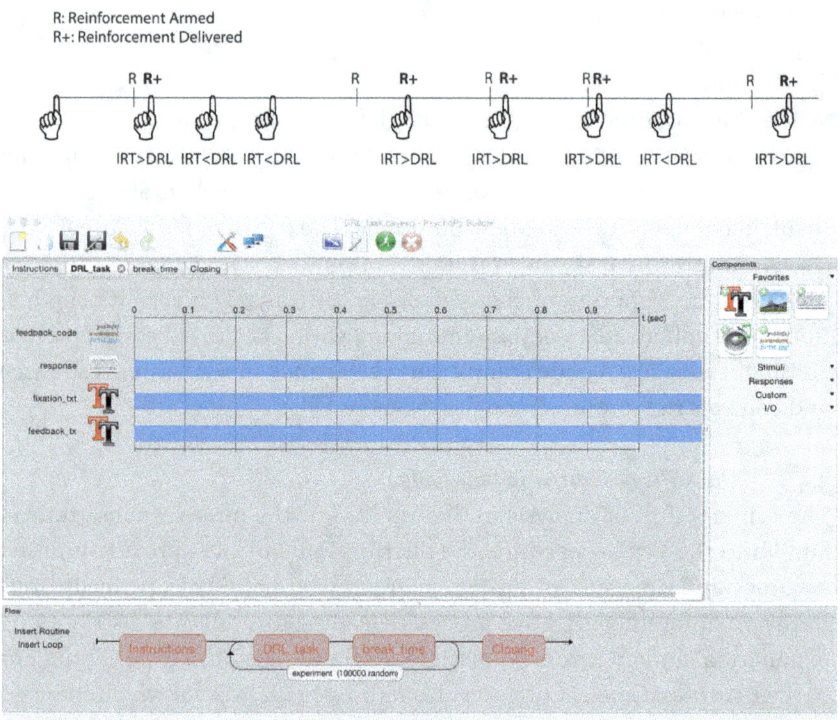


FIGURE 6.8 *The* DRL *Task. Top: A schematic of the* DRL *task. Bottom: A screenshot of the PsychoPy builder used to run the task.*

on the level of the timing precision (e.g., Çavdaroğlu et al. 2014; Freestone, Balcı, Simen, & Church, 2015). To a first approximation, the worse the timer, the later they should aim.

The reinforcement rate in this task is the the probability of reinforcement divided by the time to reinforcement. That is, it's the probability that an inter-response time is later than the DRL schedule divided by the average inter-response time. $(p(r;\mu,\gamma\mu)/\mu)$. As the participants aim to wait longer, the numerator of this ratio – the probability of a reinforcement – increases. The denominator – the time cost – also increases. The longer they wait, the more likely they are to be reinforced, but the longer they have to wait for reinforcement. The reinforcement rate maximizing (i.e., optimal) inter-response time finds the balance between these two time varying quantities. The tutorial accompanying this chapter allows the reader to conduct the optimality analysis of data gathered from the DRL task.

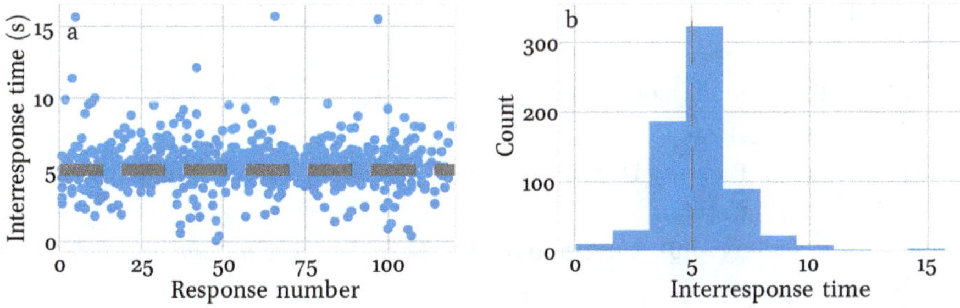


FIGURE 6.9 *Example inter-response times for a single human participant on the* DRL *task. The left panel shows every inter-response time in a scatter plot, and the right panel shows the histogram.*

## 4.6 *Analysis Code*

The analysis code on our Github page is called "DRL_Analysis.ipynb." It walks the user through the analysis up to assessing optimality in the participants, and generating figures. A detailed mathematical treatment of the optimal solution accompanies its implementation.

# 5 Conclusions

In this chapter, we have introduced three different timing procedures that are equally applicable to humans and animals. The PI procedure is a widely used timing task that has provided valuable information regarding the nature of the generative psychological (e.g., Church, Meck & Gibbon, 1994; Gibbon & Church 1990), and neurobiological processes (e.g., Meck 2006) that underlie interval timing. For instance, the "break-run-break" pattern of responding in individual trials (as well as other derived quantities such as the middle time and spread) provide insights regarding the possible sources of noise in the timing behavior. The switch procedure allows the experimenters to track the evolving temporal belief state of the individuals in each trial. This procedure is particularly well-suited to study how quanties like probabilities and payoffs change timing behavior. When it is coupled with the participant's level of timing precision, one can apply statistical decision theory to obtain the optimal response against which the actual data can be compared. Finally, the DRL task is ideal for the study of timed inhibitory control. Performance on this task can also be compared against the optimal response times, and reinforcement-rate maximizing wait time is determined purely by timing precision.

## References


Allan, L.G., & J. Gibbon (1991). Human bisection at the geometric mean. *Learning and Motivation, 22,* 39–58.

Balcı, F., & Simen P. (2014). Decision processes in temporal discrimination. *Acta psychologica, 149,* 157–168.

Balcı, F., & P. Simen (2016). A decision model of timing. *Current Opinion in Behavioral Sciences, 8,* 94–101.

Balcı, F., E.B. Papachristos, C.R. Gallistel, D. Brunner, J. Gibson, & G.P. Shumyatsky (2008). Interval-timing in the genetically modified mouse: A simple paradigm. *Genes, Brain, and Behavior, 7*(3), 373–384.

Balcı, F., D. Freestone, & C.R. Gallistel (2009). Risk assessment in man and mouse. *Proceedings of the National Academy of Sciences, 106*(7), 2459–2463.

Balcı, F., D. Freestone, P. Simen, L. deSouza, J.D. Cohen, & P. Holmes (2011). Optimal temporal risk assessment. *Frontiers in Integrative Neuroscience, 5,* 1–15.

Balcı, F., & Simen, P. (2014). Decision processes in temporal discrimination. *Acta psychologica, 149,* 157–168.

Balcı, F., M. Wiener, B. Cavdaroglu, & B.H. Coslett (2013). Epistasis effects of dopamine genes on interval timing and reward magnitude in humans. *Neuropsychologia, 51*(2), 293–308.

Catania, A.C. (1970). Reinforcement schedules and psychophysical judgments: A study of some temporal properties of behavior. In Schoenfeld, W.N. (ed.), *The theory of reinforcement schedules*. New York: Appleton-Century-Crofts.

Çavdaroğlu, B., M. Zeki, & F. Balcı (2014). Time-based reward maximization. *Philosophical Transactions of the Royal Society of London B: Biological Sciences, 369*(1637), 20120461.

Church, R.M., & M.Z. Deluty (1977). Bisection of temporal intervals. *Journal of Experimental Psychology: Animal Behavior Processes, 3*(3), 216.

Church, R.M., W.H. Meck, & J. Gibbon (1994). Application of scalar timing theory to individual trials. *Journal of Experimental Psychology: Animal Behavior Processes, 20,* 135–155.

Doughty, A.H., & J.B. Richards (2002). Effects of reinforcer magnitude on responding under differential-reinforcement-of-low-rate schedules of rats and pigeons. J Exp Anal Behav. 78,(1) 17–30.

Drew, M.R., B. Zupan, A. Cooke, P.A. Couvillon, & P.D. Balsam (2005). Temporal control of conditioned responding in goldfish. *Journal of Experimental Psychology: Animal Behavior Processes, 31,* 31–39.

Ferster, C.B., & B.F. Skinner (1957). *Schedules of reinforcement.* New York: Appleton-Century-Crofts.

Freestone, D.M., F. Balcı, P. Simen, & R.M. Church (2015). Optimal response rates in humans and rats. *Journal of Experimental Psychology: Animal Learning and Cognition, 41*(1), 39.

Freestone, D.M., M.L. MacInnis, & R.M. Church (2013). Response rates are governed more by time cues than contingency. *Timing & Time Perception, 1*(1), 3–20.

Gallistel, C.R., S. Fairhurst, & P. Balsam (2004). The learning curve: implications of a quantitative analysis. *Proceedings of the national academy of Sciences of the United States of America, 101*(36), 13124–13131.

Gibbon, J., & R.M. Church (1990). Representation of time. *Cognition, 37*, 23–54.

Kheifets, A., & C.R. Gallistel (2012). Mice take calculated risks. *Proceedings of the National Academy of Sciences, 109*(22), 8776–8779.

Kheifets, A., D. Freestone, & C.R. Gallistel (2017). Theoretical implications of quantitative properties of interval timing and probability estimation in mouse and rat. *Journal of the Experimental Analysis of Behavior, 108*(1), 39–72.

Machado, A., & R. Keen (2003). Temporal discrimination in a long operant chamber. *Behavioural Processes, 62*, 157–182.

Meck, W.H. (2006). Neuroanatomical localization of an internal clock: A functional link between mesolimbic, nigrostriatal, and mesocortical dopaminergic systems. *Brain research, 1109*(1), 93–107.

Paterson, N.E., F. Balcı, U. Campbell, B. Olivier, & T. Hanania (2011). The triple reuptake inhibitor DOV216,303 exhibits limited antidepressant-like properties in the differential reinforcement of low-rate 72-sec responding assay, likely due to dopamine reuptake inhibition. *Journal of Psychopharmacology, 25*(10), 1357–1364.

Pavlov, I. P. (1927). Conditional reflexes: An investigation of the physiological activity of the cerebral cortex (V Anrep, Trans). Martino Fine Books.

Platt, J.R., & E.R. Davis (1983). Bisection of temporal intervals by pigeons. *Journal of Experimental Psychology: Animal Behavioral Processes, 9*, 160–170.

Ploberger, W., & W. Krämer (1992). The CUSUM test with OLS residuals. *Econometrica: Journal of the Econometric Society*, 271–285.

Rakitin, B.C., J. Gibbon, T.B. Penney, C. Malapani, S.C. Hinton, & W.H. Meck (1998). Scalar expectancy theory and peak-interval timing in humans. *Journal of Experimental Psychology: Animal Behavior Processes, 24*, 15–33.

Roberts, S. (1981). Isolation of an internal clock. *Journal of Experimental Psychology: Animal Behavior Processes, 7*(3), 242.

Schneider, B. A. (1969). A two-state analysis of fixed interval responding in the pigeon. *Journal of the Experimental Analysis of Behavior, 12*(5), 677–687.

Simen, P., F. Balcı, L. deSouza, P. Holmes, & J.D. Cohen (2011). A model of interval timing by neural integration. *Journal of Neuroscience, 31*(25), 9238–9253.

Simen, P., F. Rivest, E.A. Ludvig, F. Balcı, & P. Killeen (2013). Timescale invariance in the pacemaker-accumulator family of timing models. *Timing & Time Perception, 30,* 159–188.

Stoddard, L. T., Sidman, M., & Brady, J. V. (1988). Fixed-interval and fixed-ratio reinforcement schedules with human subjects. *The Analysis of verbal behavior, 6*(1), 33–44.

Wearden, J.H. (1990). Maximizing reinforcement rate on spaced-responding schedules under conditions of temporal uncertainty. *Behavioural processes, 22*(1), 47–59.

Wearden, J. H. (1991). Human performance on an analogue of an interval bisection task. *The Quarterly Journal of Experimental Psychology, 43*(1), 59–81.

Wearden, J.H. (2002). Travelling in time: A time-left analogue for humans. *Journal of Experimental Psychology: Animal Behaviour Processes, 28,* 200–208.

Wickham, H. (2011). The split-apply-combine strategy for data analysis. *Journal of Statistical Software, 40*(1), 1–29.

CHAPTER 7

# Towards a Process Model of Temporal Generalization

*Florian Klapproth*

## 1 Introduction

There are just a few studies concerned with response time in stimulus generalization. For example, Pierrel and Murray (1963) showed that when people compared a given standard with stimuli differing from the standard only in weight magnitude, the decision time increased the smaller the difference was between the standard and the comparison stimulus. Desiderato (1964) presented participants with stimuli repeatedly after a 12-s interstimulus interval in the training phase. The participants were required to release a button as soon as they experienced the stimuli. In the subsequent testing phase, the same stimuli were presented after intervals that were either shorter than, longer than, or the same as the initial 12-s interstimulus interval. The time it took to release the finger from the button increased as a function of the difference between the standard and the comparison intervals. The task that was applied by Desiderato is commonly called a temporal-generalization task.

In temporal generalization, the participants are usually presented with a standard duration after which a series of stimuli of different durations follows, and the participant's task is to judge whether each of the successive durations matches the standard duration or not, by making a *same* or *different* response. A typical result is that on most of the trials when it is presented the standard duration is reliably identified (that is, a *same* response is given). On some trials, however, people will make a *same* judgment even though a non-standard duration had been presented. Hence, people are likely to generalize a *same* judgment to stimuli that are shorter or longer than the standard, and the probability of making a *same* response after the presentation of a non-standard comparison stimulus increases as the difference between the non-standard comparison and the standard itself becomes smaller. When plotting the proportions of *same* responses against the stimulus durations presented a gradient of generalization will result. The steeper the gradient is, the more accurately the participants have identified the standard amongst the other durations of the series. Usually, the gradient is slightly asymmetrical, with more *same*

 | DOI 10.1163/9789004280205_008

responses occurring at stimuli longer than the standard (e.g., Wearden, 1992; Wearden, Denovan, Fakhri, & Haworth, 1997; Weisman et al., 1999).

When comparing the standard duration with a comparison duration in a temporal-generalization task, people are supposed to relate the durations in a way suggested by Wearden (1992). Wearden assumed that people decide for the sameness of both intervals if $abs(s\text{-}t)\ /t < b$, with $s$ being the standard duration, $t$ being the comparison duration, and $b$ representing a decision threshold. Wearden found that a decision according to that rule fitted empirical data to a high degree of precision, when $s$ and $b$ were variables with random error, but with fixed mean and standard deviation, and $t$ was an error-free variable, changing its value only depending on the comparison duration.

The selection of the parameters used in the Wearden's (1992) model was inspired by the *scalar timing theory* (Gibbon & Church, 1984; Gibbon, Church, & Meck, 1984), which offers an information processing model of the timing processes in humans and animals. The model comprises a clock stage, a memory stage, and a comparison stage. It assumes that when the duration of a stimulus has to be judged, a clock-like mechanism produces pulses, and these pulses are added and temporarily stored in an accumulator. The number of pulses summed in the accumulator represents the duration to be judged. When an interval is regarded as important (like the standard in a temporal generalization task), it should not only be stored temporarily, but for a longer period (at least for the duration of the task at hand). To account for this, the scalar timing theory posits a longer-term (or reference) memory. When making a comparison between the standard and a just-presented duration, the scalar timing theory presumes that a sample of the standard represented in reference memory enters the comparison stage and is related to the current comparison interval. The representation of the standard is assumed to be noisy, so that it may vary from trial to trial due to variations in encoding the standard or as a result of transferring the standard to the comparison stage (Gallistel & Gibbon, 2000; Gibbon, 1991).

The scalar timing theory specifies the stages that temporal information has to pass through before a decision can be made. However, the model does not describe or predict the time one needs to make a temporal decision. An attempt to characterize the time course of temporal decisions was made by Klapproth and Müller (2008). In their experiments, participants carried out a temporal generalization task, where one group was encouraged to make a temporal judgment as quickly as possible, whereas the other group was not instructed to make a fast judgment. Klapproth and Müller were interested in investigating the relationship between the duration of the comparison stimuli and the timing of the participants' responses, and they expected that when the speed of responses is emphasized, it would result in reliable estimates of the time needed to judge the duration of a stimulus.

## 2 A Theoretical Model for Response Times in Temporal Generalization

Klapproth and Müller (2008) suggested a model of the relationship between $abs(s - t)/t$, the response time (RT), and the stimulus duration ($t$), which is depicted in Figure 7.1. Response time is defined as the interval between stimulus onset and the participant's response.

The time of the decision, and, hence, the response time (RT) for a correct decision, can be predicted by the relation of the ratio $abs(s - t)/t$ to the criterion $b$. The ratio $abs(s - t)/t$ decreases as the duration of the stimulus presentation ($T$) increases, and reaches a minimum at the instant $T_S$, where $t = s$. A further increase of the stimulus duration $T$ would cause an increase of $abs(s - t)/t$ which eventually approximates 1, since the longer t is, the more similar will be the numerator and the denominator of the ratio. As can be seen in Figure 7.1, the function $f(T) = abs(s - t)/t$ crosses the threshold $b$ twice. The interval between the two points of intersection, $(T_1; b)$ and $(T_2; b)$, marks the duration values for which a *same* response is expected to occur

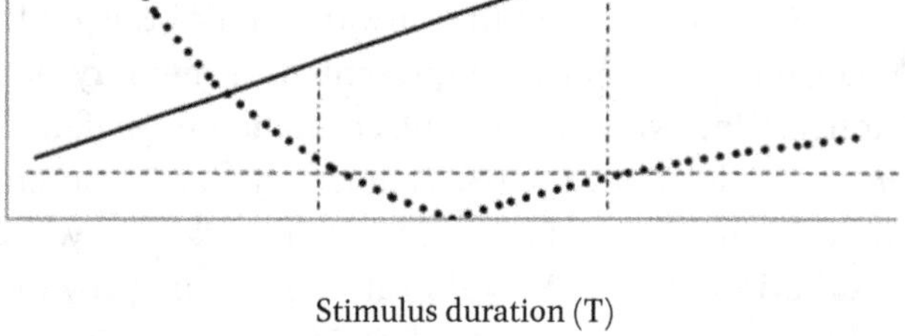


FIGURE 7.1 *The relation between abs(s – t)/t, response time (RT), and stimulus duration (T). The dotted curved line represents abs(s – t)/t as a function of T, the solid line represents the response times as a function of T, and the vertical lines represent the time values T1 and T2. The horizontal dotted line shows b.*
MODIFIED AFTER KLAPPROTH AND MÜLLER (2008).

Response times for *same* responses should increase with increasing stimulus duration, as the participants cannot make a *same* decision until after the stimulus has ended, because the stimulus might always go on for longer. For *different* responses, this is true up until the standard duration (plus some additional time) has been arrived at, then response times should flatten out, as the participant may have enough information to make the response before the entire stimulus has ended.

Based on these assumptions, Klapproth and Müller (2008) proposed a temporal decision rule according to which a decision for the *sameness* of two intervals is expected if $T_1 \leq T \leq T_2$, where $T$ is the duration of the stimulus just presented. A decision for the *difference* between two intervals is expected if $T < T_1$ or $T > T_2$. At $T_1$, $t$ is smaller than $s$, so that $t(T_1) = s/(1 + b)$, where $t(T_1)$ is the mental representation of $T_1$. The reverse relation between $t$ and $s$ can be observed at $T_2$, where $t > s$, so that $t(T_2) = s/(1 - b)$.

The response times should be linearly related to the stimulus duration for stimulus durations that are shorter than or equal to $T_2$, whereas it was presumed that constant response latencies should occur for durations that are longer than $T_2$. Similar assumptions have been made, for instance, by Balcı and Simen (2014)

## 2.1 *Gradients of Temporal Generalization*

In the study conducted by Klapproth and Müller (2008) both proportions of *same* responses as well as response times were recorded and analyzed. The gradients depicted below (Figure 7.2) reflect the proportions of *same* responses obtained at different stimulus durations in the first experiment of this study. The standard stimulus duration was 1000 ms. When the participants responded quickly, a leftward shift of the gradient occurred, compared to the gradient obtained from the participants who were not urged to respond quickly.

Klapproth and Müller (2008) presumed that the leftward shift of the gradient was the result of a change of the standard represented in memory due to the truncation of the longest intervals of the series at hand, and referred to the *adaptation-level theory* (Helson, 1948, 1964; Thomas, 1993) as a potential source of explanation of the gradient's shift. According to this theory, the mean value of all stimulus durations presented within an experiment provides a frame of reference for making a decision about the duration of each comparison stimulus. In temporal-generalization tasks normally used, the mean of all durations equals (or is near to) the standard duration. In the speed conditions of the Klapproth and Müller (2008) study, however, the participants may not have perceived the full duration of the longest comparison stimuli of the series

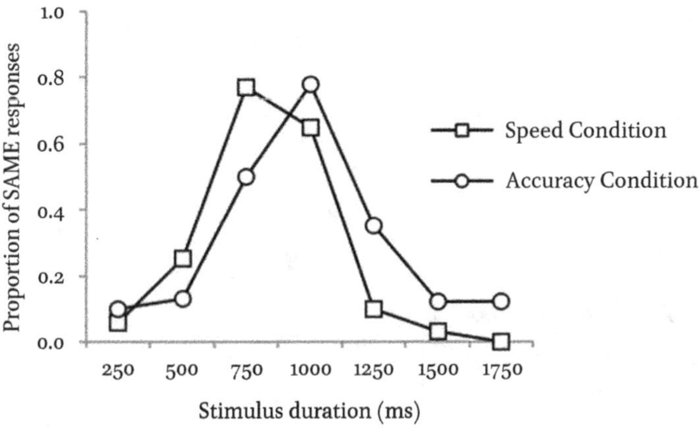


FIGURE 7.2 *Temporal generalization gradients obtained from Experiment 1 of the Klapproth and Müller (2008) study.*
MODIFIED AFTER KLAPPROTH AND MÜLLER (2008).

because they were able to make a response prior to the cessation of these long stimuli. Therefore, the mean of the durations experienced by the participants may have been smaller in the speed conditions than in the control conditions. Klapproth and Müller concluded that the participants in the speed conditions related the comparison durations not to the standard itself but to the mean of all experienced durations, a value which apparently was smaller than the nominal standard duration. Hence, the gradient of the speed conditions had a leftward shift, whereas the gradient of the control conditions was not shifted.

### 2.2 *Response Times*

The response times were much shorter in the speed groups than in the control groups, and the time taken for *same* responses increased linearly with stimulus duration, whereas the time taken for *different* responses grew proportionally with stimulus duration but then remained approximately constant at stimulus durations longer than the standard. Figure 7.3 shows the standardized response times obtained from the Klapproth and Müller (2008) experiments. Standardized response times were response times minus the intercept of the linear regression of response times, divided by the standard duration. In the Klapproth and Müller study, the standard durations used were 750, 1000, and 1250 ms. The standardization has therefore resulted in response times that were close to the identity line (which is, in Figure 7.3, the dotted diagonal), and response times would have been larger without standardization.

However, some unexpected deviations from the predicted RT model occurred. First, *same* responses at the longest durations were given faster than

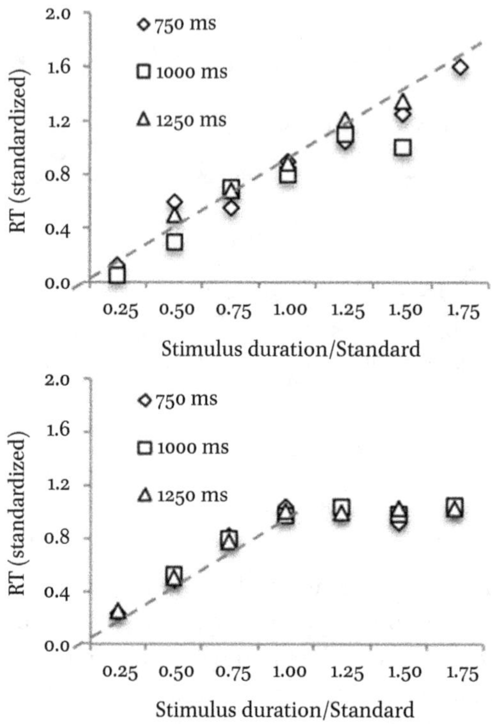


FIGURE 7.3 *Standardized response times, plotted against comparison stimulus duration divided by the corresponding standard. Data are shown separately for the conditions with the 750, 1000, and 1250 ms standard. The dotted grey line in the background represents the expected standardized* RT. *Upper panel: response times for "same" responses; lower panel: response times for "different" responses.*
MODIFIED AFTER KLAPPROTH AND MÜLLER (2008).

*same* responses at shorter durations. This might have been done because these responses were merely "fast guesses". Moreover, the number of *same* responses given at the longest durations was quite small, so that the data points at these durations were rather unreliable (and, therefore, might not reflect a stable trend). Second, *different* responses were somewhat slower than *same* responses before the turning point ($T_2$) occurred. According to the literature pertinent to this phenomenon (see, for example, Farrell, 1985), I will call this result a *fast-same effect.*

### 2.3 The Fast-same Effect

How could *same* responses be faster than *different* responses? This is an "old" question since there are quite a lot of results and models being related to this question, of which some have been proven to be more successful than others

(Farell, 1985). For example, fast-same responses have been discussed as a matter of *encoding facilitation* (Nickerson, 1975; Posner, Klein, Summers, & Buggie, 1973; Posner & Snyder, 1975). The time taken to process a stimulus might depend not only on the stimulus presented, but also on preceding stimuli. In particular, stimulus repetition might facilitate the encoding of latter occurrences of that stimulus, perhaps by priming neural pathways. Since in temporal generalization (and, in particular, in the experiments we conducted) the target stimulus (i.e., the standard) had been presented more frequently than other stimuli, encoding facilitation might be a cause for the fast-same effect observed. However, in our experiments fast-same responses were given mainly at intervals shorter than the initial standard, and repetition of the standard would not have facilitated the encoding of stimuli shorter than the standard.

According to Krueger's noisy-operator model (Krueger, 1978), comparing a test stimulus with a target might involve counting the mismatches between a number of stimulus attributes. Due to noise in the comparison process, some mismatch counts are indecisive, leading to a rechecking process. The probability of rechecking is supposed to be greater for *different* responses than for *same* responses, thus *different* response times will on average be longer *same* response times. However, in temporal generalization, only one feature (the duration) is relevant for comparison.

Despite there being some approaches aiming at explaining the fast-same effect, there seems to be no explaining model that especially covers this phenomenon with timing data. My approach to resolve this problem is based on fairly simple assumptions. The first assumption is that response time depends on the subjective probability for an event to occur in choice tasks. The more likely an event is to occur, the more "prepared" should a participant be, the higher should be his or her "readiness" to respond to that event.

Figure 7.4 illustrates the probability model. Suppose there are two alternatives a participant has to select from, for example, the choice between a *same* response and a *different* response. In temporal generalization, the initial (subjective) probability for the occurrence of the standard (and, hence, for responding with *same*) is approximately the same as the probability for a *different* response, for the frequency of standard and non-standard stimuli presented to the participant is more or less equal. As time goes by, that is, as the stimulus is being presented, the probability for a *same* response increases whereas it decreases for a *different* response. The reason is as follows: short durations are no longer subject of choice when the stimulus presentation has passed those durations. Hence, the number of non-standards that still can be chosen reduces as the stimulus presentation prolongs. The probability for a *same* response grows to its maximum when the presented duration equals the

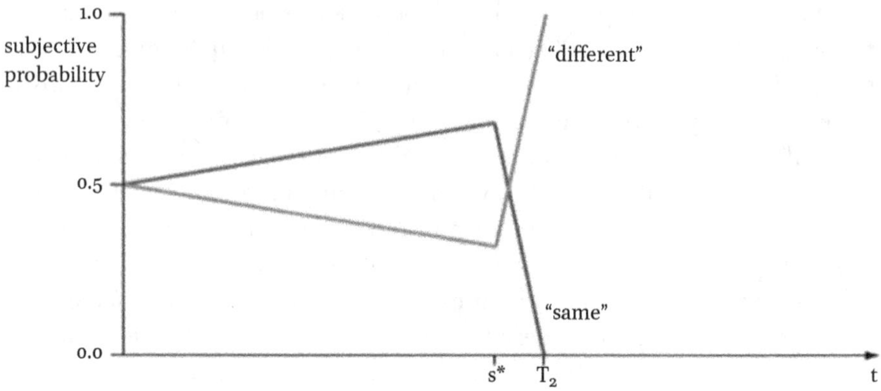


FIGURE 7.4 *A model of subjective probability of same or different responses in temporal generalization, dependent on the duration of the comparison interval t.*

(subjective) standard duration, $t = s^*$. After that instant, the probability for a *same* response decreases rapidly and reaches its minimum ($p = 0$) shortly after passing the threshold $T_2$.

With respect to subjective probability, the speed of responding with *same* should increase the closer the current duration is to $s^*$. Accordingly, the speed of responding with *different* should decrease. As can be inferred from Figure 7.4, *same* responses ought to be faster than *different* responses for intervals shorter than the (subjective) standard. Whereas they should be fairly equal at very short durations, our data suggest that even at very short intervals (e.g., 100 ms), the same-different difference occurred. How could this finding be explained?

The answer may lie in the "dynamic nature" of $s^*$, which leads to my second assumption (see Figure 7.5). The scalar timing theory suggests that *same* responses are made when $t = s^*$. If $s^* < s$, *same* responses will be made at durations shorter than the standard. That is, *same* responses made at intervals shorter than the standard (even at very short intervals) might be done *because* in those trials the subjective value of the standard was very small, too. *Different* responses, however, do not "need" a change of the value of $s^*$ to be executed at short intervals. They can be made even if $s^* = s$.

According to the probability model, both *same* responses and *different* responses can be sped due to the increased "preparedness" of the participant. Consider first the case that a *same* response is given which should happen when $t = s^*$. The line indicating *same* responses in Figure 7.5 shows that the subjective probability for a *same* response is higher than that for a *different* response. Thus, response time will be relatively short. Now imagine that the participant's memory sample of the standard is the same but the current interval is longer than $s^*$. As can be seen in Figure 7.5, the subjective probability for

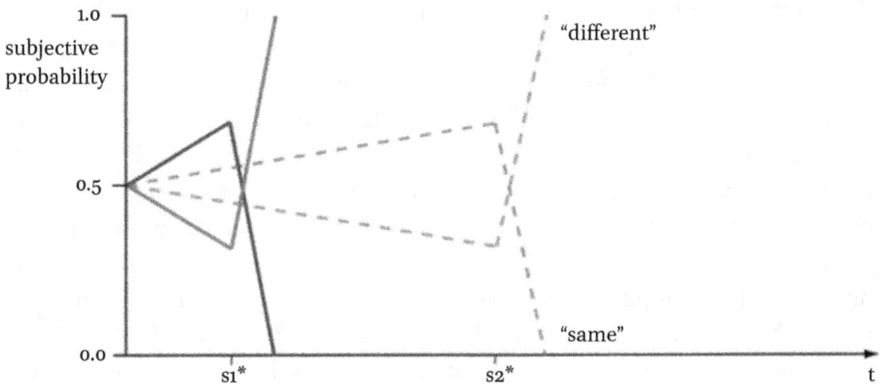


FIGURE 7.5 *The development of subjective probability of same or different responses in temporal generalization, dependent on the mental representation of the standard duration s*.*

a *same* response will decrease and—accordingly—will increase for a *different* response. In that case, a *different* response will occur which is made rapidly because it is considered to be very likely. (This would also be the case when intervals longer than the subjective standard are truncated.) Now suppose a third case in which the current interval is shorter than the subjective standard $s^*$. A *different* response would occur which takes more time to be executed than a *same* response would do at the same interval, for a *same* response should be expected to a higher degree than a *different* response.

This example illustrates that there might be three classes of responses: fast-same responses, fast-different responses, and slow-different responses. Fast-same responses should be made when $t = s^*$, fast-different responses should be made when $t > s^*$, and slow-different responses should be made when $t < s^*$. To yield a fast-same effect, fast-same responses must overweigh fast-different responses (i.e., *same* responses must *on average* be faster than *different* responses given at a certain stimulus duration).

Compared to *different* responses, *same* responses were observed to need more time to be given at intervals larger than the standard. This was presumably the case because *different* responses could be made without completely processing stimulus duration of larger intervals whereas *same* responses could be made only after the whole stimulus was experienced. Therefore, the fast-same effect was observed only for intervals shorter than the standard.

### 2.4 *Another Experiment*

In an experiment conducted by Klapproth and Wearden (2011), a temporal-generalization task was used where the standard duration was not at the center of all durations presented, but either the smallest or the largest value of the durations. The participants were assigned to four conditions, half of which were speed conditions, the remaining half control conditions. It was expected that only durations longer than the standard were truncated when participants were urged to respond quickly. Therefore, a shift should occur only in the speed condition with long durations and not in the speed condition with short durations. Figure 7.6 shows the gradients obtained from the Klapproth and Wearden experiment (dark-grey: speed conditions; light-grey: accuracy conditions).

As hypothesized, a shift of the gradients towards shorter durations only occurred when the participants were presented with stimulus durations that were either the same as or longer than the standard duration. However, when comparison durations were shorter than or equal to the standard, no shift was observed. Therefore, experiencing stimulus durations that are longer than the standard appears to be necessary to produce the shift in response gradients during temporal generalization under time pressure. Figure 7.7 shows the response times obtained from this experiment.

As Figure 7.7 shows, response times for *same* responses increased linearly with stimulus duration, with some deviations from linearity occurring at the shortest and longest stimulus duration which might be considered outliers since each of these data points represented only a few responses (1 or 3, respectively). Like with the response times of the Klapproth and Müller (2008) experiment, *different* responses increased up to a certain point and then maintained their amount. What is striking here is that response times of the "short" condition were apparently longer at the 1000 ms duration than response times of the "long" condition. This was true for both *same* responses and *different* responses. In regard to the *same* responses, however, the difference in response times

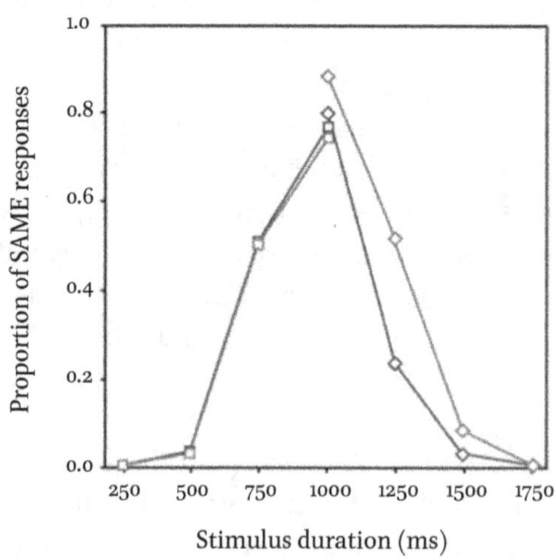


FIGURE 7.6
*Temporal-generalization gradients obtained from Experiment 2 of the Klapproth and Wearden (2011) study.*
MODIFIED AFTER KLAPPROTH AND WEARDEN (2011).

between both conditions was not statistically significant, $t(28) = 1.93$, $p = .07$, $d = 0.81$. Yet, the difference between the "short" and the "long" condition with respect to *different* responses was much more substantial, $t(26) = 3.73$, $p < .001$, $d = 1.41$. The turning point at which response times changed from increasing to keeping their value constant occurred earlier in the "short" condition than in the "long" condition. This might be attributed to a difference in $s^*$ between both conditions. Remember that in the "long" condition longer intervals had been experienced than in the "short" condition (although the longest might have been truncated). According to the adaptation-level theory, the reference value for comparisons with current duration values should be larger in the "long" condition than in the "short" condition. Therefore, the instant to make a *different* response at intervals larger than the subjective standard should be later in the "long" condition than in the "short" condition.

Furthermore, the fast-same effect observed in the Klapproth and Müller (2008) experiment did not occur in this experiment (except for the few data points at the shortest duration). On the contrary, *different* responses in the "long" condition were on average faster than *same* responses, whereas in the "short" condition, same and different responses were approximately of the same speed. This finding might be explained by assuming that intervals longer than the standard were truncated and therefore less time was needed to make a *different* response than to make a *same* response. For the "short" condition,

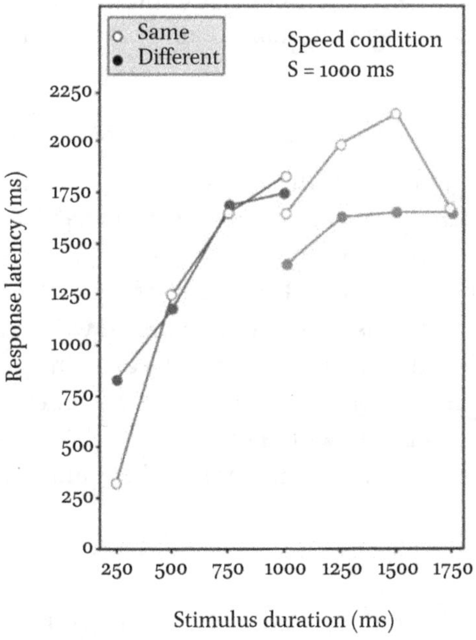


FIGURE 7.7
*Mean latencies of same and different responses in Experiment 2 of the Klapproth and Wearden (2011) study. In the "short" condition, intervals ranging from 250 to 1000 ms were presented, in the "long" condition, intervals ranging from 1000 to 1750 ms were presented.*
MODIFIED AFTER KLAPPROTH AND WEARDEN (2011).

however, this explanation is not feasible since there was no interval longer than the standard. However, the probability model seems to be suitable for explaining the data from the "short" condition quite well. In this condition, $s^*$ should vary only between $s$ and zero, that is, it is not supposed to be larger than $s$. Consider now the case that a *different* response is made when the standard is presented. This would only be possible if $s^*$ is smaller than $s$. According to the model, fast-different responses are given when $t > s^*$, hence, *different* responses at $t = s$ should always be fast responses. The next case that is worth to be looked at is when a *different* response is made at an interval shorter than the standard. In that case, $s^*$ is either longer than the current interval or shorter. If it is longer, a slow-different response would occur, if it is shorter, a fast-different response would occur. Although the former case is more likely than the latter one, a mixture of fast and slow responses is expected which might contribute to a rather small discrepancy between the latencies of *same* and *different* responses.

## 3 Challenges for Scalar Timing Theory?

According to the scalar timing theory, a *same* response will be given if $abs(s - t) / t < b$. The results of the speeded temporal-generalization experiments put forward the following question: How do the values of the representation of the standard duration, $s^*$, the representation of the presented duration, $t$, and the decision threshold, $b$, change when the participants make temporal judgments as quickly as possible?

### 3.1 *Changes of s*

From the experiments my colleagues and I conducted, it was inferred that the peak shift in temporal generalization under time pressure was presumably altered by the truncation of the longest stimuli of the series of stimulus durations. It was suggested that this truncation lowered the reference value of the standard duration. Thus, the $s$-value that was used for comparison with the current duration was not the same than the initial value of $s$. The experiments therefore suggest that $s$ cannot be simply regarded as the standard duration that is presented prior to the test trials or even within every test trial, but instead $s$ must be seen as a value that is strongly dependent on other durations that are experienced by the subject.

### 3.2 *Changes of t*

The $t$-value corresponds to the current duration, which is compared with the standard duration. However, it was shown that only durations smaller than or equal to the standard duration were fully considered for comparison. Comparison durations longer than the standard had presumably been truncated through premature responding. Therefore, the value of *t* is not the same as the duration of the current interval, as long as the duration is larger than the standard. Instead, the maximum value of *t* is supposed to be somewhat larger than the standard (according to the Klapproth and Müller, 2008, model: $t_{max} = s/1\text{-}b$). Since $t_{max}$ is determined by the standard, and since intervals will be truncated after they had been presented for some time longer than the standard duration, the truncation in turn affects the value of $t_{max}$, which is supposed to decrease when long stimuli are not experienced in their full length.

### 3.3 *Changes of b*

It has also been suggested that there is a special relationship between the *b*-value and response time. In one of their experiments, Klapproth and Wearden (2011) increased the difficulty of the generalization task by decreasing the stimulus spacing. As a result, the participants improved their discrimination performance and produced steeper gradients. According to the scalar timing theory, this should have been done through adapting a strict decision criterion, which corresponds to a rather low value of *b*. Moreover, in this experiment the participants not only made their decisions more precisely, but made them also more quickly. This seems counterintuitive at first glance, but is suitable to the model proposed for decision times in temporal generalization (Figure 7.1). In this model, the time to make a *different* response at long intervals is directly related to *b*: the larger the value of *b*, the longer participants will need to make a *different* decision. This would also imply that large values of *b* will allow larger intervals to be completely perceived than will small values of *b*. Therefore, the larger *b* is, the smaller the truncation effect will be, and hence, the smaller will be the gradient's shift. Figure 7.8 illustrates the interdependency of premature responding and the alterations of the main parameters of the scalar timing theory.

## 4 Conclusion

What can we conclude from the findings reported in this chapter? First of all, the results imply that when intervals have to be judged within an experiment, the standard *s* that is used for the comparative judgments seems to be not a fixed value, but rather dynamic. Its dynamic character results from all stimulus durations presented within the experiment, from which the mean of durations is supposed to form the reference duration. This "dynamic" representation of the reference duration has several implications for the judgment of subsequent

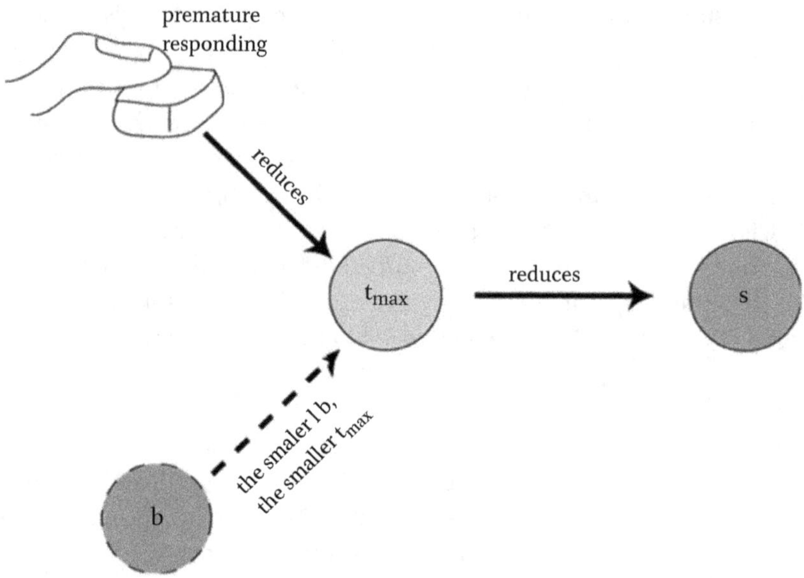


FIGURE 7.8 *Synopsis of the effects of speeded duration judgment in temporal generalization on the timing parameters* t, s, *and* b.

(comparison) durations, and also for the scalar timing theory which often serves as a theoretical framework for comparative duration judgments. According to the scalar timing theory, duration judgments depend on at least three components of the judgment process, which are the reference (or standard) duration *s*, the to-be-judged (or comparison) duration *t*, and a decision threshold *b*. Since the value of *s* depends on durations that were previously encountered within or across experimental trials, comparison durations would be judged as being larger, the shorter the previously encountered durations are. When individuals are encouraged to respond as quickly as possible, truncation of long comparison durations is likely to occur (resulting in a "shortening" of the encountered durations) so that subsequent comparison durations will be judged as being longer than they would be if the individuals take all time they need to make their judgments. As the standard *s* depends on previous comparisons, and subsequent comparisons depend on the value of *s*, each single temporal comparison, where *s* and *t* were related to one another, depends on former comparisons and affects later comparisons. This exactly is the dynamic nature of the comparison process. The decision process and its outcomes are further affected by the decision threshold *b*, according to which a presented interval is categorized as either the standard duration or not the standard duration. If an individual applies a "strict" threshold, meaning that only small differences between the standard and comparison intervals are accepted for judging a comparison interval as equaling the standard, it is likely that long comparison durations are truncated. The truncation of long intervals, however, will result in lowering the value of the standard and hence in overestimation of subsequently presented comparison intervals. However, if the decision threshold is rather lax, truncation of comparison intervals is less likely, so that the respective standard is rather long, making overestimation of comparison intervals also less likely.

## References

Balcı, F., & P. Simen (2014). Decision processes in temporal discrimination. *Acta Psychologica, 149*, 157–168.

Desiderato, O. (1964). Effect of anxiety and stress on reaction time and temporal generalization. *Psychological Reports, 14*, 51–58.

Farell, B. (1985). Same-different judgments: A review of current controversies in perceptual comparisons. *Psychological Bulletin, 98*, 419–456.

Gallistel, C.R., & J. Gibbon (2000). Time, rate, and conditioning. *Psychological Review, 107*, 289–344.

Gibbon, J. (1991). Origins of scalar timing. *Learning and Motivation, 22*, 3–38.

Gibbon, J., & R.M. Church (1984). Sources of variance in information processing models of timing. In Roitblat, H.L., T.G. Bever, & H.S. Terrace (Eds.), *Animal cognition* (pp. 465–488). Hillsdale, NJ: Lawrence Erlbaum Associates.

Gibbon, J., R.M. Church, & W.H. Meck (1984). Scalar timing in memory. In Gibbon, J. & L.G. Allan (Eds.), *Timing and time perception* (pp. 52–77). New York, NY: New York Academy of Sciences.

Helson, H. (1948). Adaptation-level as a basis for a quantitative theory of frames of reference. *Psychological Review, 55*, 297–313.

Helson, H. (1964). *Adaptation-level theory*. New York, NY: Harper & Row.

Klapproth, F., & M. Müller (2008). Temporal generalization under time pressure in humans. *Quarterly Journal of Experimental Psychology, 61*, 588–600.

Klapproth, F., & J.H. Wearden (2011). Why do temporal generalization gradients change when people make decisions as quickly as possible? *Quarterly Journal of Experimental Psychology, 64*, 1646–1664.

Krueger, L.E. (1978). A theory of perceptual matching. *Psychological Review, 85*, 278–304.

Nickerson, R.S. (1975). Effects of correlated and uncorrelated noise on visual pattern matching. In Rabbitt, P.M.A. & S. Domic (Eds.), *Attention and performance* (pp. 655–668). London: Academic Press,.

Pierrel, R., & C.S. Murray (1963). Some relationships between comparative judgment, confidence, and decision-time in weight-lifting. *American Journal of Psychology, 76*, 28–38.

Posner, M.I., & C.R.R. Snyder (1975). Facilitation and inhibition in the processing of signals. In Rabbitt, P.M.A. & S. Domic (Eds.), *Attention and performance* (pp. 669–682). London: Academic Press.

Posner, M.I., R. Klein, J. Summers, & S. Buggie (1973). On the selection of signals. *Memory and Cognition, 1*, 2–12.

Thomas, D.R. (1993). A model for adaptation-level effects on stimulus generalization. *Psychological Review, 100*, 658–673.

Wearden, J.H. (1992). Temporal generalization in humans. *Journal of Experimental Psychology: Animal Behavior Processes, 18*, 134–144.

Wearden, J.H., L. Denovan, M. Fakhri, & R. Haworth (1997). Scalar timing in temporal generalization in humans with longer stimulus durations. *Journal of Experimental Psychology: Animal Behavior Processes, 23*, 502–511.

Weisman, R., L. Brownlie, A. Olthof, M. Njegovan, C. Sturdy, & D. Mewhort (1999). Timing and classifying brief acoustic stimuli by songbirds and humans. *Journal of Experimental Psychology: Animal Behavior Processes, 25*, 139–152.

CHAPTER 8

# Reaction Time Analysis for Interval Timing Research

*Patrick Simen*

## 1 Introduction: A Space of Timing Models

Reaction times are a rich form of data that has been widely used to understand how humans and other animals make simple, two-alternative perceptual discriminations, and how they time intervals. However, statistical techniques for analyzing reaction time data have been developed more extensively in the case of non-temporal, perceptual decision research than in timing research. This chapter describes how to apply reaction time (RT) analysis techniques from non-temporal decision research to the temporal domain, and it describes Matlab code that can be used to implement these analyses (see book's GitHub repository).

In order to motivate the application of these techniques for interval timing research, it helps first to consider a simple class of computational models of timing, the pacemaker-accumulator (PA) models (Creelman, 1962; Gibbon & Church, 1984; Killeen & Fetterman, 1988; Treisman, 1963). Simulating these models is interesting in its own right. For the purpose of this chapter, though, simulation is merely a tool that allows researchers to generate "fake" reaction time data that they can compare against their empirical data, test their intuitions, and debug their analysis code.

Importantly, starting with a simple, classic family of timing models also provides an example of how RT analysis can be used to select the "best" model within a class. Model selection is currently a topic of great interest in non-temporal decision research. Furthermore, starting with simulations may also help explain why the techniques used in non-temporal decision research have so far not reached the same level of use in timing research. That is, in many cases the sophistication of these techniques may seem to be wasted on the data in timing. Timing data, after all, typically has fewer degrees of freedom than decision data, as we will describe shortly. However, some examples from the literature will serve to illustrate the point that RT analysis can, contrary to this view, be extremely helpful in assessing models of timed behavior (e.g., Balci & Simen, 2014; Simen, Vlasov, & Papadakis, 2016).

DOI 10.1163/9789004280205_009

### 1.1 *Fixed Clock-speed Pacemaker-accumulators*

We now consider two different types of timing model and ways to simulate them, thereby producing fake RT data. We can then apply the RT analysis methods described in the final section of this chapter to test whether these methods can properly determine which model best fits the fake data. This is an important step to undertake before applying the methods to real RT data from an experiment. If these methods cannot accurately determine which model generated the data even when we already know the answer, then they will not be useful in the real world. For this reason, model simulation should always be the first step in vetting a model-fitting procedure. We will restrict our attention in this chapter to the most basic and most classical type of timing model: the PA.

PAs are just about the simplest possible model of timing. They operate like a stopwatch, which counts up regular or irregular clock ticks until hitting some threshold either for making a decision (such as, whether the current interval is longer or shorter than another interval in memory) or for producing a response (such as pressing a lever to earn a reinforcement). We will refer to these two possibilities as decision and production tasks respectively. There are many possible variations on the basic PA idea, however, and different variations make different predictions, especially about production times. RT analysis techniques will therefore be useful in teasing these models apart.

Creelman (1962) developed an early PA model in which a source of clock pulses emits pulses randomly, at a fixed rate. The time between pulses was exponentially distributed, making the process of counting them up into a Poisson counting model. Since this basic process is central to several of the models for which we provide Matlab code, we begin with it. *Poisson_countermodel.m* contains the code in its entirety (see book's GitHub repository). It generates exponentially distributed random inter-pulse intervals, then adds them up, and checks when their sum has exceeded a threshold.

An important variation on this model allows for changes in pacemaker rate across trials, as well as variability in the pulse count threshold across trials. Gibbon and Church (1990) considered how these variations could account for a problem with Creelman's model, which is its conflict with a widely observed phenomenon in timing known as 'scalar invariance'. This benchmark phenomenon in interval timing research is one in which the standard deviation of remembered interval durations appears to grow linearly with the timed duration. This pattern yields a constant ratio of the standard deviation divided by the mean – a ratio known as the coefficient of variation (CV). Indeed, in production tasks, the production time distributions for different durations frequently superimpose when the data from different duration-conditions are divided by their means. Creelman's model, in contrast, predicts decreasing CVs, and, thus, increased relative precision, as durations increase. Gibbon (1992) referred to this pattern as 'Poisson timing' (though as we shall see below, a Poisson timer that works on different principles yields perfect scalar invariance).

Treisman (1963) developed an early variant of this PA model that did not make specific assumptions about the distribution of clock pulses except for a key constraint on their statistics. The constraint is that the inter-pulse intervals are mostly relatively short in some trials, compared to the global average over trials, and mostly relatively long in others. That is, the clock speed varies from trial to trial (but always around a fixed average). Another way to say it is that, across trials, the *i*th inter-pulse interval following the clock-start is correlated with the *j*th inter-pulse interval in the same trial, for all intervals $i$ and $j$ occurring prior to the end of the timed duration. Simulations of this model demonstrate that it takes only a remarkably small amount of such correlation (i.e., variation in the pacemaker rate) across trials to recover the pattern of constant CVs. *Treisman63.m* contains code (see book's GitHub repository) that allows any desired distribution of inter-pulse intervals to be simulated while still observing the constraints. Part of what the code produces is a correlation matrix for selecting random inter-pulse intervals with the specified level of correlations. As users can see for themselves, correlations can be reduced nearly to 0 in a Creelman-style PA model but still obtain constant CVs.

### 1.2 *Variable Clock-speed Pacemaker-accumulators*

The models in the preceding section used a pacemaker that emits random pulses at a constant rate on average, and stores different pulse-count totals to encode different durations $T$. In contrast, the models in this section use a fixed pulse-count threshold (call it $\theta$) and a variable pacemaker rate, $A = \theta/T$. When the pulses are purely excitatory, the model is equivalent to Killeen and Fetterman's (1988) Behavioral Theory of Timing (BeT) model (although the "pulses" in that case are interpreted as transitions between states of behavior). When we add negative pulses emitted at a rate proportional to the positive pulse emission, the resulting process closely approximates a process known as a diffusion process (Simen et al., 2013). Such processes are idealized, one-dimensional Brownian motion systems, in which a particle (equivalent to a pulse-count) drifts upward while being continuously perturbed by noise at every moment. This kind of perturbation (in three dimensions) accounts for the diffusion of particles in a liquid or gas over time. One-dimensional drift-diffusion models (DDMs) are leading models of two-alternative perceptual decision making, where the "particle position" represents something like the log odds ratio of one hypothesis vs. the alternative. In either case, BeT or the time-adaptive, opponent Poisson DDM (TopDDM) of Simen et al. (2011, 2013), scalar invariance falls out of the models automatically. The code in *opponent_Poisson_Appendix3.m* explicitly compares the simulation of actual pulse-counting to a diffusion approximation of pulse-counting (specifically, the Euler-Maruyama method, in which spike counting *per se* does not occur). The primary computational difference between these approaches is that in the former, a sequence of pulse-times is generated and then summed; in the latter, properly scaled deterministic and random noise increments are added to a running sum at every one of a sequence of time steps. The primary theoretical difference is that the pulses are finite in number and spaced out at separate points in time, in the former, while they are idealized as occurring infinitely often at every moment in the latter. Since the latter theoretical assumption of the diffusion model is not likely to be true in reality, it is worth noting that this assumption allows some very simple approximate mathematical expressions to be used to fit RT distributions.

## 2 Reaction Time Analysis Methods

Built-in Matlab functions that are particularly useful for reaction time analysis include sorting functions (sort), plotting functions (plot, hist, ksdensity), and many of the features of the Matlab Statistics Toolbox (this toolbox is an add-on product that comes with its own licensing fee). These functions are incorporated into several of the functions accompanying this chapter.

Growing numbers of researchers these days are also opting for alternatives to Matlab, such as the free, open-source, Python programming and scripting language, or the statistical programming language, R, in addition to open-source versions of Matlab such as Octave. Python, for example, operates in many ways like Matlab, particularly when the NumPy and SciPy numerical and scientific computing packages and the Matplotlib graphics packages are imported (see, e.g., Anaconda, for a complete development environment for scientific computing in Python). Here, however, we restrict our attention to programs in Matlab. The primary virtue of Matlab, in the author's opinion, is that, although it must be purchased and renewed yearly for updates, the software is stable and platform-independent across Mac, Windows, and Linux computing platforms; documentation for each function is generally trustworthy and well organized; and there are fewer of the problems that seem to accompany open-source software (e.g., the confusion surrounding multiple, slightly different parallel versions – e.g., Python 2 vs. Python 3 – and updates that frequently break functioning software until workarounds are developed). Whether those features are worth the cost is a matter of personal opinion.

### 2.1 *Using Moments to Characterize the Data and Evaluate Conformance to the Scalar Invariance Pattern*

The shape of a probability distribution can be characterized by its moments: that is, by the mean, variance, skewness, kurtosis, etc. Mathematically, the *n*th central moment is defined as the expected value of the *n*th power of the data minus the mean (when such expectation exists). In the case of skewness and kurtosis, the third and fourth central moments are normalized through division by the standard deviation. In Matlab, these moments can be computed with a single function call. For this, we will assume that the reaction time values are contained in a vector having n elements in the variable rt. With such a variable, we can compute:

```
mean(rt), var(rt), std(rt), skewness(rt), kurtosis(rt)
```

Documentation for any of these can be easily obtained at the Matlab command prompt by typing, e.g.:

```
doc mean
```

In the literature on timing, the most famous phenomenon that can be observed by measuring RTs is the constancy of relative temporal precision (Gibbon, 1977), that is, the constant ratio between the first two central moments of production time distributions. In studies that require animals to learn to press a lever after a delay of *T* seconds from a stimulus to obtain a reward, production times are found to be typically accurate: the average of the production time distribution is close to *T*. The precision of the production time across trials, however, decreases as *T* increases. Such a relationship can be captured by computing the CV, which is the standard deviation (square root of the variance) divided by the mean of the response times contained in the vector variable rt. In Matlab, the CV can be obtained with:

```
CV = std(rt)/mean(rt)
```

CVs are often found to be roughly constant as a function of *T* across groups of participants. Violations of CV constancy are often observed if multiple durations *T* are mixed in a single experiment, with both human and non-human participants (Bizo, Chu, Sanabria, & Killeen, 2006).

### 2.2 *Skewness and Model Selection*

Skewness, the third standardized moment, captures how symmetric a distribution is. Gaussian distributions (a shape that is often used to fit production time data) have zero skewness. Positive skewness, in which the mean is greater than the mode, and the tail on the right is heavier than on the left, is widely observed in two-alternative perceptual decision RTs, but is less often observed in timing tasks, which typically yield skewness levels closer to 0. Nevertheless, skewness may provide a fairly robust test of the predictions of a timing model, because models that predict a Gaussian distribution of RTs can be tested by verifying whether skewness equals 0. Simple Poisson counter models, such as BeT, predict skewness equal to twice the CV. Diffusion models such as Top-DDM predict skewness equal to three times the CV (see Simen et al., 2011, 2013, for discussion of proofs; these properties can be verified by running the model simulation code provided in the GitHub).

It should be noted that when the CV is small, skewness is, therefore, also expected to be small. Thus, if conditions can be created that generate large CVs, such as by driving attentional resources away from the primary task, then skewness should be expected to increase for models such as BeT and DDM, but not for models that predict a normal distribution of RTs, such as the information processing implementation (Gibbon, Church, & Meck, 1984) of scalar expectancy theory (SET; Gibbon 1977).

In summary, skewness is worth computing as it helps in model selection. Matlab has a built-in skewness function that can be used in this way. The Matlab Statistics Toolbox also includes helpful functions for assessing normality or deviations from normality in a set of RT data in variable rt (e.g., `normplot(rt)`).

### 2.3 *Quantiles and Timescale Invariance*

In Section 2.1, we have discussed how the CV tends to remain the same for different base intervals. Another famous feature of timing data that implies constant CVs, but is a stronger form of invariance, is called scalar invariance or timescale invariance. This is a phenomenon in which the entire distribution of RTs can be rescaled, through division by the mean, so that the shape of the RT distributions with different means superimpose on each other perfectly after rescaling.

A good test of this form of invariance is to examine a feature of the RT data that is widely used in 2AFC research. To obtain a compact empirical description of the distribution, the RTs are divided into quantiles. For example, after the RTs are sorted from fastest to slowest, it is possible to calculate the 10th, 30th, 50th, 70th, and 90th percentile by taking the 10% fastest, 30% fastest, 50% fastest, 70% fastest, and 90% fastest RT, respectively. If there is scalar invariance, the plot of the quantiles from one duration condition should line up with the corresponding quantiles from another condition.

The following Matlab commands will compute the quantiles by sorting the reaction time data in the vector variable rt :

```
rt_sorted=sort(rt);
```

determine how many trials there are

```
lrt=length(rt);
```

and compute the quantiles of reaction time data stored in the Matlab variable rt:

```
quantiles1=rt_sorted(floor(lrt*[0.10.30.50.70.9]));
```

Supposing that there are quantiles from a different data set stored in the variable quantiles2, the two RT distributions can be examined for time scale invariance by plotting them against each other:

```
plot(quantiles1,quantiles2,'o');
```

These plot points should form nearly a straight line, with slope equal to the ratio of the average of the timed durations in the two different conditions. Additional analyses based on linear regression can be computed and the best fitting line can be superimposed on the data (a quick but incomplete way of doing this is to go to the Tools menu of the figure window in Matlab, select the Basic Fitting menu option, and then select "linear" in the popup dialog box that appears).

### 2.4 *Maximum Likelihood Fitting of Distributions to* RT *Data*

The most sophisticated method for testing model predictions is by fitting model parameters to empirical RT distributions. In two-alternative perceptual decision research, this approach is powerful and widely used. Despite having a similar goal, different fitting methods greatly vary in terms of their computational speed, their robustness (resilience from data not actually produced by the process under investigation, but instead delayed by distraction, for example), and the amount of data they require for fitting accurately. For this reason, a large amount of work is devoted to comparing the fitting methods and trying to find new ones.

Maximum likelihood fitting is a standard technique for fitting models to data in many areas. The *likelihood function* is the point-by-point product of a candidate probability density function's values evaluated at each of the observed RTs. For example, a Gaussian density has two parameters, a mean, and a variance. To calculate the likelihood of the RT data for a given value of mean and variance, we multiply the probability density value at each RT point to get the overall likelihood of the data, given the particular parameter values used. If a new set of values for the mean and variance parameters gives a higher likelihood, then we prefer that set of parameters to the previous set. Intuitively, those parameters that make the data appear most likely are themselves likely to be the parameters that generated the data in reality. Thus, an iterative procedure can be implemented, wherein a parameter value is chosen (or multiple values are chosen, for models with more than one parameter), the likelihood of the data is computed, and then a new parameter is chosen to see if it gives a higher likelihood. This could be done purely randomly, but search procedures in Matlab's Statistics Toolbox, and also in its Optimization Toolbox, can be used to do much more intelligent searching over parameter space. The code for implementing such a search is shown in *fit_models.m* (see book's GitHub repository). It is relatively concise, because the fitdist function hides a large degree of complexity.

The maximum likelihood method is particularly useful in timing research, because PA models produce RT distributions with simple, closed-form expressions, such as the normal, gamma, or inverse Gaussian distributions. The most common distributions are built in to Matlab's Statistics Toolbox, allowing samples to be drawn easily. Moreover, because of the closed-form expression, the fit to the data can be done by simply modifying the parameters of the distribution and finding the best fit rather than by sampling the data and simulating the model output as in the case of models with no closed-form expression. Instead, a computation-intensive process is required to evaluate the likelihood function for the data of models without a known, closed-form RT distribution, so that maximum likelihood methods are frequently outperformed by methods such as the chi-square method (Ratcliff & Tuerlinckx, 2002). Furthermore, new methods such as hierarchical Bayesian methods (e.g., Wiecki, Sofer, & Frank, 2013) are becoming increasingly interesting to researchers, given their ability to fit small amounts of data and to infer population-level parameters efficiently. For example, they can infer a set of parameter values that represents patients with attention deficit hyperactivity disorder versus a set of values that represents neurotypical control participants. We do not go into those methods here, since they are at the forefront of development in the two-alternative decision domain, and have not yet (to my knowledge) been widely used in fitting timing data.

### 2.5 *Model Complexity*

Maximum likelihood fitting offers one other useful property, which is that likelihood methods can easily be adapted so as to penalize for model complexity. Occam's Razor is the principle that the simplest explanation of real data should be preferred, all else being equal. When one statistical model has more parameters than another, it has more flexibility to fit a wider range of data patterns. In the extreme, a model that has as many parameters as data points is likely to fit the data perfectly. However, such a model is not at all likely to generalize well to data that has not been fit. Such models are said to overfit observed data, at the risk of failing to fit unobserved data. To combat this risk, tests such as the Akaike Information Criterion (Akaike, 1974), and the Bayesian Information Criterion (Schwarz, 1978), can be used to rule out overly complex models. Both of these methods simply add a penalty to the logarithm of the computed likelihood function, with the amount of the penalty depending on the number of parameters. Thus, a model A that fits the data less well, but with fewer parameters than model B, may in the end have a higher likelihood score. In such a case, we select model A over model B.

If the log likelihood of the data is computed by the fitting methods described in the previous section and stored in variable ll, it can be adjusted for parameter penalties as follows, where the number of parameters in the model is k (for example, Creelman's model has a pulse rate parameter and a pulse-count threshold parameter, so k = 2):

$$AIC = 2^* k - 2^* ll;$$

Given that the log likelihood is subtracted from the parameter penalty in the standard formulation of the AIC, the goal is to select the model with the minimum AIC score.

The Bayesian Information Criterion is also used for model selection and tends to implement a stronger penalty for parameters:

$$BIC = k^* \log\left(\text{length}(rt)\right) - 2^* ll;$$

### 2.6 *Outlier Treatment*

One factor that bedevils RT research is the presence of contaminated data. If a participant does not pay attention during a trial of an experiment, they may issue a response that is far later, or far earlier, than would normally occur. There is a host of different approaches to removing outliers from data, though none can be assured of doing it correctly (see, e.g., Ratcliff, 1993). After all, a very long RT may just be ... a very long RT. Still, the presence of just one unusually long RT that results from inattention can throw off maximum likelihood fitting methods, because the likelihood of such a data point is so low according to the true parameter values.

Here is a simple technique for eliminating outliers that is by no means guaranteed to work perfectly, but in any case can be adapted by users to be more or less conservative as they see fit. It is included here primarily to emphasize an incredibly useful technique for indexing RT data that is outside an outlier cutoff range.

We can eliminate unusually long RTs by computing the standard deviation of the data, and then keeping only those data that are smaller than some number of standard deviations above the mean, for example 3:

```
rt=rt(rt<=mean(rt)+3*std(rt));
```

The syntax inside the outer parentheses creates a vector of logical 1s and 0s. Only those elements of the rt array that have a logical 1 in the corresponding array created by the <= operation will be assigned to rt. The result is that any RT greater than 3 standard deviations above the mean will be deleted from the array rt.

## 3 Conclusion

The Matlab code that accompanies this chapter (see book's GitHub repository) is intended to help researchers who are new to RT analysis to begin analyzing their data with the aim of model selection. Arguably the best thing about Matlab is its extensive, easily searchable Help documentation (for Matlab version 2014a or later, type doc at the Matlab command prompt to bring up the documentation viewer). This author learned Matlab simply by progressing through the Matlab documentation and trying out the examples, which are provided in nearly every help topic at the command prompt. The MathWorks website also includes tutorial videos and forums for getting help from other users. After consulting the Matlab tutorial documentation section (called Getting Started With Matlab in version 2014b), I recommend that new users try out some of the functions provided with this chapter, using the debugger in Matlab to step through lines of code one by one to see how variables in memory are changing as a script or function is executing, and to learn how particular Matlab built-in commands are used for RT analysis. Experienced users may wish to use this code as a stepping-off point for investigating methods more widely used in the two-alternative decision making literature, some of which, such as hierarchical Bayesian model fitting, have yet to be used widely in timing research.

I have not addressed the very useful technique of fitting RTs in retrospective timing tasks, as was done for example in Balci and Simen (2014). In such tasks, estimates of time intervals are used as the inputs to a decision process. Balci and Simen (2014) applied this technique to data from a temporal bisection task, in which intervals are presented and the participant must categorize them as being either closer to a short reference interval, or closer to a long reference interval. Choice probabilities for long and short choices and corresponding RTs in this case are not necessarily directly related to the mechanism of time estimation, but the RT data here provide important information about the mechanism by which temporal discriminations are made after an interval is over. Because the literature on non-temporal, two-alternative decision making covers this type of RT analysis extensively, and because fitting two-choice RT data is more complicated than fitting "1-choice" production times, I do not address such methods here. However, they are really just extensions of the approaches described here. Ratcliff and Tuerlinckx (2002) offer a comprehensive discussion of how model fitting is done, as just one example, and a number of fitting algorithms and tutorials exist in the two-choice perceptual decision domain.

## References

Akaike, H. (1974). A new look at the statistical model identification. *IEEE Transactions on Automatic Control, 19*, 716–723.

Balci, F., & P. Simen (2014). Decision processes in temporal discrimination. *Acta Psychologica, 149*, 157–168.

Bizo, L.A., J.Y.M. Chu, F. Sanabria, & P.R. Killeen (2006). The failure of Weber's law in time perception and production. *Behavioral Processes, 71*, 201–210.

Creelman, C.D. (1962). Human discrimination of auditory duration. *The Journal of the Acoustical Society of America, 34*, 582–593.

Gibbon, J. (1977). Scalar expectancy theory and Weber's law in animal timing. *Psychological Review, 84*, 279–325.

Gibbon, J. (1992). Ubiquity of scalar timing with a Poisson clock. *Journal of Mathematical Psychology, 35*, 283–293.

Gibbon, J., & R.M. Church (1984). Sources of variance in an information processing theory of timing. In Roitblat, H.L., T.G. Bever, & H.S. Terrace (Eds.), *Animal Cognition* (pp. 465–488). Erlbaum.

Gibbon, J., & R.M. Church (1990). Representation of time. *Cognition, 37*, 23–54.

Gibbon, J., R.M. Church, & W.H. Meck (1984) Scalar timing in memory. In Gibbon, J. and L.G. Allan (Eds.), *Annals of the New York Academy of Sciences: Timing and Time Perception,* Vol. 423 (pp. 52–77), New York Academy of Sciences.

Killeen, P.R., & J.G. Fetterman (1988). A behavioral theory of timing. *Psychological Review*, *95*(2), 274–295.

Ratcliff, R. (1993). Methods for dealing with reaction time outliers. *Psychological Bulletin, 114*, 510–532.

Ratcliff, R., & F. Tuerlinckx (2002). Estimating parameters of the diffusion model: Approaches to dealing with contaminant reaction times and parameter variability. *Psychonomic Bulletin and Review*, *9*(3), 438–481.

Schwarz, G. (1978). Estimating the dimension of a model. *The Annals of Statistics*, *6*(2), 461–464.

Simen, P., F. Balci, L. deSouza, J.D. Cohen, & P. Holmes (2011). A model of interval timing by neural integration. *Journal of Neuroscience, 31*, 9238–9253.

Simen, P., F. Rivest, E.A. Ludvig, F. Balci, & P.R. Killeen (2013). Timescale invariance in the pacemaker-accumulator family of timing models. *Timing & Time Perception, 1*, 159–188.

Treisman, M. (1963). Temporal discrimination and the indifference interval: Implications for a model of the 'internal clock'. *Psychological Monographs, 77*, 1–31.

Wiecki, T., I. Sofer, & M.J. Frank (2013). HDDM: Hierarchical Bayesian estimation of the drift-diffusion model in Python. *Frontiers in Neuroinformatics*, 7, 14.

CHAPTER 9

# Analysing Multi-person Timing in Music and Movement: Event Based Methods

*Mark T. Elliott, Dominic Ward, Ryan Stables, Dagmar Fraser, Nori Jacoby and Alan M. Wing*

## 1 Introduction

Accurate timing of movement in the hundreds of milliseconds range is a hallmark of human activities such as music and dance. Its study requires accurate measurement of the times of events (often called responses) based on the movement or acoustic record. This chapter provides a comprehensive overview of methods developed to capture, process, analyse, and model individual and group timing.

In a classic paper on sensorimotor timing, Stevens (1886) used a combination of paced and unpaced tapping over a range of tempos to describe what we would now recognise as characteristic attributes of movement timing. Participants tapped with a metronome set to various tempo values for a number of repetitions and then tapped on their own to reproduce the metronome tempo as accurately as possible. Stevens presented his results graphically as time series of intervals between successive responses. He showed that timing is highly adjustable but is subject to variability in produced intervals, which increases as the target interval lengthens. Moreover, he observed that the variability is not purely random but has a characteristic patterning. This includes distinct tendencies to short-term alternation between shorter and longer intervals (at faster tempos) and longer term drift around the target interval (at slower tempos).

Many papers (e.g., see Repp & Su (2013), for review) subsequent to Stevens (1886) have examined paced and unpaced finger tapping. The goals of the research include characterising influences on timing accuracy in terms of mean and variability and also understanding the nature of patterns in the variation. Although the majority of these studies has focused on individual performance, recently there has been growing interest in the relation between the timing of multiple individuals attempting to synchronise their joint performance, with the goal of achieving coherent ensemble timing (see Elliott, Chua, & Wing, 2016, for a review of this emerging area in the context of mathematical

 | DOI 10.1163/9789004280205_010

models). Where previously the theoretical interest focused on understanding component sources of variance in the individual (i.e., timer, memory, attention, input, and output delays), the new paradigms raise questions about forms of timing linkage, including feedback correction and anticipatory adjustments, that keep participants moving together.

Stevens (1886) collected data using Morse code signal set transmission key presses (see next section for further detail). More recently movement timing study methods have ranged from switching devices such as computer keyboard keys, push button switches, resistive and capacitive contact switches to sensors such as force transducers and motion tracking cameras capable of characterising the dynamics as well as the timing of the movements. A subset of sensorimotor timing studies often involves research around timing in musical production. This research can also involve a variety of input devices, each with a unique set of methodological constraints. When using acoustic instruments, for example, additional data capture devices need to be considered, along with methods of extracting onset locations from the musical signal. Similarly, when using Musical Instrument Digital Interface (MIDI; a universal interface to a wide range of electronic musical instruments) devices, variability and latency in the system can cause issues when relaying the device's output to the participant in real-time.

Studies of timing in western music have largely focussed on the use of a piano (Repp 1995; Shafer, 1984), largely due to the simple relation between movement, note sounded, and the possibility of mechanical measurement. Similarly, these experiments are confluent with finger-tapping studies given that expert pianists tend to exhibit particularly strong timing capabilities (Keele et al., 1985; Loehr & Palmer, 2007). The piano also supports research into a range of synchronisation types such as two players following each other (Goebl & Palmer, 2009), a single player following an external stimulus (Goebl & Palmer, 2008), and a single player using both hands (Goebl et al., 2010). With both upright and grand pianos, sensors or microphones can be placed inside the instrument (Palmer & Brown 1991; Shafer, 1984) in order to record the moments at which the hammer strikes the string. More recently, electric pianos tend to be more widely used (Goebl & Palmer, 2008; Henning, 2014) due to their ability to output MIDI messages and to modify musical parameters such as playback time and timbre.

Other research has considered a broad spectrum of instrument types, each bringing challenges in terms of capturing the acoustics and defining movement timing events. De Poli et al. (1998) analysed expressivity in solo violin performances, whereas Rasch (1979), Wing et al. (2014), and Stables et al. (2014) present models for interpersonal synchronisation in small string ensembles, namely trios and quartets. Similarly, Ellis (1991) and Friberg and Sundström

(2002) investigate swing ratios in solo saxophone and percussion performances respectively.

This chapter is structured in five main sections, as follows. We start with a review of data capture methods, working, in turn, through a low cost system to research simple tapping, complex movements, use of video, inertial measurement units, and dedicated sensorimotor synchronisation software. This is followed by a section on music performance, which includes topics on the selection of music materials, sound recording, and system latency. The identification of events in the data stream can be challenging and this topic is treated in the next section, first for movement then for music. Finally, we cover methods of analysis, including alignment of the channels, computation of between channel asynchrony errors and modelling of the data set.

## 2 Data Capture

### 2.1 *Capturing Movement*

Early studies into sensorimotor synchronisation focused on a very simple motor action in the form of a finger tap (Repp, 2005). Not only is this a simple action for most participants to perform, it is also an easy event to record. When people produce a finger tap action, there is an asymmetry in the flexion and extension segments of the movement (Balasubramaniam, Wing, & Daffertshofer, 2004). This results in a short impact time of the finger onto the surface, generating strong tactile feedback (Balasubramaniam et al., 2004; Elliott, Welchman, & Wing, 2009a) that participants align with the external beat. By recording the impact time of the finger, researchers subsequently have an accurate event onset time of each finger tap. This is how one of the earliest known sensorimotor synchronisation experiment was implemented (Stevens, 1886). Participants tapped their finger on a Morse code key with the electrical contact recorded on a smoked drum kymograph. On a kymograph the timing is measured from distances between pulse marks on the surface of a drum rotating at constant velocity.

The modern equivalent of Stevens' (1886) approach is to use some form of touch sensor connected to a computer. The times between movements are determined by reference to distinct events registered by the sensor. Force sensitive resistor (FSR) materials are particularly useful for registering finger taps (e.g., Elliott, Wing, & Welchman, 2010; Schultz & Vugt, 2015). In addition to being very low cost, the sensors come in the form of a thin membrane, meaning that there is no 'travel' when the finger hits the surface (as might be the case if one used a button press or keyboard to record events).

Similar devices include piezo-electric sensors and the more recent capacitive sensing technology (as used on modern touch-screens). While low cost and practical for recording the impulse response of the tap, the aforementioned sensors tend not to be sufficiently linear for measuring the amplitude or shape of the signal. In scenarios where these parameters are of interest, a force sensor (e.g., ATI Industrial Automation; http://www.ati-ia.com/) can be used (Elliott et al., 2009a).

Interfacing these sensors to a PC for recording responses usually requires a data acquisition card (DAQ). These devices capture the analogue signal from the sensor and convert them into a digital value for import into Matlab or similar software. DAQs, such as those from National Instruments, Measurement Computing and Labjack have a wide price range, depending on number of channels, maximum sampling rate and the number of functions the device has. A key advantage is that the devices can be used to output the external cues and also trigger any other external devices, so all data is both output and recorded with a common time base, i.e., synchronised. Time resolution depends on the sampling rate, but it is possible to achieve very reliable and consistent event timings from these devices.

The close relationship of sensorimotor synchronisation research to musical contexts has meant that often MIDI equipment has been used to record participant responses. In particular, drum-pads have been used as an effective tapping sensor (Manning & Schutz, 2013; Pecenka & Keller, 2011), providing a large surface area and no movement in the surface itself. Keyboards have also been used (Goebl & Palmer, 2008; Keller, Knoblich, & Repp, 2007), however the time difference between the finger hitting the key and the key travelling down to hit the sensor adds an uncertainty as to when the event onset actually occurred. There is also a level of time lag and variability in MIDI communications between devices and the computer software. This has been identified as a small but not insignificant amount of delay (Repp & Keller, 2008; Schultz & Vugt, 2015) and, hence, should be characterised and accounted for when using this interface for timing experiments.

### 2.2 *Example of a Simple, Low Cost System for Recording Finger Taps to Auditory Cues*

Both the sensors and hardware for collecting data from tapping studies can range from very high-cost (e.g., force sensors with a high specification data acquisition card) to low-cost (simple impulse detecting sensor, with sound card input). Figure 9.1 provides an example of a simple solution that can be applied in fieldwork to record one or more participants performing a tapping experiment.

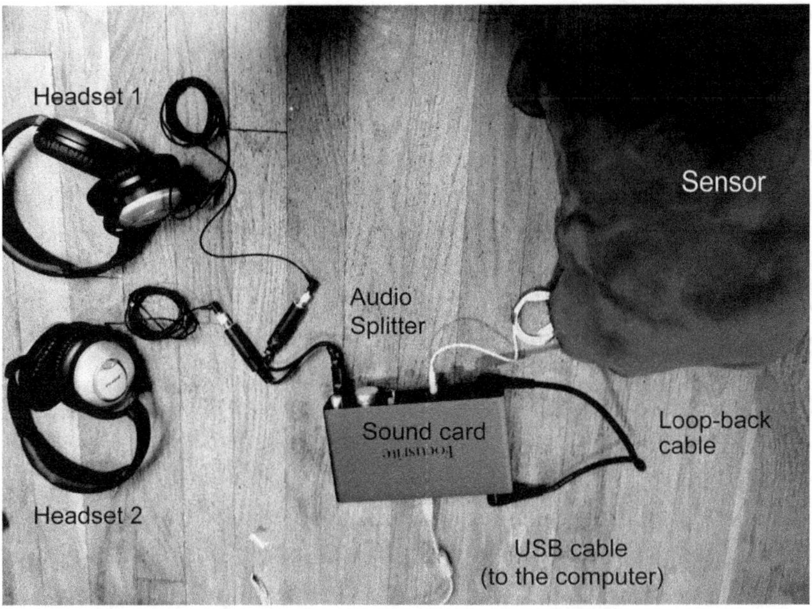


FIGURE 9.1 *Example of a simple, low-cost experimental setup for capturing finger tap responses from group timing experiments.*

One or more sensors are connected to a sound card. Sensors can be as simple as a wooden box or a soft pad in which earbuds are installed. These may be used as very cheap but low-sensitivity microphones that are well suited to record a direct touch on the surface to which they are attached, and are insensitive to external noises. This setup is able to detect even a light touch, if the soundcard is set to a high gain.

An external sound card is connected to a computer via USB. The computer sends an auditory stimulus (such as a metronome) through Digital Audio Workstation (DAW) software such as Cubase or through designated software such as MatTAP (Elliott, Welchman, & Wing, 2009b; see Section 2.6). A loopback cable is installed so that stimulus and output are recorded with zero latency, and the responses and loopback stimulus are recorded on two or more separate channels. One or more high-quality headsets are connected to an audio splitter so that participants can hear the stimulus. If participants synchronise only to one another, the headsets can be removed and sensors can be made of hard resonating material such as a wooden box.

### 2.3 *Complex Movements*

It is clear why finger tapping became the *de facto* task for early sensorimotor synchronisation experiments: simple equipment setups can be used to accurately measure the event onsets for timing analysis. More recently, researchers have been interested in increasingly complex movements. For larger groups performing more complex movement interactions, data involving temporal and spatial dimensions must be recorded. This could be upper limb movements such as choreographed ballet movements (Honisch, Roach, & Wing, 2009), or lower limb movements such as walking or bouncing (Georgiou, Racic, Brownjohn, & Elliot, 2015). There are two ways of typically capturing these movements. The first is using marker-based motion capture and the second is using inertial measurement units (IMU). Here, we provide some recommendations in their use in the context of movement timing experiments.

### 2.4 *Video Motion Capture*

3D motion capture systems such as Vicon (Vicon Motion Systems Ltd, UK) and Qualisys (Qualisys AB, Sweden) are considered to provide the gold standard in terms of accuracy. Movements can typically be captured with an error of <0.5mm at hundreds of samples per second. On the negative side, systems tend to require calibration over a specified capture volume at the beginning of each session and reflective markers must be accurately placed on the bony landmarks of the participant's body. Furthermore, post-processing can be a tedious task in terms of labelling markers for each trial, such that trajectories can be identified for analysis. While most software packages associated with these systems have an 'auto-label' feature to identify markers, this is liable to fail. Labelling can be become particularly complex for multi-person studies. At the start of an analysis, the researcher is presented with a cloud of unlabelled markers in 3D space. Markers in successive frames must be linked to define trajectories, which can be identified with body segments of each participant. To help identify individual participants in a group it is often advantageous to add extra markers (not used for tracking) somewhere on the body that is a unique formation for each group member. For example, marking out the corners of a small square on the back of Participant A, versus the corners of a triangle on Participant B can help identify which person is which during labelling. There are 'active marker' systems, where the marker itself is electronic and hence can be pre-assigned a label or ID. An example of this type of system is the Polhemus Liberty (Polhemus, USA), which uses active markers in a magnetic field to track motion.

For event based timing analysis, one is often only interested in the temporal aspects, even for complex movements. Therefore, a small number of markers can be used, rather than a full body marker set (the Vicon Plug-In-Gait marker set is in the region of 40 markers per person). It is important to choose a marker location that will provide the primary trajectory for analysis. This might be a marker on the finger for upper limb movements, or the heel for lower limb movements. However, the marker must always be in good view of the cameras, minimising the chances of occlusions. Additional markers can be applied to other areas of the same limb or body segment, both to aid identification and also as a secondary trajectory source in case there are problems with the primary trajectory data (a consistent trajectory location across participants should be maintained however).

Trajectories for each marker can usually be output from the software as a text file, with each marker having an individual X, Y, and Z coordinate at each time sample. We provide example code for parsing these text files, using a representative output from Vicon Nexus software (see this book's GitHub repository).

### 2.5 *Inertial Measurement Units*

IMUs consist of two or three sensing devices. Two-sensor devices consist of an accelerometer, that measures acceleration (in units of $m/s^2$ or g) and a gyroscope that measures rate of angular rotation (in units of radians/second or degrees/second). Three sensor devices have an additional magnetometer included (units of Tesla or Gauss). Recent devices output values from each sensor in 3 axes. IMUs use a local coordinate system, so it is not easily possible to infer the location of a device in global coordinates. That is, if there were two devices attached to a person, it would not be possible to directly calculate the relative distance between those devices (unless the starting positions were known). Additional data fusion algorithms allow the advantages of three sensors to be combined such that accurate motion analysis can be achieved. Without these algorithms, trying to infer the positional trajectory of movement from accelerometer data alone (by integrating the data twice) will result in drift and inaccuracy from the true position. However, for measuring timing of movements (rather than position) the associated drift is not such a big issue as the timing in the data remains intact. With some initial alignment of the data with video, it is possible to identify the peaks and troughs in the acceleration data that relate to key parts of the movement cycle (e.g., walking). Alternatively, integrating to velocity can produce a clean, and easier to interpret signal, by applying both low and high-pass filters to the data.

An IMU's main advantage is that the participant is free to move around without restriction. There is no capture volume as with video motion capture and occlusions are not an issue due to the sensors being within the device. Participants 'wear' one or more of the devices on the body and are then free to move naturally. This is particularly useful for gait analysis: in a video motion capture gait lab, only a small number of gait cycles can be recorded within the capture volume. With IMUs, the participant can complete a long walk or even be recorded over a full day, dependent only on the on-board memory of the device. A number of IMU companies have seen the potential for this and developed software alongside their devices that includes the algorithms to do a full or partial gait analysis. Current examples include APDM (APDM Inc., US; www.apdm.com), XSens (XSens Technologies B.V., NL; www.xsens.com), GaitSmart (ETB Ltd; www.gaitsmart.com), Shimmer3 (Shimmer Sensing, www.shimmersensing.com), and BTS (BTS S.p.A, IT; www.btsbioengineering.com). All but BTS use multiple devices worn on the body. The actual gait parameters provided vary amongst the software packages and many still struggle to provide accurate step length measures due to the drift issues mentioned above.

A particularly useful feature of the APDM and Shimmer devices is that they are wireless, time-synchronised devices. They come as a set of IMUs to be fitted on different body segments of a single individual. However, the software also allows raw data access such that each device could instead be fitted to separate individuals, with their activity recorded wirelessly. Given the resulting data is time-synchronised, this is ideal for group timing studies (Georgiou et al., 2015).

### 2.6 *Dedicated Sensorimotor Synchronisation Software*

The main challenge with setting up the data acquisition and cue presentation for both single- and multi-person sensorimotor synchronisation experiments is in minimising timing uncertainty. Multi-tasking operating systems, such as Microsoft Windows, imply that executing commands is an asynchronous process. That is, you might run a segment of code which outputs a cue stimulus every 500 ms, but the operating system will not necessarily execute that command immediately if it is busy dealing with another application in the background. This can create jitter in the cue generation, so that a stimulus that should occur exactly every 500 ms might instead execute on average every 500 ms, with actual intervals produced varying around that value (e.g., 490, 515, 516, 502 ms etc.). If the standard deviation of these intervals becomes relatively large then the impact on the analysed movement timing results will be significant (Repp, 1999). Interval variance will increase as participants correct their movements to remain in time with the varying beat. Asynchrony variance will also be artificially inflated as both the variance in the movement and the cue sum together. On the other hand, controlled manipulation of cue jitter can be effective for investigating cue reliability effects (Elliott et al., 2010; Elliott, Wing, & Welchman, 2014).

Similar issues occur with capturing responses. If a participant is required to tap a key on a standard PC keyboard in time with the beat, it is difficult to reliably record the onset time due to lags in the operating system servicing the event. Therefore, when designing an experimental setup, minimising lag time and jitter in the cue and response signals are a key consideration. Because of this, a number of software toolboxes have been developed to assist with data capture in sensorimotor synchronisation (SMS) experiments. These include, FTAP (Finney, 2001), one of the earlier toolboxes written for Linux. This toolbox interfaced directly with MIDI instruments and allowed accurate control of cue timing and recorded finger tap responses. Max/MSP (Cycling '74, USA) is a commercial software package that provides a visual programming interface with full MIDI support. This has often been used in timing experiments with the visual programming interface allowing relatively simple onset detection and analysis to be set up with minimal coding skills. As previously mentioned, MIDI suffers from lag and some jitter. These are substantially smaller and less significant than trying to use traditional programming approaches and PC hardware interfaces, which must still be characterised. A recent study (Schultz & Vugt, 2015) has characterised both Max/MSP and FTAP experimental setups for a sensorimotor synchronisation experiment, finding the mean lag (with respect to an FSR sensor response) to be 15.8 ms and 14.6 ms, respectively. The standard deviation (jitter) of the lags was 3.4 ms and 2.8 ms, respectively. The study contrasted the two MIDI setups to a novel hardware setup using a low cost embedded controller (Arduino, www.arduino.cc). By using the Arduino device (taking the signal processing away from the PC), the lag was reduced to 0.6 ms with a jitter of 0.3 ms.

It is clear, therefore, that moving the signal processing away from the PC to dedicated hardware such as an embedded controller or a data acquisition card (e.g., National Instruments, Measurement Computing) is a good way to get an accurate cue presentation and corresponding response times. This philosophy was used to develop another sensorimotor synchronisation toolbox. MatTAP (Elliott et al., 2009b) uses data acquisition hardware interfaced to the MATLAB programming environment to provide a comprehensive toolbox that offers virtually no lag or jitter in the signal output and response capture. By using a loop-back method (see Figure 9.2), both the output signal and response can be sampled under a common clock at very high sampling rates (e.g., 10kS/s) allowing highly accurate measures of asynchrony (see Section 5.2). The toolbox further uses a graphical user interface that allows the user to accurately control cue presentation, store data and run analyses. We have successfully interfaced the toolbox with both accelerometer devices (APDM Opal) and video motion capture (Qualisys) to allow accurate measures of group movements to an auditory metronome or visual cue. The downside to this high level of accuracy is increased expense, with both the hardware (data acquisition) and Matlab (with appropriate toolboxes) adding up to a relatively high cost compared to other solutions. Regardless, much of the code we provide with this chapter has

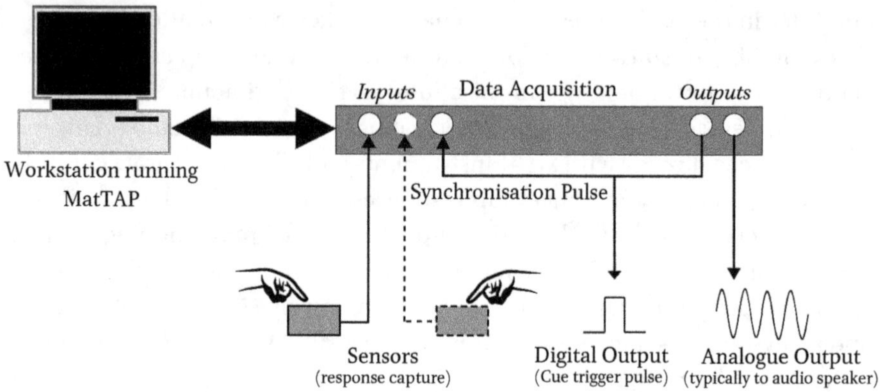


FIGURE 9.2 *A typical hardware configuration for cue generation and response capture using MatTAP (reproduced from Elliott et al., 2009b). The toolbox uses data acquisition hardware to achieve a high level of timing accuracy. (1) Two outputs are generated to drive stimuli. One is an analogue waveform, typically used to drive an audio speaker, but can also be applied to haptic or visual devices. The second output is a digital square pulse, which can be used to trigger bespoke stimuli equipment. To increase accuracy further, this pulse is fed back into the system to compensate for processing delays in the hardware. (2) Up to two sensors can be utilised to capture responses. Any sensor that produces an analogue or digital voltage, typically in the range ±5 V can be used to record events. (3) The output signals and corresponding responses are captured and stored automatically in individual, sequentially numbered .mat format files.*

evolved from MatTAP, generated from the requirements of new experiments within our labs.

## 3 Audio Capture

In musical timing research, both single and multi-agent studies generally adopt a similar methodology. Based on the study's objectives, selections are made for instruments, number of players and source material. Environmental constraints such as visual and auditory occlusion are then set and trials are implemented to acquire data. The performance data can be captured in a number of formats ranging from acoustic waveforms, MIDI messages, sensor-data taken from IMUs (see Section 2.5), movement data from a video motion capture device (see Section 2.4), or a combination of these systems. In this section we concentrate on the capture of the audio signals, rather than the movements of the agents producing the music. This involves recording the acoustic waveform and configuring the system to effectively derive the relevant events.

### 3.1 *Material Selection*

The selection of performance material is generally based on the extent to which the piece enforces performance characteristics. In synchronisation experiments, long passages of concurrent isochronous events (equally spaced notes performed at the same time) are desirable, often limiting the pool of recognised works. For this reason, Moore and Chen (2010) opted to use an excerpt from Shostakovich's String Quartet Op. 108, No. 7, which included 260 events performed in quick succession by two members of a string quartet. Furthermore, all notes are generated by individual bow-strokes, and are rhythmically partitioned into groups of four. Similarly, both Wing et al. (2014) and Stables et al. (2014) used an excerpt from Haydn's String Quartet Op. 74 no. 1, which consists of 48 x 8th notes performed continuously by almost all members of the ensemble. For studies investigating phase relationships such as Shaffer (1984), multiple voices with independent subdivisions are desirable, leading to the selection of an excerpt from Chopin's Trois Nouvelles Etudes. Specially composed pieces are also commonly used in timing studies, typically when there is a requirement for tractable context and specific musical conditions. This is the case in Goebl & Palmer (2009), where the content is easy to perform and subdivisions vary between players. This allows trained musicians to easily perform the experiment with no pre-requisite knowledge of the content.

### 3.2 *Sound Recording*

In some cases, it can be impractical to capture event-based performance data such as MIDI due to the acoustic properties of the instrument, or the physical restrictions that controllers impose on a participant. An acoustic violin, for example, produces notes with legato (i.e., in a smooth continuous manner, without breaks between notes) and has a small area of sound propagation. This means it is difficult to incorporate a MIDI device into the instrument without restricting the movement of the musician. This often introduces a requirement for audio recording, followed by post-processing to perform onset detection in order to derive a symbolic representation from the captured acoustic data. For music listening, instruments are typically recorded by placing microphones at acoustically relevant locations around the source and surrounding environment, with the intention of achieving a desired aesthetic. This can differ from analytical recordings where the aim is to isolate signals and derive an accurate representation of the performer's onset locations via further signal decomposition. For well recorded monophonic signals (e.g., solo instruments) comprising homogeneous fragments of sound, timing data can be extracted more easily when compared to polyphonic signals (e.g., multiple instruments played by a group) or those contaminated by noise. For this reason, close-miking techniques in which the microphones are placed near the sound source (e.g., the strings) are generally preferred to room or ambient miking in order to obtain a higher signal-to-noise ratio; in the case of multi-person performance, at least one microphone is used per source, thereby minimising acoustic bleed. The multi-microphone setup is also advantageous in that further separation of the sources can be achieved through cross-analyses. Here, the sound pressure level of each source will vary across microphones, allowing for the attenuation of other instruments in the recording. In the case of polyphony, it may be necessary to first employ source separation algorithms such as those proposed by Vincent (2006), and then perform onset detection on the resulting streams. In the latter case, noise-reduction can be used as a pre-processing step.

For close-miking, clip-on condenser microphones such as the AKG C519 range (similar to those used in Polak & London, 2014), provide high sensitivity and greater frequency and transient responses than dynamic microphones. Whilst omnidirectional microphones can be used for this task (sound from all directions is captured with equal sensitivity), localised polar patterns such as cardioid and hypercardioid (sound in front of the microphone is recorded with higher sensitivity) are preferable as they mitigate sound capture from external sources. During this process, careful miking techniques are necessary to gain proximity to the instrument and therefore achieve a high signal-to-noise ratio. When using the close-miking technique, the microphones tend to be placed on the areas of the instrument that don't dampen the sound or prevent any free-flowing movement. In string instruments such as violins, violas and cellos, the clips are often located on the bridge of the instrument. For percussive instruments such as drums and cymbals, the microphones are clipped to the rim or to stands, so as not to interfere with the skin or plate. For brass instruments it is difficult to avoid the resonant surface of the instrument, so microphones tend to be clipped to the bell, with the microphone located inside or near to the opening.

An alternative to close-miking instruments is to use a vibrational pickup, in which a transducer reacts to vibrations of the instrument's surface material. These tend to be less common as they have poorer transient responses, but can be used when close-miking is not plausible or susceptible to noise, such as in large ensembles.

For field recordings such as those in Polak and London (2014), portable recording devices such as the Roland R-4 or the Tascam DP006 can be interfaced with microphones to capture the signal. These generally record data to an onboard hard disk or portable storage device, which can later be transferred to another machine for analysis in an uncompressed format. The main benefit of these devices is that they can record multiple channels (usually up to 6) without the need for additional computational hardware, however they have limited auditory feedback options and recording length is often dependent on battery life. More commonly, an external soundcard with a desktop or laptop computer is used, in which the soundcard connects to a host machine via USB or Firewire. Soundcards support varying numbers of inputs and outputs and generally have assignable sampling frequencies (often set by default to 44.1kHz) for use during analogue to digital conversion. In this case, a software interface is also required to record the inputs to disk, which can be done using a DAW such as Logic Pro, Ableton Live, Audacity, Reaper or Cubase, all of which share a similar multi-track interface, with varying levels of control over the audio signal.

### 3.3 *System Latency*

Due to the computational overhead involved in reading, writing and processing a large number of samples each second, audio processing systems incur a time lag, known as latency, at numerous points throughout the processing chain. Furthermore, this latency is shown to exhibit high variability and information loss (Wang et al., 2010) when systems are subject to high processing loads (e.g., when multiple channels are being used to record a large ensemble), thus leading to unreliable playback. For this reason, it is generally not recommended to feed the system output back to participants via headphones when musical timing is being measured, as latency will create negative recurrent effects on the performer. In isochronous rhythmic sequences, the threshold for perception of delay is observed by Friberg and Sundberg (1995) to be around 6 ms for tones with relatively short intervals, and periodic timing correction to the delayed stimuli is observed to occur at time lags of as little as 10 ms (Thaut et al., 1998). Further to this, the standard deviation of inter-onset intervals (IOI, time between consecutive onsets) in performed rhythmic sequences is widely accepted to increase with auditory delay time (Pfordresher & Palmer, 2002). This suggests that even minimal system latency (observed by Wang et al., 2010 to be around 19 ms for Audacity with Mac OS X 10.6, when running under low computational load) is likely to impact the validity of results. If no other options are available, the signal path can often be configured to route the analogue signal directly to the headphone output, bypassing the processing chain and minimising latency caused by play-through.

If processed auditory feedback is unavoidable, such as in experiments where participants will be played manipulated versions of their input signals, the buffer size of the host software should be reduced in order to reduce the latency time in the system. This limits the time allocated to the system to process the audio samples, thus allowing the signal to reach the playback device in a shorter time period. The buffer size can often be controlled via the DAW, and can be set experimentally between 32–1024 samples. Whilst lowering the buffer size reduces latency, it also puts strain on the computer's CPU as the processor is then required to complete more operations in a shorter period of time. Negative effects of excessively lowering the buffer size include the addition of audible noise to the signal path caused by loss of information. The input/output latency of the system also has implications for the use of external stimuli during experiments. Given the time delay incurred during playback, it is recommended to record audio signals generated by the computer back into the system whilst recording performers. This means any pre-recorded accompaniment or metronome tracks should be captured using headphones and a microphone in order to limit computational asynchronies caused by the variation in system delay at multiple points in the processing chain.

## 4 Onset Detection and Analysis

> A bold onset is half the battle.
>
> GIUSEPPE GARIBALDI

One of the key challenges in post-processing for event-based analysis (both for movement and audio signals) is accurate onset detection. This step needs to be applied to both the cue signal and the participant's responses. In this section, we cover the key approaches used to achieve accurate onset detection. The three main stages of onset detection are shown in Figure 9.3 and are common in both movement and music data analysis. However, completion of each stage often requires a specific approach, based on the origin of the signals.

### 4.1 *Extracting Movement Onsets*

In movement, onsets correspond with physical events, (e.g., the peak pressure applied to a point, a finger tap on a surface, or a sudden change in motion as measured by position, velocity or acceleration). Reliable onset detection is vital for analysing sensorimotor responses (Elliott et al., 2009b) by allowing accurate measurement of the asynchrony between the cue and the corresponding motor response.

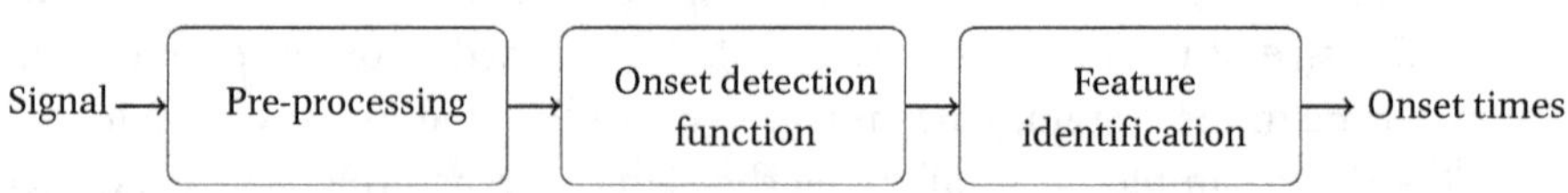


FIGURE 9.3 *Key stages involved in onset detection.*

In the example shown (Figure 9.4), a finger tap onto a surface is captured by a force sensor, converting the force into a voltage output. This shows the *baseline,* which represents the signal prior to the finger making contact with the sensor. A rise in the signal from *baseline* identifies the *onset of attack*. The *attack* represents the rise of energy in the system from the prior state, i.e., the initial impact of the finger onto the sensor. *Peak attack* occurs when the finger reaches maximum force onto the sensor. The *onset of decay* indicates the beginning of *evanescence*, i.e., the return to *baseline* as the finger begins to lift again, off the sensor surface. For movement onset detection, it is usually the *onset of attack* or the *peak attack* that is identified as the onset time of the signal.

### 4.1.1 Movement Data Pre-processing

Pre-processing is the transformation of raw data to facilitate processing by the onset detection function (ODF). The first step in pre-processing is experimental design; facilitating the optimal capture of data and encoding the movement. The experimental hardware must have a sampling rate of sufficient magnitude to capture the movement without aliasing. The sensor's rise time and evanescent should be an order of magnitude faster than the movement. The magnitude of the onset sought should be readily distinguishable from that of the noise of the recording system, and distinguishable from common artefacts. The experiment should ideally offer a dedicated input channel for each element of participant response of interest, i.e., one touch sensor per finger, or a marker for each limb.

Algorithmic pre-processing addresses practical flaws in movement already captured. Low frequency human motion (below 10–50Hz) is generally

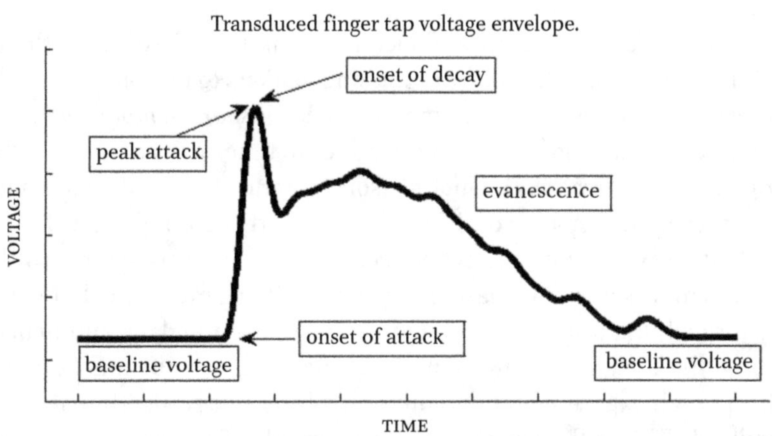


FIGURE 9.4 *Example sensor output signal resulting from a single tap.*

contaminated with higher frequency noise. This noise comes from such sources as non-ideal sensors, the environment (e.g. 50/60Hz mains hum, participant's heartbeat) and the harmonics of movement itself. The frequency components of interest are commonly emphasised via low-pass filtering which attenuates all frequency components of a given signal above a cut off value, optimally without adding delays or distortion. Classically, a low-pass filter known as a zero lag 4th order Butterworth practically implements these requirements (Winter, 2009). These filters can be implemented algorithmically in software such as Matlab, rather than requiring a hardware implementation to be used. Following the application of the filter, signal peaks of interest should remain prominent whilst noise peaks should be reduced. For onset detection, a heuristic cut off frequency can be determined by visual inspection and iteration. Note that by filtering, relevant information in the signal can be altered or lost. For temporal studies in particular, it is important to use a zero-phase filter. This is a special case of a linear-phase filter which avoids any frequency-dependent lag. The *filtfilt* command in MATLAB applies a filter both forwards and backwards, cancelling out any phase effects of the filtered signal.

To facilitate filtering, data recovered from non-ideal sensors must be sanitised. Such data should be continuous, machine-readable and exhibit values that readily allow for computation (e.g., numeric values within the maximal and minimal machine accuracy limits). *Numerical sensor artefacts* such as those arising from sensor dropouts, misconfigured apparatus etc. may return numerical error codes, missing values (e.g., empty set [] or NaN) or default values (e.g., zero). These values have no useful relation to the effect being measured and must be excluded to maintain the integrity of any analysis. Hence, data exploration and visualisation, i.e., a check upon the sanity of the data, should always be a first step.

For systems such as the Qualisys and Vicon motion tracking software, there are explicit functions that allow data for missing markers to be approximated. For less integrated systems, MATLAB functions such as *isnan, isempty, isnumeric* can be used to find invalid, non-numeric values in time-series data.

*Numerical sensor aberrations* include sensor saturation (where the recorded movements exceed the capacity of the sensor to report), sensor drift, warmup trends and battery exhaustion, power bounce, and other artefacts of the recording. The values may have some relation to the effect being measured, but have been transformed in a fashion not shared by the rest of the data, and hence may decrease the integrity of any analysis. These effects can be ameliorated by initialising each experimental session with a brief test run with real time sensor feedback. This will reveal aberrant values, allowing action (replacing batteries, adjusting sensor gain). As noted previously, thoughtful experimental design is the most essential pre-processing step.

The experimental artefacts, listed above, are distinct from *participant artefacts*, in which the participant offers responses outside of those anticipated but still within the scope of measurement; ambiguous touches, mistaken taps and involuntary movements etc. These should not be removed in pre-processing, which attempts to faithfully relay participant action to the ODF. Participant outlier artefacts are treated with rigour in Section 5.3.

### 4.1.2 Onset Detection Functions (ODFs)

The ODF renders clearly the presence of attacks within the original signal. In musical onset detection this is often called the *Reduction step*, where the sound signal is traditionally downsampled to a 'low' sample rate (e.g., hundreds of hertz (Dixon, 2006)). However movement ODFs typically eschew downsampled.

There are many varieties of ODF: time and frequency domain, probabilistic and machine learning (Bello et al., 2005; Dixon, 2006; Eyben et al., 2010). In the context of movement, we focus on the time domain methods. In many sensorimotor studies the end of the attack, i.e. the peak of expressed force, can be considered the *intentional* onset of response. In sensorimotor timing, onsets might include peak velocity (Pelton, Wing, Fraser, & van Vliet, 2015), acceleration (Honisch, Elliott, Jacoby, & Wing, 2016), or even higher derivatives such as jerk (Balasubramaniam et al., 2004; Elliott et al., 2009a).

For attacks that are obvious to an annotator, i.e., large increase in voltage amplitude, such as transduced force in tapping experiments (see Figure 9.3 above) a simple envelope follower can be used to algorithmically extract the peaks of attacks (Eq. 1).

$$E_{(n)} = \frac{1}{N} \sum_{m=-N/2}^{(N/2)-1} \left| x(n+m) \right| w(m) \tag{1}$$

in which $w(m)$ is an N-point smoothing kernel centred at $m = 0$. This can be extended to use of the derivative which marks abrupt rises in energy with narrowed peaks (Bello et al., 2005; Eq. 2):

$$E_{(n)} = \frac{1}{N} \sum_{m=-N/2}^{(N/2)-1} \left[ x(n+m) \right]^2 w(m) \tag{2}$$

A direct method of detecting onsets arises from the derivative of the signal (1st or higher), which illuminate periods of change in the movement. The *onset of attack* would be the beginning of the periods where the 1st derivative is positive. The *onset of decay* corresponds with the end of the *attack*, in this example, marked by the beginning of the period where the 1st derivative is negative. However, this naïve approach is still susceptible to the presence of noise (non-ideal sensors), overlapping participant responses (i.e., no return to baseline) or competing sources of spurious onsets (e.g., movement artefacts, movement harmonics).

#### 4.1.3 Event Detection

Peak Picking involves a decision about candidate onsets (which are normally local maxima), resulting from the previous stage(s). If the ODF has been sufficiently well constructed, or the pre-processed data itself is suitable due to experimental design, this final stage is often simple thresholding. That is, candidate onsets that have a peak value above a certain threshold are considered to be movement onsets. This can be readily hand tuned in well-formed movement experiments. Other domain specific knowledge can be added, such as the expected recurrence of onsets within specific durations, a minimal/maximal duration etc.

We provide in the accompanying code, *peakdet*, one of the more robust peak detection algorithms written by Eli Billauer.[1] However, even the best algorithms are likely to have false or missed detections, again due to noise on the signal from imperfections in the sensor or due to human artefacts such as false movements. Therefore, onset detection methods are typically complemented with manual visual checks to ensure any errors are removed. We have further written a Matlab based graphical user interface to visually check the peak onsets extracted using the *peakdet* code, which accompanies this chapter.

To measure the effectiveness with each change made in the process, we need measures of performance. If we consider that merely capturing all of the movement onsets is not sufficient, we must also reject non-movement onsets, which gives rise to two measures: Sensitivity and Specificity. We define Sensitivity, also known as the true positive rate, as:

Sensitivity (Recall) = Correct Movement Onset Detections / Total True Movement Onset Detections

In which Total True Movement Onset Detections are the total number of true movement onsets detected + the number of missed onsets. We define the Positive Predictive Value (PPV), as:

PPV = Correct Movement Onset Detections / Total Movement Onset Detections

1 http://www.billauer.co.il/peakdet.html

In which Total Movement Onset Detections are the total number of true movement onsets + false onsets detected.

Trivially, we could have a Sensitivity 1, by setting the threshold below the lowest peak. All true movement onsets would be captured by this threshold. This would unfortunately minimise PPV, i.e., permit a maximal number of non-movement onsets to pass the threshold and be labelled falsely as movement onsets. There is thus a trade-off between the two values.

### 4.1.4 Dimensionality Reduction, Clustering and Machine Learning

The algorithms presented perform similarly to expert annotators' subjective agreement of the incidence of onsets in single channel data. Multiple marker systems can result in onset complexes, in one or more channels, coincident with a true movement onset. Onset complexes in isolation would correspond with an onset in a single channel system. These onset complexes require a further stage to evaluate when they become multi-channel features. Simple stages include considering one channel of data as representative of the whole (identical to prior mono channel approaches), or a sum of coincident onset complexes across channels compared to a threshold. Such a threshold may not capture the expert appraisal of multichannel cues that give rise to effective subjective onset labelling (e.g., in electromyography (EMG) contiguous channel onset complexes may result from electrodes associated with one muscle vs. multi-channel artefacts such as heartbeat contamination). Consistent labelling of multichannel onset complexes can be facilitated by: dimensionality reduction strategies (such as principal component analysis) and/or machine learning (clustering with an additional classification stage).

Principal component analysis (PCA) is a linear method of data re-expression which returns a set of $n$ components, where $n$ is equal (or less) to the dimension of the original data. These components are ordered by their explicative power of the variance, of the original signal. If the underlying movement is the greatest source of variance, then the principal component will be a single channel representative of the underlying movement. By focusing on that principal component, mono-channel strategies can be re-employed. Other methods of dimensionality reduction include independent component analysis (ICA) and multidimensional scaling. When lower dimensional expressions do not collapse to one obvious channel, i.e., suggesting multiple and/or non-linear underlying generators, machine learning methods can be applied (e.g., self-organising maps, generative topographic mapping).

Whilst Matlab has a Neural Network Toolbox at an extra cost, these approaches can be implemented in Matlab using the excellent NETLAB toolbox

(Nabney, 2002; http://www.aston.ac.uk/eas/research/groups/ncrg/resources/netlab/). NETLAB is also largely compatible with Octave.

### 4.2 *Extracting Onsets from Audio Recordings*

An audio signal contains distinct events pertaining to one or multiple acoustic sources. Examples include a sequence of musical notes, a chain of percussive hits, and consonant and vowel segments comprising continuous speech. Event timing information is conveyed through variation in some physical property of the source. These changes are detected by the listener and registered as distinct events that are often inter-connected at a higher contextual level. The task of extracting timing information about events embedded within an audio signal involves estimating perceptually important points of change. In particular, we are interested in detecting the presence of new acoustic events and annotating associated temporal information, (e.g., start time, end time and event duration). Most research in the field of audio signal processing targets the automatic detection of event onsets. Onset detection is highly relevant when studying the synchronisation in music performance where accurate measurement of response time is imperative.

As with movement (Section 4.1), the term *onset* is generally used to denote the earliest time at which a signal evolves quickly (Bello et al., 2005). This definition relates to the physical properties of the source and thus does not necessarily correlate with the perceived start of an event (Von & Rasch, 1981) or the perceived *attack* time which refers to the moment of rhythmic emphasis for a musical tone (Gordon, 1987; see Collins, 2006, for a review of modelling perceptual attack time and associated problems therein). Nevertheless, most recent work on music onset detection takes a pragmatic approach by tuning and assessing detection algorithms using hand-labelled datasets. Such ground-truth data is typically generated by experienced individuals who combine critical listening with spectro-temporal analysis using state-of-the-art software to best identify the beginning of acoustic events that satisfy the requirements of many practical applications.

Most onset detection algorithms deal with a monophonic signal corresponding to a *single* acoustic stream. The onset detection process follows the same principle as that described for movement onset detection (Section 4.1): Pre-processing, ODF transformation and finally event/feature extraction.

#### 4.2.1 Signal Feature Based Detection Functions

The success of the system is fundamentally dependent on the reduction stage and so most effort has been on developing and evaluating different detection functions (Bello et al., 2005; Böck et al., 2012b; Collins, 2005a, Dixon, 2006). Perhaps the simplest of approaches to onset detection are those based on the amplitude envelope (Masri, 1997; Schloss, 1985). The general idea is that the onset of a new sound leads to a sharp rise in the envelope of the waveform. The local energy of the signal can also be followed rather than the amplitude (Bello et al., 2004), for example by applying a running sum low-pass filter to the square of the signal. It is common to use the time derivative of the envelope such that significant changes in amplitude (or energy) are transformed to sharp peaks that are easily detected by thresholding the resulting detection function.

ODFs based on temporal features are generally adequate for percussive sounds and provide good temporal resolution and have low computational demand. Klapuri (1999) suggested taking the logarithm of the envelope prior to differencing to minimise spurious local maxima after the physical onset of the sound and emphasise lower intensity onsets. A further refinement is to incorporate spectral information since transients tend to introduce energy at high frequencies. The short-time Fourier transform (STFT) is commonly used for this purpose, although auditory filter banks have also been employed (Klapuri, 1999). Masri (1997) used the STFT to focus the local energy measurement towards high frequencies, a technique useful for emphasising the percussiveness of a sound. This high frequency content (HFC) detector can, however, be problematic for low-pitched and non-percussive instruments (Bello et al., 2005). In order to incorporate changes in the distribution of spectral energy over time, Masri (1997) proposed the spectral flux detector. Rather than summing the weighted magnitudes prior to differencing, the algorithm first sums over all positive changes in magnitude in each frequency bin between consecutive analysis frames generated by the STFT. Because changes in magnitude are measured across different frequency bands, the detection function is more reliable compared to one based solely on the temporal envelope.

Additional spectral methods make use of the phase spectra to enhance subtle tonal variations in the signal, and are less dependent on changes in energy (Bello et al., 2004). The idea is that during the steady-state portion of the signal, differences between the (unwrapped) phase of consecutive spectral frames will be constant. The phase deviation, defined as the second difference of the phase, i.e. the change in instantaneous frequency, can then be used to signify changes in the stationarity of the signal; large deviations are more probable during the attack region of a transient. Although methods incorporating phase information are better suited for sounds with soft onsets, one of the shortcomings of the phase deviation detector is its susceptibility to phase distortion and noise in low-energy components. Refined techniques include the weighted phase deviation and variations of the complex domain method, the latter combining both phase and magnitude information (Dixon, 2006; Duxbury et al., 2003). Finally, Collins (2005b) used the constant-Q pitch estimator (Brown & Puckette, 1993) as the primary feature driving an onset detection algorithm targeting pitched non-percussive instruments. The algorithm incorporates vibrato suppression to better emphasise note transitions, outperforming the phase deviation algorithm of Bello et al. (2004)

Böck and Widmer (2013a) also proposed an onset detector with vibrato suppression, based on the common spectral flux method. The detector, called SuperFlux, uses a maximum filter applied to a logarithmic-frequency scaled spectrogram to better track spectral trajectories. The performance of SuperFlux outperformed the pitch-based detector of Collins (2005b) and another specialised detector targeting pitched non-percussive sounds (Schleusing et al., 2008). A second algorithm, the ComplexFlux, also based on differences in magnitude spectra was later developed (Böck & Widmer, 2013b) to suppress both vibrato and tremolo in solo pitched instruments. Figure 9.5 shows the temporal waveform and spectrogram of a recording of a violinist, playing with a détaché bowing style, from which four ODFs have been extracted. The signal was pre-processed by applying a 3rd order Butterworth high-pass filter to remove low-frequency noise picked up by the clip-on microphone. The simplest of detectors, which we have found to work well on signals with well-defined disconnected notes, is the Log HFC, obtained by applying the first-order difference to the logarithm of the frequency-weighted energy.

#### 4.2.2 Classification Based Onset Detection

In recent years, machine learning techniques have been employed to overcome the issue of source-dependent onset detectors (Zhu et al., 2014) as well as establishing more sophisticated detection functions by learning directly from the human annotated datasets traditionally used to evaluate the aforementioned heuristic approaches (Davy & Godsill, 2002; Eyben et al., 2010; Lacoste & Eck, 2007; Marchi et al., 2014; Marolt et al., 2002; Toh et al., 2008). In general, the task is treated as a classification problem where spectral frames extracted from the audio signal are classified as being onsets or non-onsets. Supervised machine learning techniques such as Support Vector Machines (SVM) and Gaussian Mixture Models (GMM) have been employed (Kapanci & Pfeffer, 2004; Toh et al. 2008) to handle pitched non-percussive instruments such as the singing voice where "soft" onsets often occur between smooth pitch transitions and tend to be accompanied by complex modulations in pitch and amplitude.

Neural networks have proven successful in automatically locating onsets in a range of musical signals and define the current state-of-art (Böck et al., 2012a; Eyben et al., 2010; Marchi et al., 2014; Schluter & Böck, 2014). These methods use features such as cent-scaled magnitude spectrograms and linear prediction errors derived from multi-resolution spectra as inputs to a neural network which has been trained using binary labelled features to discriminate between

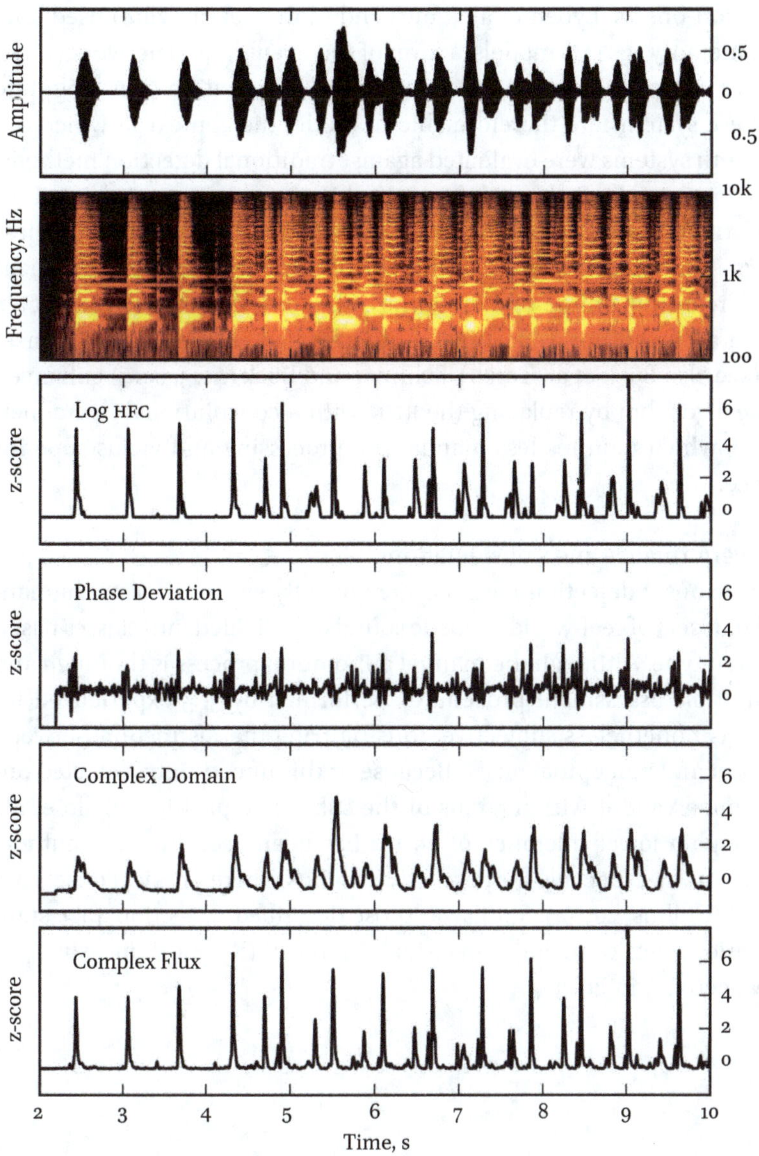


FIGURE 9.5 *Temporal waveform, log magnitude spectrogram and four onset detection functions (ODF) extracted from a violin recording. The ODFs have been standardised by setting their means to zero and standard deviations to one.*

onsets and non-onsets. Eyben et al. (2010) and Marchi et al. (2014) used long short-term memory (LSTM) models, a form of recurrent neural network (RNN) that provides complete access to past and future information over long time periods. These systems are therefore able to model the context in which onsets occur. Both systems were evaluated against traditional detection methods, (e.g., those presented by Bello et al. (2005) and Dixon (2006)), and showed superior performance with respect to F-measure (see Section 4.2.3), regardless of onset type and are therefore considered to be robust. Böck et al. (2012a) developed a real-time online version of the offline RNN method (Eyben et al., 2010), which although not as accurate, outperformed existing non-ML onset detectors (see also Böck et al., 2012b). Schluter and Böck (2013, 2014) enhanced their offline algorithm by replacing the RNN with a convolutional neural network (CNN), which requires less manual pre-processing and yields superior performance.

### 4.2.3 Performance and Considerations

Today's music onset detection methods are typically evaluated using human-annotated datasets of real-world acoustic sounds subdivided into classes based on instrument type. Although the manual annotation process is thorough and involves multiple assessment procedures performed by 3–5 experienced individuals, it is nonetheless subjective, thus blurring the distinction between physical onset and perceptual onset. Because of this uncertainty, detected onsets are deemed valid if within 50 ms of the subjective position (Bello et al., 2005), although a lower tolerance of 25 ms has been used by some authors (Böck et al., 2012b), especially for percussive sounds where physical onsets are well-defined (Collins, 2005a). Similar to those described in section 4.1.3, standard evaluation metrics include precision (P), recall (R) and F-measure (F), defined respectively in Eq. 3–5.

$$P = \frac{O_c}{O_c + O_{fp}} \tag{3}$$

$$R = \frac{O_c}{O_c + O_{fn}} \tag{4}$$

$$F = 2\frac{PR}{P + R} \tag{5}$$

Here, $O_c$ is the number correctly detected onsets, $O_{fp}$ is the number of false positives and $O_{fn}$ is the number of false negatives. In offline settings one might favour high recall over precision, since there is greater chance that the detector is simply reacting to noise or modulations in the signal unrelated to onsets. In this respect, it is less subjective to manually remove data points than to add them.

Table 9.1 gives the average F-measures by instrument category for four of the best onset detectors submitted the 2015 onset detection contest run by the Music Information Retrieval Evaluation eXchange[2] (MIREX). Both Universal (Böck et al., 2015; Eyben et al., 2010) and Fusion (Chen, 2015) algorithms use probabilistic methods, whereas SuperFlux (Böck & Widmer, 2013a) and ComplexFlux (Böck & Widmer, 2013b) are refined versions of two classic spectral-based algorithms. As with the MIREX results, The detectors are ranked by the average of their class means, though we have omitted the third best ranking algorithm as it is an online version of the Universal detector (Böck et al., 2012a). For this dataset, the four techniques perform similarly over all classes (around 80%) but there are clear differences between the algorithms within each category. For example, the probabilistic methods outperform the simpler flux algorithms in the majority of classes with a few exceptions (e.g., solo singing voice). This may be attributable to a lack of training data and/or because the flux detectors – especially ComplexFlux – were designed to better handle instruments with strong vibrato and tremolo. For all algorithms, performance appears to deteriorate for the voice, sustained strings and wind instruments,

TABLE 9.1 *Summary of average F-measures for four state-of-the-art onset detectors submitted to 2015 MIREX audio onset detection contest.*

| **Class** | **Universal** | **Fusion** | **Complex-Flux** | **Super-Flux** | **Mean (SD)** |
|---|---|---|---|---|---|
| Complex | 79.4 | 79.5 | 75.7 | 77.5 | 78.0 (1.8) |
| Poly pitched | 94.1 | 93.9 | 91.7 | 91.6 | 92.8 (1.4) |
| Solo bars & bells | 100.0 | 100.0 | 96.5 | 96.7 | 98.3 (2.0) |
| Solo brass | 82.1 | 77.0 | 75.3 | 76.6 | 77.8 (3.0) |
| Solo drum | 93.1 | 93.1 | 93.1 | 92.4 | 92.9 (0.3) |
| Solo plucked strings | 90.9 | 91.5 | 89.7 | 89.8 | 90.5 (0.8) |
| Solo singing voice | 52.1 | 55.3 | 60.4 | 60.6 | 57.1 (4.1) |
| Solo sustained strings | 72.9 | 66.9 | 57.5 | 58.8 | 64.0 (7.2) |
| Solo winds | 74.0 | 72.2 | 74.6 | 68.6 | 72.3 (2.7) |
| Mean (SD) | 82.1 (13.8) | 81.0 (13.9) | 79.4 (13.4) | 79.2 (13.5) | |

2 http://www.music-ir.org/mirex/wiki/2015:Main_Page

which might be explained by difficulties in detecting softer onsets and/or that the human annotations were more in line with perceptual attack time rather than the physical onsets picked up by the algorithms. The average recall (and precision) across all test stimuli for the four algorithms was: Universal – 87.9% (86.2%); Fusion – 86.2% (87.0%); ComplexFlux – 84.3% (86.8%); Spectral-Flux – 85.6% (85.4%).

The choice of algorithm for detecting sonic events is evidently dependent on the both source type of the technical requirements of a given application. When studying synchronisation in musical performance, the measurement of player timing must be sufficiently accurate to reflect the tempo of the piece and capture salient asynchronies between the note onsets of each player and those of an external auditory stimulus such as a metronome and/or the note played by respective partners. For example, capturing small asynchronies in timing is imperative when studying how performers correct for deviations from an external beat (Vorberg & Wing, 1996) or from fellow musicians (Rasch, 1979, Wing et al., 2014), or how players utilise asynchrony for expressive purposes (Palmer, 1996). In general, methods based on amplitude envelope following provide the highest temporal resolution and are computationally efficient compared to frequency-domain and especially ML approaches. The latter are more suitable for acoustic sources with soft attacks and complex modulations following the onset, such as those produced by bowed string instruments, flute or the singing voice. When using frequency-domain methods, it is important to consider the parameters used to configure the time-frequency decomposition, such as window length and window hop size in the case of the STFT. For example, reducing the window hop size improves temporal precision at the cost of increasing the workload and smoothing variations in the resulting detection function. The choice of window size, which defines the temporal resolution, is signal-dependent and therefore multi-resolution analysis is more favourable in the case of complex signals.

In short, it is preferable to employ an offline onset detector, which, along with the peak-picker, can be tuned for maximum accuracy. With sufficient training data, probabilistic multi-resolution methods are robust, but one should be cautious of the quality of the subjective data used to train the classification. For more objective measures of onset, the flux methods can be chosen and combined with other detectors to increase the likelihood of capturing new events based on changes across multiple signal features. Most state-of-the-art onset detectors output onset times to text files which, along with the audio signal, can be imported to audio analysis software such as Sonic Visualiser (Cannam et al., 2010) for cross-validation using displays of spectrograms and other signal features, and vari-speed playback.

## 5 Analysis and Modelling

### 5.1 *Alignment*

Once the challenge of accurate onset or event detection has been achieved, the standard measures of analysis are relatively simple. However, to analyse synchrony between the events of two or more independent sources (e.g., the heel strikes of a group of people walking or the auditory onsets of a string ensemble) the onsets need to be aligned to understand which event from one source is temporally related to an event from the other source(s). Here we describe an approach to pairwise-align response onsets to the cue onsets (Elliott et al., 2009b). Note, that typically for multi-person alignment, there is ideally a common vector of 'cue' onsets to which all the response onsets from each group member can be aligned to. This may be an external metronome stimulus or the movements of the lead person in the group, for example.

In this approach we use a dynamic programming method to find the shortest distance between response and cue onsets (Figure 9.6). Starting with two vectors, one containing the 'cue' onsets, ***m*** (this could be another person's movement onsets, or a fixed stimulus such as a metronome); the other containing the response onsets, ***t***. The length of ***m*** and ***t*** do not need to be equal. We subsequently make a matrix of squared distances, ***d***, between each cue onset and each response onset. Alignment occurs by matching up each response onset to the closest (i.e., shortest squared distance) cue onset. If a cue onset

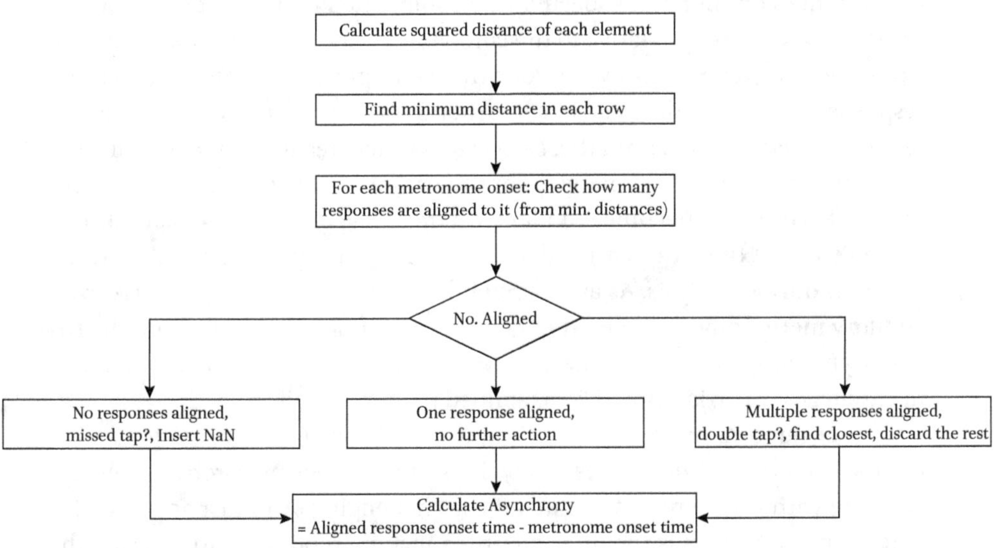


FIGURE 9.6 *Flowchart of the algorithm used to align onsets between two sources (taken from Elliott et al., 2009b).*

has no matching response, then a *NaN* is inserted. If a cue onset has multiple matching responses, we first check to see if the previous onset is empty (*NaN*). If so, we assign the earlier response to the previous cue onset. Of the remainder we assign the response, which has the shortest distance to that cue onset.

### 5.2 *Calculating Asynchrony and IOI/ISI*

Once the onsets of the participant(s) have been aligned to either an external cue or other participant onsets, calculating the time difference or asynchrony ($A$; Eq. 6) between related onsets and the IOIs (also labelled as inter-stimulus intervals, *ISI* for cue onset intervals; Eq. 7) is relatively trivial.

$$A_k = t_k - m_k \quad (6)$$

$$IOI_k = t_{k+1} - t_k \quad (7)$$

where $t_k$ is the $k$ th response onset and $m_k$ is the $k$ th cue onset.

### 5.3 *Participant Outliers*

Sensorimotor synchronisation analysis can be very sensitive to outliers. Outliers will generally emerge in the latter stages of analysis, where the IOI or the asynchronies have been calculated. For example, a missed movement onset to a cue stimulus with an interval of 500ms, will suddenly introduce an IOI of 1000 ms. Another common issue, occurs when someone's movement onset occurs very late, or very early, relative to the comparative cue onset. This will result in a phase wrapped asynchrony (e.g., one that is assigned, via alignment, as a late response to the preceding cue onset rather than an early response to the current cue, or vice-versa). Both these occurrences will result in large within-trial standard deviations (SD) emerging for the IOI and asynchronies, respectively. In fact, it is useful to become familiar with the range of IOI and/or asynchrony SD you would expect from a 'good' trial. This helps to spot potentially erroneous trials during analysis. As an example, for a simple finger tapping task to an auditory metronome with an ISI of 500 ms, one would expect the both the IOI and asynchrony SD to be in the range of 15–30 ms. Values far exceeding this range suggest the trial should be examined in more detail.

Outlier removal must be dealt with methodologically and consistently. Using the IOIs to find outliers is often the simplest and most robust method. Working with asynchronies is much more challenging. A robust approach for detecting IOI outliers is the inter-quartile range method. In Matlab, using the *median* command to find the median IOI of the trial data and then the *iqr* command to find the inter-quartile range, the threshold for upper and lower outlier values is:

$$\text{threshUpper} = \text{median}(\text{IOI}) + \text{N}^{*}\text{iqr}(\text{IOI});$$
$$\text{threshLower} = \text{median}(\text{IOI}) - \text{N}^{*}\text{iqr}(\text{IOI});$$

Where *IOI* is a vector of IOIs calculated from a trial. *N* defines how 'strong' the outlier detection function is. *N*=3 should be the minimum and will heavily cleanse the data, while *N*=6 will be more conservative and only remove extreme outliers.

Matlab's *find* command can subsequently be used to locate values exceeding either the upper or lower threshold and can be removed from the IOI vector. Removal will usually consist of replacing the value with *NaN* and henceforth using *nanmean, nanstd* is required to calculate the mean and SD of the cleansed IOIs. However, if cross-correlation or other calculations relying on a continuous series are to be applied, then an alternative replacement method should be used (e.g. replacing with the average (median) value or similar).

The removal/replacement of IOIs should be reflected in the corresponding Asynchrony vector. Assuming *IOI* and *A* are calculated as defined in Eq. 6–7, then removal of $IOI_k$ should result in removal of $A_{k+1}$.

Identifying outliers from asynchrony data is more challenging. Recall, the alignment process will allocate a response onset to a cue onset within the range –ISI/2 to ISI/2, where ISI is the inter-stimulus interval of the cue. Therefore, given all asynchronies will be bound within this range, there are no outliers as such. However, if phase wrapping occurs there will be sign changes where the onsets go from being large and negative to large and positive, or vice versa. This corresponds to drift (see Figure 9.7) where the participant is not synchronising with the cue and therefore asynchronies become increasingly negative until they hit the lower bound and subsequently the next response is closer to previous cue onset but with a positive asynchrony.

There is little that can be done with linear analyses in these scenarios. The SD becomes very large when these discontinuities occur. And given that typically the presence of drift suggests the participant isn't synchronising to the cue, it is often a case of discarding trials where this occurs. There are occasions where phase-wrapping is likely and of interest (e.g., analysing data where the response and cue have differing tempos). In these cases, it is recommended that circular statistics be used to analyse the data. Circular mean and SD can be used as an alternative without being susceptible to the phase wrapping discontinuities. Further details of circular statistics

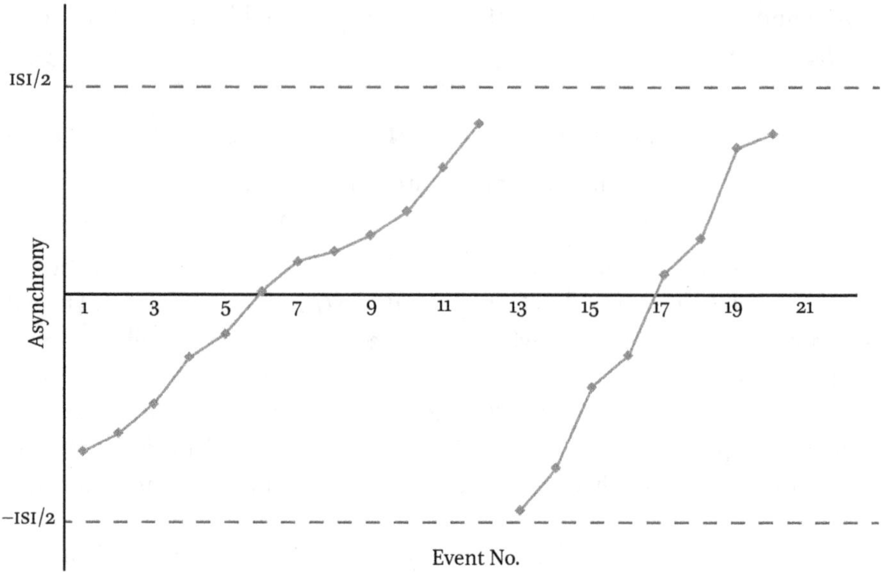


FIGURE 9.7 *Typical pattern of asynchronies when participant is exhibiting drift – i.e. not synchronising with the cue. Note the phase wrapping occurs around +/– ISI/2 and can result in highly inflated asynchrony variance.*

are beyond the scope of this chapter but an excellent free toolbox for Matlab is available (Berens, 2009; http://www.mathworks.com/matlabcentral/fileexchange/10676-circular-statistics-toolbox--directional-statistics-).

### 5.4 *Cross- and Auto-Covariance and Event Based Synchronisation Models*

It is often useful to measure the auto-covariance of the IOIs and asynchronies, or the cross-covariance between two IOI / asynchrony time series. The auto-covariance shows the dependency between current and past time-series values. Wing and Kristofferson (1973) proposed a model for tapping without an external stimulus that predicts that finger tapping intervals have a lag -1 dependence (resulting in a short-long-short-long pattern). The model follows the hypothesis that tapping is based on two internal processes: time keeping (that maintain a temporal interval) and a motor action (that is a result of the execution of a given motor command). The model can be written as:

$$IOI_k = T_k + M_{k+1} - M_k \tag{8}$$

Where, $IOI_k$ is the $k$th IOI (see Eq. 7) and $T_k$ and $M_k$ are the timekeeper interval and motor delay respectively.

The model predicts that:

$$\gamma_{\mathrm{IOI}}(1) = -\sigma_T{}^2 \tag{9}$$

$$\gamma_{\mathrm{IOI}}(0) = \sigma_T{}^2 + 2\sigma_M{}^2 \tag{10}$$

Where $\gamma_I(k)$ is the lag $k$ auto-covariance, $\sigma_T{}^2$ is the timekeeper variance, $\sigma_M{}^2$ is motor variance.

The model has become a highly efficient tool to characterise tapping (Wing, 2002). Empirical results indicate several intriguing relations between timekeeper variance, motor variance, and tempo. Namely: (a) Motor noise remains constant when the base tapping tempo is changed, but timekeeper variance increases with tempo; (b) Motor noise is smaller than timekeeper noise $\left(\sigma_M{}^2 < \sigma_T{}^2\right)$

The success of the model led to its generalisation to the case of tapping to an external metronome. Vorberg and Wing (1996) proposed a revised model which included a correction gain parameter to describe the process of synchronising to an external cue. The correction gain, α (often also referred to as phase correction), explains how much of the previous error (asynchrony) is corrected for in the next movement.

$$IOI_k = \alpha A_k + T_k + M_{k+1} - M_k \tag{11}$$

The gain is stable in the range, $0 \leq \alpha \leq 2$, where $\alpha = 1$ is full correction, $\alpha > 1$ is overcorrection, and $\alpha < 1$ is undercorrection. In most cases, empirical estimates of $\alpha$ are usually in the range of 0.5 to 1.

In the case of a relatively stable metronome (no significant tempo changes), the correction gain can be deduced simply by calculating the cross-covariance between the cue and response intervals, *if* the cue intervals do not have zero variance (i.e., an isochronous metronome has zero interval variance). The relationship between the covariance and the correction gain is as follows:

$$\gamma\_CI(j) = \alpha(1-\alpha){}^\wedge(j-1)\sigma\_C{}^\wedge 2, j \geq 0 \tag{12}$$

where, $\gamma_{CI(j)}$ is the cross-covariance function between the stimulus response intervals of lag j and $\sigma_c{}^2$ is the variance of the stimulus intervals. The model of Eq. 11 can be generalised to ensemble synchronisation:

$$IOI_k = \sum_i \alpha_i A_{k,i} + T_k + M_{k+1} - M_k \tag{13}$$

Where $A_{k,i}$ are the asynchronies between the studied player all other players, and $\alpha_i$ is the phase correction parameter associated with adapting to a specific player $i$.

Note that this model is a generalisation of an ensemble synchronisation model proposed by (Wing, Endo, Bradbury, & Vorberg, 2014) that has been used to study synchronisation within string quartets. Their model is identical to Eq. 13, but with the assumption that the parameter $\sigma_M{}^2 = 0$, and therefore $M_{k+1} - M_k$ in equation Eq. 13 does not play a role.

It is possible to generalise the model of equation Eq. 11 to the case where there are substantial tempo changes (Schulze, Cordes, & Vorberg, 2005). Here it is often assumed that an additional period correction process occurs (Repp & Keller, 2008):

$$\tau_k = \tau_{k-1} - \beta A_k \tag{14}$$

Where $\tau_k = mean(T_k)$ and $\beta$ represents the period correction constant.

The model can be cast as a standard Autoregressive Moving Average (ARMA) model (Diedrichsen, Ivry, & Pressing, 2003)

$$IOI_k - IOI_{k-1} = -(\alpha + \beta)A_k + \alpha A_{k-1} + T_k - T_{k-1} + M_{k+1} - 2M_k + M_{k-1} \tag{15}$$

Here $T_k$ and $M_k$ are two independent random variables with a fixed mean.

### 5.5 *Bounded General Least Squares (bGLS) Method for Parameter Estimation*

While the cross- and auto-correlation approaches to parameter estimation of the linear phase correction model are relatively simple to compute, their application is limited to one participant with small tempo variations. In the case of ensemble synchronisation, the estimation procedure based on the autocovariance function requires a slow iterative model fitting approach.

Moreover, recent work (Jacoby, Keller, Repp, Ahissar, & Tishby, 2015) showed that the structure of the models described above generates an inherited dependency in the accuracy of estimating the parameters $\alpha$, $\sigma_M{}^2$ and $\sigma_T{}^2$. Since the parameters are inherently interdependent, they cannot be jointly estimated by the autocovariance method or by *any other method* without using further assumptions. Therefore, directly applying the autocovariance method or standard linear estimation techniques such as the Matlab *armax* command on data will often lead to unreliable estimations. While the problem exists also for single participant synchronisation with a metronome with small tempo changes, it becomes much more notable in the case of ensemble synchronisation or when there are large tempo changes. Fortunately, there is a simple solution to this problem. As mentioned above it has been empirically observed that the motor noise is smaller than timekeeper variance. If this additional simple constraint is taken into account in the estimation process the interdependency problem is practically resolved (Jacoby et al., 2015). Moreover, the group further proposed an algorithm, the bounded general least squares (bGLS) method that can estimate the parameters in case of a single person and ensemble synchronisation as well as tempo changing sequences (Jacoby, Tishby, Repp, Ahissar, & Keller, 2015).

The method has been applied in music related studies, re-analysing the earlier quartet study by Wing et al., (Jacoby et al., 2015) and investigating metrical structure in Malian jembe drumming (Polak, Jacoby, & London, 2015). In addition, it has been applied to a group sensorimotor synchronisation task in order to estimate changes in correction along a chain of individuals moving in time with each other (Honisch et al., 2016).

The mathematical derivations of the method are fully explained in the two aforementioned publications (Jacoby et al., 2015) and, hence, won't be reiterated here (for a short overview of the method see Elliott, Chua & Wing, 2016). However, the bGLS Toolbox for Matlab is provided with example code for this chapter (see book's GitHub repository).

## 6 Conclusion

We have presented methods for collecting, conditioning and analysing the timing of movements, ranging from simple finger tapping where response events can be captured by switches, force transducers or motion capture systems to the complexities of music performance where the data commonly requires acoustic recording, or in some cases, motion capture data. Regardless of the particular technology for capturing timing data, our goal has been to maximise the measurement accuracy in order to better characterise, not only the accuracy of timing in terms of mean and variability, but also the form of variability, in order to reveal the underlying mechanisms that are so often key to the skilled performance of complex sequential activities such as music and dance.

## References

Balasubramaniam, R., A.M. Wing, & A. Daffertshofer (2004). Keeping with the beat: Movement trajectories contribute to movement timing. *Experimental Brain Research*, *159*(1), 129–134.

Bello, J.P., L. Daudet, S. Abdallah, C. Duxbury, M. Davies, M.B. Sandler, & S. Member (2005). A tutorial on onset detection in music signals. *IEEE Transactions on Speech and Audio Processing*, *13*(5), 1–13.

Bello, J.P., C. Duxbury, M. Davies, M. Sandler, & S. Member (2004). On the use of phase and energy for musical onset detection in the complex domain. *IEEE Signal Processing Letters*, *11*(6), 553–556.

Berens, P. (2009). CircStat: A MATLAB toolbox for circular statistics. *Journal of Statistical Software*, *31*(110).

Böck, S., & G. Widmer (2013a). Maximum filter vibrato suppression for onset detection. In *16th International Conference on Digital Audio Effects (DAFx-13)*.

Böck, S., & G. Widmer (2013b). Local group delay based vibrato and tremolo suppression for onset detection. In *Proceedings of the 13th International Society for Music Information Retrieval Conference (ISMIR)*.

Böck, S., A. Arzt, F. Krebs, & M. Schedl (2012a). Online real-time onset detection with recurrent neural networks. In *15th International Conference on Digital Audio Effects* (DAF*x*).

Böck, S., F. Krebs, & M. Schedl (2012b). Evaluating the online capabilities of onset detection methods. In *13th International Society for Music Information Retrieval Conference (ISMIR)* (pp. 49–54).

Böck, S., F. Krebs, F. Korzeniowski, & G. Widmer (2015). *MIREX 2015 Submissions.* MIREX Audio Onset Detection. http://nema.lis.illinois.edu/nema_out/mirex2015/results/aod/.

Brown, J.C., & M.S. Puckette (1993). A high resolution fundamental frequency determination based on phase changes of the fourier transform. *Journal of the Acoustical Society of America*, *94*(2).

Cannam, C., C. Landone, & M. Sandler (2010). Sonic visualiser: An open source application for viewing, analysing, and annotating music Audio Files. In *Proceedings of the ACM Multimedia 2010 International Conference* (pp. 1467–1468).

Chen, C. (2015.). *An improved onset detection algorithm by ODF fusion.* MIREX Audio Onset Detection. http://nema.lis.illinois.edu/nema_out/mirex2015/results/aod/.

Collins, N. (2005a). A comparison of sound onset detection algorithms with emphasis on psychoacoustically motivated detection functions. In *Proceedings of the 118th Audio Engineering Society Convention.*

Collins, N. (2005b). Using a pitch detector for onset detection. In *Proceedings of the International Symposium on Music Information Retrieval* (pp. 100–106).

Collins, N. (2006). Investigating computational models of perceptual attack time. In *Proceedings of the 9th International Conference on Music Perception & Cognition (ICMPC)*.

Davy, M., & S. Godsill (2002). Detection of abrupt spectral changes using support vector machines an application to audio signal segmentation. In *Proceedings of IEEE International Conference on Acoustics, Speech, and Signal Processing* (*ICASSP*) (pp. 1313–1316).

De Poli, G., A. Roda, & A. Vidolin (1998). Note-by-note analysis of the influence of expressive intentions and musical structure in violin performance. *Journal of New Music Research*, *27*(3), 293–321.

Diedrichsen, J., R.B. Ivry, & J. Pressing (2003). Cerebellar and basal ganglia contributions to interval timing. In Meck, W.H. (Ed.), *Functional and neural mechanisms of interval timing*. CRC Press.

Dixon, S. (2006). Onset detection revisited. In *Proceedings of the International Conference on Digital Audio Effects* (DAFx).

Duxbury, C., J.P. Bello, M. Davies, & M. Sandler (2003). Complex domain onset detection for musical signals. In *Proceedings of the 6th International Conference on Digital Audio Effects* (DAFx).

Elliott, M.T., W.L. Chua, & A.M. Wing (2016). Modelling single-person and multi-person event-based synchronisation. *Current Opinion in Behavioral Sciences*, *8*, 167–174.

Elliott, M.T., A.E. Welchman, & A.M. Wing (2009a). Being discrete helps keep to the beat. *Experimental Brain Research*, *192*(4), 731–737.

Elliott, M.T., A.E. Welchman, & A.M. Wing (2009b). MatTAP: A MATLAB toolbox for the control and analysis of movement synchronisation experiments. *Journal of Neuroscience Methods*, *177*(1), 250–257.

Elliott, M.T., A.M. Wing, & A.E. Welchman (2010). Multisensory cues improve sensorimotor synchronisation. *European Journal of Neuroscience*, *31*(10), 1828–1835.

Elliott, M.T., A.M. Wing, & A.E. Welchman (2014). Moving in time: Bayesian causal inference explains movement coordination to auditory beats. *Proceedings of the Royal Society B: Biological Sciences*, *281*(1786), 20140751.

Ellis, M.C. (1991). An analysis of swing subdivision and asynchronization in three jazz saxophonists. *Perceptual and Motor Skills*, *73*(3), 707–713.

Eyben, F., S. Böck, B. Schuller, & A. Graves (2010). Onset detection with bidirectional long short-term memory neural networks. In *Proceedings Annual Meeting of the MIREX 2010 community as part of the 11th International Conference on Music Information Retrieval* (pp. 589–594).

Finney, S.A. (2001). FTAP: A Linux-based program for tapping and music experiments. *Behavior Research Methods, Instruments, & Computers*, *33*, 65–72.

Friberg, A., & A. Sundströöm (2002). Swing ratios and ensemble timing in jazz performance: Evidence for a common rhythmic pattern. *Music Perception: An Interdisciplinary Journal*, *19*(3), 333–349.

Georgiou, L., V. Racic, J.M.W. Brownjohn, & M.T. Elliot (2015). Coordination of groups jumping to popular music beats. In Caicedo, J. & S. Pakzad (Eds.), *Dynamics of Civil Structures, Volume 2* (pp. 283–288). Springer International Publishing.

Goebl, W., & C. Palmer (2008). Tactile feedback and timing accuracy in piano performance. *Experimental Brain Research, 186*(3), 471–479.

Goebl, W., & C. Palmer (2009). Synchronization of timing and motion among performing musicians. *Music Perception: An Interdisciplinary Journal, 26*(5), 427–438.

Goebl, W., S. Flossmann, & G. Widmer (2010). Investigations of Between-hand Synchronization in Magaloff's Chopin. *Computer Music Journal, 34*(3), 35–44.

Gordon, J.W. (1987). The perceptual attack time of musical tones. *The Journal of the Acoustical Society of America, 82*(1), 88–105.

Hennig, H. (2014). Synchronization in human musical rhythms and mutually interacting complex systems. *Proceedings of the National Academy of Sciences, 111*(36), 12974–12979.

Honisch, J.J., M.T. Elliott, N. Jacoby, & A.M. Wing (2016). Cue properties change timing strategies in group movement synchronisation. *Scientific Reports, 6.*

Honisch, J.J., N. Roach, & A.M. Wing (2009). Movement synchronization to a virtual dancer: How do expert dancers adjust to perceived temporal and spatial changes whilst performing ballet versus abstract dance sequences. In *Proceedings of the ISSP 12th World Congress of Sport Psychology.*

Jacoby, N., P.E. Keller, B.H. Repp, M. Ahissar, & N. Tishby (2015a). Lower bound on the accuracy of parameter estimation methods for linear sensorimotor synchronization models. *Timing & Time Perception, 3*(1–2), 32–51.

Jacoby, N., N. Tishby, B.H. Repp, M. Ahissar, & P.E. Keller (2015b). Parameter estimation of linear sensorimotor synchronization models: Phase correction, period correction, and ensemble synchronization. *Timing & Time Perception, 3*(1–2), 52–87.

Kapanci, E., & A. Pfeffer (2004). A hierarchical approach to onset detection. In *Proceedings of the International Computer Music Conference.*

Keele, S.W., R.A. Pokorny, D.M. Corcos, & R. Ivry (1985). Do perception and motor production share common timing mechanisms: A correlational analysis. *Acta Psychologica, 60*(2), 173–191.

Keller, P.E., G. Knoblich, & B.H. Repp (2007). Pianists duet better when they play with themselves: On the possible role of action simulation in synchronization. *Consciousness and Cognition, 16*(1), 102–111.

Klapuri, A. (1999). Sound onset detection by applying psychoacoustic knowledge. In *Proceedings of The IEEE International Conference on Acoustic, Speech and Signal Processing (ICASSP).*

Lacoste, A., & D. Eck (2007). A supervised classification algorithm for note onset detection. *EURASIP Journal on Advances in Signal Processing.*

Lago, N.P., & F. Kon (2004). The quest for low latency. In *Proceedings of the International Computer Music Conference (pp. 33–36).*

Loehr, J.D., & C. Palmer (2007). Cognitive and biomechanical influences in pianists' finger tapping. *Experimental Brain Research, 178*(4), 518–528.

Manning, F., & M. Schutz (2013). "Moving to the beat" improves timing perception. *Psychonomic Bulletin & Review, 20*(6), 1133–1139.

Marchi, E., G. Ferroni, F. Eyben, L. Gabrielli, S. Squartini, & B. Schuller (2014). Multi-resolution linear prediction based features for audio onset detection with bidirectional LSTM neural networks. *2014 IEEE International Conference on Acoustics, Speech and Signal Processing* (*ICASSP*).

Marolt, M., A. Kavcic, M. Privosnik, & S. Divjak (2002). On detecting note onsets in piano music. In *Proceedings of the 11th IEEE Mediterranean Electrotechnical Conference* (pp. 385–389).

Masri, P. (1997). Computer modelling of sound for transformation and synthesis of musical signals. Ph.D Thesis. University of Bristol.

Moore, G.P., & J. Chen (2010). Timings and interactions of skilled musicians. *Biological Cybernetics, 103*(5), 401–414.

Nabney, I. (2002). *NETLAB: Algorithms for Pattern Recognition*. Springer Science & Business Media.

Palmer, C. (1996). On the assignment of structure in music performance. *Music Perception*, 14, 23–56.

Palmer, C., & J.C. Brown (1991). Investigations in the amplitude of sounded piano tones. *The Journal of the Acoustical Society of America, 90*(1), 60–66.

Pecenka, N., & P.E. Keller (2011). The role of temporal prediction abilities in interpersonal sensorimotor synchronization. *Experimental Brain Research, 211*(3–4), 505–515.

Pelton, T.A., A.M. Wing, D. Fraser, & P. Vliet (2015). Differential effects of parietal and cerebellar stroke in response to object location perturbation. *Frontiers in Human Neuroscience, 9*.

Pfordresher, P., & C. Palmer (2002). Effects of delayed auditory feedback on timing of music performance. *Psychological research, 66*(1), 71–79.

Polak, R., & J. London (2014). Timing and meter in mande drumming from Mali. *Music Theory Online, 20*(1).

Polak, R., N. Jacoby, & J. London (2015). Ensemble entrainment in jembe music from Mali. *15th Rhythm Production and Perception Workshop, Netherlands, July 6–8, 2015*.

Rasch, R.A. (1979). Synchronization in performed ensemble music. *Acta Acustica united with Acustica, 43*(2), 121–131.

Repp, B.H. (1995). Quantitative effects of global tempo on expressive timing in music performance: Some perceptual evidence. *Music Perception: An Interdisciplinary Journal, 13*(1), 39–57.

Repp, B.H. (1999). Detecting deviations from metronomic timing in music: Effects of perceptual structure on the mental timekeeper. *Perception & Psychophysics, 61*(3), 529–548.

Repp, B.H. (2005). Sensorimotor synchronization: A review of the tapping literature. *Psychonomic Bulletin & Review, 12*(6), 969–992.

Repp, B.H., & P.E. Keller (2008). Sensorimotor synchronization with adaptively timed sequences. *Human Movement Science, 27*(3), 423–456.

Repp, B.H., & Y.-H. Su (2013). Sensorimotor synchronization: A review of recent research (2006–2012). *Psychonomic Bulletin & Review, 20*(3), 403–452.

Rosao, C., R. Ribeiro, & D.M. De Matos (2012). Influence of peak selection methods on onset detection. In *ISMIR* (pp. 517–522).

Schleusing, O., B. Zhang, & Y. Wang (2008). Onset detection in pitched non-percussive music using warping-compensated correlation. In *Proceedings of the IEEE International Conference on Acoustics, Speech and Signal Processing* (*ICASSP*).

Schluter, J., & S. Böck (2013). Musical onset detection with convolutional neural networks. In *Proceedings of the 6th International Workshop on Machine Learning and Music* (*MML*).

Schluter, J., & S. Böck (2014). Improved musical onset detection with convolutional neural networks. *IEEE International Conference on Acoustic, Speech and Signal Processing* (*ICASSP*).

Schultz, B.G., & F.T. Vugt. (2015). Tap Arduino: An Arduino microcontroller for low-latency auditory feedback in sensorimotor synchronization experiments. *Behavior Research Methods*, 1–17.

Schulze, H.H., A. Cordes, & D. Vorberg (2005). Keeping synchrony while tempo changes: Accelerando and ritardando. *Music Perception, 22*(3), 461–477.

Shaffer, L.H. (1984). Timing in solo and duet piano performances. *The Quarterly Journal of Experimental Psychology, 36*(4), 577–595.

Stables, R., S. Endo, & A. Wing (2014). Multi-Player microtiming humanisation using a multivariate markov model. In *International Conference on Digital Audio Effects* – DAF*x 14* (pp. 109–114).

Stevens, L.T. (1886). On the time sense. *Mind, 11*, 393–404.

Thaut, M.H., B. Tian, & M.R. Azimi-Sadjadi (1998). Rhythmic finger tapping to cosine-wave modulated metronome sequences: Evidence of subliminal entrainment. *Human Movement Science, 17*(6), 839–863.

Toh, C.C., B. Zhang, & Y. Wang (2008). Multiple-feature fusion based onset detection for solo singing voice. In *Proceedings of the 9th International Conference on Music Information Retrieval* (*ISMIR*) (pp. 515–520).

Vincent, E. (2006). Musical source separation using time-frequency source priors. *IEEE Transactions on Audio, Speech, and Language Processing, 14*(1), 91–98.

Vorberg, D., & A.M. Wing (1996). Modeling variability and dependence in timing. In *Handbook of perception and action* (pp. 181–262). London: Academic Press.

Vos, J., & R. Rasch (1981). The perceptual onset of musical tones. *Perception & Psychophysics*, *29*(4), 323–335.

Wang, Y., R. Stables, & J Reiss. (2010). Audio latency measurement for desktop operating systems with onboard soundcards. In *Audio Engineering Society Convention 128*. Audio Engineering Society.

Wing, A.M. (2002). Voluntary timing and brain function: An information processing approach. *Brain and Cognition*, *48*(1), 7–30.

Wing, A.M., & A.B. Kristofferson (1973). Response delays and the timing of discrete motor responses. *Perception and Psychophysics*, *14*, 5–12.

Wing, A.M., S. Endo, A. Bradbury, & D. Vorberg (2014). Optimal feedback correction in string quartet synchronization. *Journal of The Royal Society Interface*, *11*(93), 20131125.

Winter, D.A. (2009). *Biomechanics and Motor Control of Human Movement*. John Wiley & Sons.

Zhu, B., J. Gan, J. Cai, Y. Wang, & H. Wang (2014). Adaptive Onset Detection based on Instrument Recognition. In *Proceedings of the 12th International Conference on Signal Processing* (ICSP) (pp. 2416–2421).

CHAPTER 10

# Measuring Temporal Preparation

*Mariagrazia Capizzi and Ángel Correa*

## 1 Introduction

Temporal preparation is a fundamental cognitive ability that enables us to anticipate the moment in time (in the millisecond to second range) at which a relevant event might occur. At an operational level, this means that when stimulus onsets can be predicted, both accuracy and response time to such stimuli will be improved compared to stimuli with unpredictable onset (Correa, 2010; Nobre, Correa, & Coull, 2007). Theoretically, temporal preparation is linked to the field of time perception since the ability to anticipate when something is going to occur does involve the capacity to compute the passage of time. It should be noted, however, that experiments on temporal preparation require *using* time rather than consciously estimating it (i.e., time estimation and reproduction tasks; e.g., Grondin, 2010; Ivry & Hazeltine, 1995; Chapters 2 and 4, this volume).

Early studies of temporal preparation focused on the behavioral effects that the manipulation of the "*foreperiod*" (timing during which a warning signal precedes a target stimulus) had on participants' performance (Woodrow, 1914). Subsequent studies investigated whether temporal preparation could be voluntarily acquired following a symbolic and predictive temporal cue (e.g., Coull & Nobre, 1998; Miniussi, Wilding, Coull, & Nobre, 1999; Nobre, 2001). More recent work has explored temporal preparation driven by the presentation of regular rhythmic patterns (e.g., Cutanda, Correa, & Sanabria, 2015; Jones, Moynihan, MacKenzie, & Puente, 2002; Lange, 2010; Sanabria, Capizzi, & Correa, 2011).

These complementary research approaches have identified the basic features of the temporal preparation effects that will be covered in the following sections of this chapter. Each section provides a brief description of one of the paradigms used to investigate temporal preparation along with an example of how to measure and interpret the behavioral data usually obtained in the studies employing such a paradigm. For a thorough review of the cognitive processes and neural mechanisms underlying temporal preparation, the reader is referred to Nobre and Coull (2010).

DOI 10.1163/9789004280205_011

## 2 Foreperiod and Sequential Effects

In the foreperiod paradigm, a warning signal precedes a target stimulus, to which the participant is to respond (see Coull, 2009; Niemi & Näätänen, 1981, for reviews). The foreperiod is the time interval between warning signal and target. It represents the period in which temporal preparation has been shown to take place. Consider, for example, the simplest case with only two possible foreperiods (e.g., 1000 ms – short vs. 3000 ms – long). The variability with which short and long foreperiods are presented in different trials will selectively influence participants' reaction times (RTs). When there is no variability and just one foreperiod is administered in a block of trials (i.e., fixed foreperiod design), it is typical to find shorter RTs with short foreperiods, a phenomenon known as the "fixed foreperiod effect" (e.g., Bausenhart, Rolke, & Ulrich, 2008; Mattes & Ulrich, 1997; Vallesi, McIntosh, Shallice, & Stuss, 2009a). Conversely, when short and long foreperiods are randomly intermixed with equal probability within a block (i.e., variable foreperiod design), shorter RTs will be associated with long foreperiods (i.e., the "variable foreperiod effect"; hereafter simply referred to as the foreperiod effect; e.g., Drazin, 1961; Mento, Tarantino, Vallesi, & Bisiacchi, 2015; Niemi & Näätänen, 1981; Woodrow, 1914). The discrepant findings between fixed and variable foreperiod designs can be accounted for by two separate factors: *time estimation* and *conditional probability of target occurrence*.

On the one hand, in a fixed foreperiod paradigm, it is likely that participants use the warning signal as a reference to estimate the time at which the target stimulus will be presented (Klemmer, 1956, 1957; Niemi & Näätänen, 1981). Given that uncertainty in time estimation increases as a function of the duration of the interval being estimated (Gibbon, 1977), it will be more difficult to anticipate the target onset in long than in short foreperiod blocks, which will lead to longer RTs in the long foreperiod blocks.

On the other hand, in a variable foreperiod paradigm, the conditional probability of target onset increases with the passage of time. Accordingly, participants learn that if the target does not appear at the short foreperiod, it will necessarily do so at the long foreperiod (Elithorn & Lawrence, 1955). Under this "strategic" view of the foreperiod effect, participants' temporal expectation would be higher for targets occurring at the long foreperiod, thus explaining the decrease in RT observed with long foreperiod trials of variable foreperiod studies.

The study by Vallesi, Lozano, and Correa (2013) will be used to explain how to measure the variable foreperiod effect. Here, we will consider only the data from the blocks with a short inter-trial interval (ITI) of Experiment 2 (simple-RT task). An adapted version of the original experiment can be found in book's GitHub repository.

In the study by Vallesi et al. (2013), the variable foreperiod design included an auditory warning signal (a 1500 Hz pure tone) and a visual target (a square or a triangle) with a foreperiod of either 1000 or 3000 ms. Each foreperiod was presented randomly with the same probability. A yellow cross, which served as the fixation point, was presented simultaneously with the warning signal and remained on the screen until the target appeared. The participants' task was to respond to the onset of the target by pressing the space bar with their index finger (see Figure 10.1A).

As can be seen in Figure 10.1B, participants' RTs were shorter for the long foreperiod than for the short foreperiod. Such a difference reflects the foreperiod effect typically found in a variable foreperiod design. Several studies have replicated the foreperiod effect across different foreperiod ranges and type of tasks (Niemi & Näätänen, 1981). Hence, an interim conclusion from this phenomenon is that our cognitive system can exploit the temporal information implicitly provided by task context to improve performance.

Another type of phenomenon that further proves the increase in sensitivity to events having predictable timing is generally termed "sequential effects" (e.g., Capizzi, Correa, Wojtowicz, & Rafal, 2015; Los & van den Heuvel, 2001; Los, 2013; Steinborn & Langner, 2012; Steinborn, Rolke, Bratzke, & Ulrich, 2008; Vallesi & Shallice, 2007; Woodrow, 1914). Sequential effects take into account the influence that the previous foreperiod duration has on preparation in the current foreperiod. Thus, following the variable foreperiod procedure, the analysis should now consider whether the previous foreperiod was short or long. To illustrate this point, we turn back to the experiment by Vallesi et al. (2013). When results are analyzed considering the previous foreperiod as a factor (pre-Short vs. pre-Long), the pattern usually conforms to the one depicted in Figure 10.1C. As shown in this figure, participants' RTs were overall shorter when the previous foreperiod was short rather than long. More importantly, a significant interaction between previous foreperiod and current foreperiod duration was also observed. This interaction reflects the "asymmetry" of sequential effects. That is, on current short foreperiod trials participants' responses were faster after a previous short foreperiod as compared to alternation from a previous long foreperiod (in Figure 10.1C, the two conditions are associated with 316 and 361 ms, respectively). Conversely, on current long foreperiod trials, participants' responses were equally fast irrespective of whether the previous foreperiod duration had been short or long (in Figure 10.1C, 305 ms for both conditions).

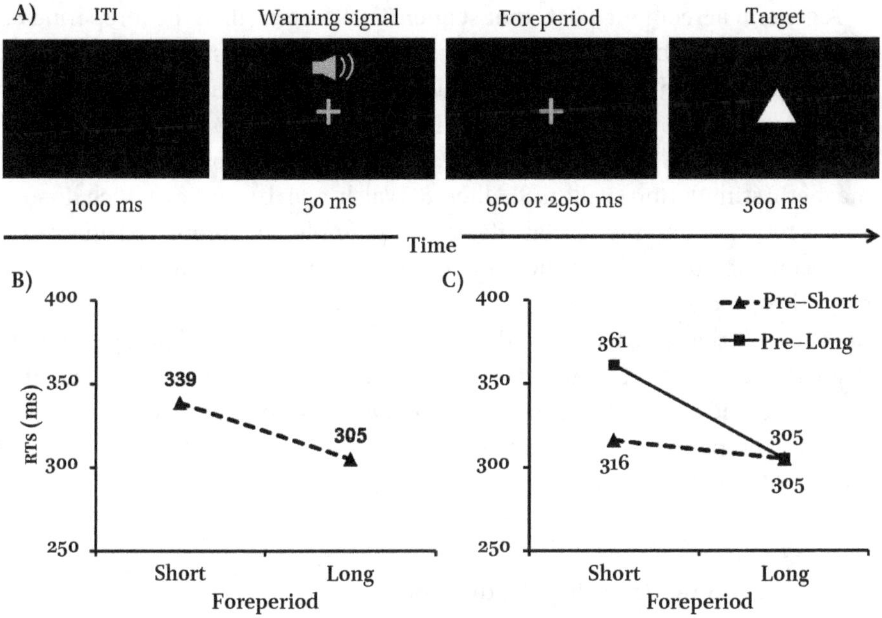


FIGURE 10.1 *(A) Experimental design. The variable foreperiod paradigm used by Vallesi, Lozano, and Correa (2013). In the original study, the cross was displayed in yellow instead of gray. Each trial started with the ITI: a black screen lasting 1000 ms. Subsequently, the auditory warning signal and the fixation cross were presented. The fixation cross remained on the screen for a variable foreperiod interval of 1000 or 3000 ms. Then, the target (a square or a triangle) appeared and participants had to respond to it by pressing the space bar. The target disappeared after either the participants' response or after 1.500 ms. (B) The foreperiod effect. Participants' RTs in ms are plotted as a function of the Foreperiod duration (short, long). The foreperiod effect is indexed by shorter RTs at the long foreperiod. (C) Sequential effects. Participants' RTs are plotted as a function of both the previous foreperiod duration (label Pre- in the figure legend) and the current foreperiod duration. The asymmetry of sequential effects is given by shorter RTs for the short-short foreperiod sequence as compared to the long-short foreperiod sequence. Procedure and data for both the foreperiod and sequential effects are adapted from Vallesi, Lozano and Correa (2013).*

Since the first demonstration of the presence of sequential effects in temporal preparation (Woodrow, 1914), there have been several replications suggesting that such effects are quite robust. A comprehensive overview of the cognitive models that have been developed to explain sequential effects is beyond the scope of this chapter (for further information, see Capizzi et al., 2015; Los & van den Heuvel, 2001; Steinborn et al., 2008; Vallesi & Shallice, 2007, and the following book chapters: Los, 2010; Vallesi, 2010).

A general agreement exists that sequential effects, unlike the foreperiod effect, are mediated by automatic processes. This claim is bolstered by converging evidence provided by behavioral (Vallesi, Arbula, & Bernardis, 2014; Vallesi et al., 2013), neuropsychological (Triviño, Correa, Arnedo, & Lupiáñez, 2010; Vallesi et al., 2007a), developmental (Vallesi & Shallice, 2007), and transcranial magnetic stimulation (Vallesi, Shallice, & Walsh, 2007b) studies, which dissociated foreperiod and sequential effects. For example, the foreperiod effect only has been shown to rely on the functioning of prefrontal structures related to executive processes, whereas it has been shown not to be the case for sequential effects (Triviño et al., 2010; Vallesi et al., 2007a). These dissociations demonstrate that temporal preparation can be flexibly built up on the basis of distinct controlled and automatic processes. A clear example of a controlled process of temporal preparation is given by "temporal orienting of attention", which is the focus of the next section.

## 3 Temporal Orienting of Attention

Coull and Nobre (1998) developed a temporal variant of the Posner's spatial orienting task (Posner, Snyder, & Davidson, 1980) to test whether attention can be voluntarily oriented in time. In a typical implementation of the temporal orienting task (see Correa, 2010), participants have to respond to the onset of a target stimulus that can appear either early in time (e.g., after a short foreperiod of 1000 ms) or late in time (e.g., after a long foreperiod of 3000 ms). A symbolic cue (e.g., a short or a long line) precedes each target presentation indicating whether the target is likely to appear early or late, respectively. On a large percentage of trials (e.g., 75%), the cue validly predicts the length of the foreperiod (valid condition). On the remaining trials, the target will appear either earlier or later than expected (invalid condition). As shown in Figure 10.2B, this design gives rise to four types of trials. The 'early cue-short foreperiod' and the 'late cue-long foreperiod' types of trials are valid, whereas the 'early cue-long foreperiod' ('delayed', cf. Coull, 2011) and the 'late cue-short foreperiod' ('premature') trials are invalid.

Figure 10.2 depicts the temporal orienting task used by Capizzi, Sanabria and Correa (2012) and their results, that is, mean RTs from the single-task condition plotted as a function of Cue Validity (valid, invalid) and Foreperiod (short, long). An adapted version of the original experiment can be downloaded from the book's GitHub repository.

As shown in Figure 10.2C, participants' RTs were shorter for valid trials as compared to invalid trials. This finding indexes the temporal orienting (or

A)

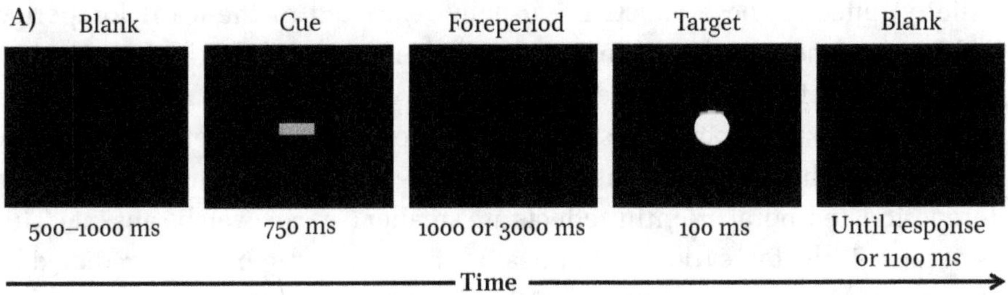


B)

| CUE | FOREPERIOD | |
|---|---|---|
| | Short | Long |
| Early | **Valid trial** (75%) | Invalid trial (Delayed) |
| Late | Invalid trial (Premature) | **Valid trial** (75%) |

C)

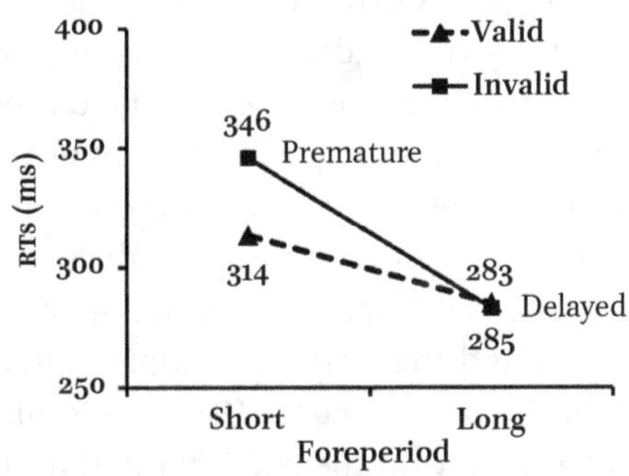


FIGURE 10.2 *(A) Experimental design. The temporal orienting task used by Capizzi, Sanabria, and Correa (2012). In the original study, the cue (here displayed in gray) could be filled with one of three colors (red, green, or blue). Each trial started with the presentation of a blank screen for a random interval between 500 and 1000 ms. The temporal cue (a short or a long line) was then displayed for 750 ms. The short line indicated that the target was likely to appear early (after a short foreperiod), whereas the long line indicated that the target would probably appear late (after a long foreperiod). Following the cue, the screen remained blank for a variable foreperiod of 1000 or 3000 ms. After the foreperiod elapsed, the target stimulus (a white dot) appeared in the center of the screen for a duration of 100 ms. Participants had to respond to the target onset by pressing the space bar with their index finger. The next trial began following the response to the target or after 1100 ms. (B) Types of trials. The matching between early cue and short foreperiod and between late cue and long foreperiod gives rise to valid trials (75% validity rate). Conversely, the combination of early cue and long foreperiod produces invalid 'delayed' trials since the target appears later than expected. Lastly, the combination of late cue and short foreperiod creates invalid 'premature' trials in which the target appears earlier than expected. (C) Temporal orienting effects. Participants' RTs are plotted as a function of Cue Validity (valid, invalid) and Foreperiod duration (short, long). Note that temporal orienting effects are significant at the short foreperiod only. This is due to the fact that when the target does not show up after the short foreperiod (as invalidly predicted by the early cue), participants can re-orient their attention to the long foreperiod, thus counteracting the negative consequences on performance of the invalid prediction. The procedure and data are adapted from Capizzi et al. (2012).*

validity) effects. These effects are usually restricted to the short foreperiod only, as demonstrated by the interaction between Cue Validity and Foreperiod. Turning to the figure, a RT benefit can be observed for validly early cued targets (early cue-short foreperiod) as compared to 'premature' targets that invalidly occur earlier than expected (late cue-short foreperiod). By contrast, at the long foreperiod, temporal orienting effects are smaller (or can even be absent). This is because if the target does not appear at the short foreperiod as predicted by the early cue, it is possible to 'infer' that it will appear later, thus enabling attention to be re-oriented to the long foreperiod (invalid 'delayed' trials in the figure) (e.g., Correa, Lupiáñez, Milliken, & Tudela, 2004; Coull & Nobre, 1998).

In the remainder of the section, we list some practical issues that should be considered when designing a temporal orienting task. One issue concerns the cue validity rate. Most studies use cue validity rates of 75 or 80%. However, it is also possible to use a 100% valid cue condition and to compare it with a neutral one in which the cue does not provide any temporal information on target onset. That is, in the neutral condition the cue simply acts as a warning signal and the target can appear after the short or long foreperiod with the same frequency (50%). Use of valid-only trials encourages participants to confidently rely on the cue information. This manipulation has been adopted in studies investigating the ability to orient attention in time in more vulnerable populations such as children (Mento & Tarantino, 2015; Mento & Vallesi, 2016) and older adults (Zanto et al., 2011).

If a researcher is interested in maximizing temporal orienting effects, it is possible to keep 'early' and 'late' temporal cues separate across different blocks of trials instead of intermixing them within trials. In that case, participants will be presented with early-blocks and late-blocks, each containing both valid and invalid trials. Specifically, the early-block will only display the cue type assigned to the early condition. On 75% of the trials, the foreperiod will match the duration predicted by the cue, whereas on 25% of the trials it will not. The same applies, but with the late cue-long foreperiod association, to the late-block condition. It is important to counterbalance the order of presentation of early and late blocks to avoid confounding any practice or boredom effects with one of the two conditions. In previous studies, we have compared between-block and within-block manipulations of temporal cues and found that the magnitude of temporal orienting effects were larger in the former case (Correa, Lupiáñez, & Tudela, 2006b). Moreover, we recently reported that temporal orienting effects elicited on a trial-by-trial basis, but not those obtained in a blocked design, were impaired in dual-task conditions that competed for common limited resources with the temporal orienting task (Capizzi et al., 2012).

A blocked manipulation of temporal cues is, thus, advisable when evaluating temporal orienting in populations that get tired and unmotivated after brief periods of time, such as frontal brain damaged patients who, notoriously, have troubles maintaining the focus of attention on the task at hand (see Triviño et al., 2011; Triviño et al., 2010).

Another factor that can affect the magnitude of temporal orienting effects includes cue encoding. It is recommended to employ intuitive cues that do not need a demanding and time-consuming decoding process. Some examples of previously used intuitive cues comprise *symbolic figures*: two concentric circles with the inner circle brightening for the short foreperiod and the outer circle for the long one (Coull & Nobre, 1998) or short and long lines (Capizzi, Correa, & Sanabria, 2013; Capizzi et al., 2012; Correa et al., 2014; Correa et al., 2004; Correa, Lupiáñez, & Tudela, 2006b; Triviño et al., 2010); *letters*: 'S' for short, 'L' for long (Zanto et al., 2011) and *words*: 'early' and 'late' (Correa, Lupiáñez, Madrid, & Tudela, 2006a).

To strengthen temporal orienting effects, it is also important to provide participants with enough practice trials to initially learn the specific cue-time interval contingencies. Moreover, participants' instructions should explicitly emphasize that using the temporal cue would help them to predict when the target will occur.

Finally, the type of task also influences temporal orienting effects. We have shown, for example, that temporal orienting effects are usually larger in a simple-RT detection task than in a choice-RT discrimination task (Correa et al., 2004; Correa, Lupiáñez, & Tudela, 2006b). However, only the latter, which imposes strong demands on perceptual analysis, boosts visual processing as revealed by event-related potential (ERP) measures. Otherwise, in simple RT-detection tasks, temporal orienting typically enhances performance through facilitation of late motor processes (see Correa, Lupiáñez, Madrid, & Tudela, 2006a, for a review). Of note, more sophisticated tasks than the originally employed detection and discrimination ones have been developed over the last years to elucidate whether temporal orienting effects can also act on processes requiring higher cognitive functioning. These studies showed that temporal orienting can modulate semantic processing (Naccache, Blandin, & Dehaene, 2002), executive control (Correa, Cappucci, Nobre, & Lupiáñez, 2010a), and visual working memory (van Ede, Niklaus, & Nobre, 2017), thus demonstrating that temporal orienting effects are robust across different tasks and cognitive demands.

The majority of the earlier studies investigating temporal orienting have used symbolic and static cues as a means of orienting attention in time.

However, since we live in a dynamic environment, which is intrinsically rhythmic, we should also be able to generate temporal expectations following the regular presentation of isochronous sequences of stimuli (i.e., rhythms). This topic will be covered in the following section.

## 4 Temporal Preparation Induced by Rhythmic Cues

According to the Dynamic Attending Theory (Ellis & Jones, 2010; Large & Jones, 1999), attention is conceived of as an endogenous oscillatory process that can be entrained by external rhythms. The repetition of a regular rhythmic pattern will, thus, synchronize participants' attentional oscillations, which improves both accuracy and response speed to a target appearing in synchrony with that pattern (e.g., Bolger, Coull, & Schön, 2014; Bolger, Trost, & Schön, 2013; Breska & Deouell, 2014; Di Luca & Rhodes, 2016; Jones et al., 2002; Lange, 2010; Rohenkohl, Cravo, Wyart, & Nobre, 2012; Sanabria et al., 2011; Sanabria & Correa, 2013). It is important to note that unlike the temporal orienting effects described above, rhythms can automatically orient attention in time. For example, participants' performance is entrained by the rhythms even if they are instructed to ignore them (Rohenkohl, Coull, & Nobre, 2011). The automaticity of temporal preparation driven by auditory rhythms has been tested in dual-task studies that showed no decrement in performance on rhythm-based preparation in the context of a concurrent memory task (Cutanda et al., 2015; de la Rosa, Sanabria, Capizzi, & Correa, 2012).

An example of how to investigate temporal preparation driven by auditory rhythms can be found in de la Rosa et al. (2012). In this chapter, we will only focus on the single-task data of Experiment 1. An adapted version of the original experiment can be downloaded from the book's GitHub repository. As shown in Figure 10.3A, the task was comprised of a sequence of six tones with a 250-ms duration each and a frequency of 700 Hz. This sequence could be temporally regular or irregular. The regular sequence was created by spacing the tones with a fixed interval of 550 ms. Conversely, in the irregular rhythm each interval duration was selected randomly among the following values: 150, 350, 550, 750, or 950 ms. Importantly, both sequences included six tones and had identical duration such that the only difference between the two concerned the isochrony of the rhythm. The target tone, which appeared after a variable foreperiod (800, 1100, or 1400 ms), was a 100-ms sound of 400 Hz. Participants had to respond to the target tone by pressing the "b" key on a computer keyboard. They were informed that the target was preceded by a sequence of sounds creating a rhythm. However,

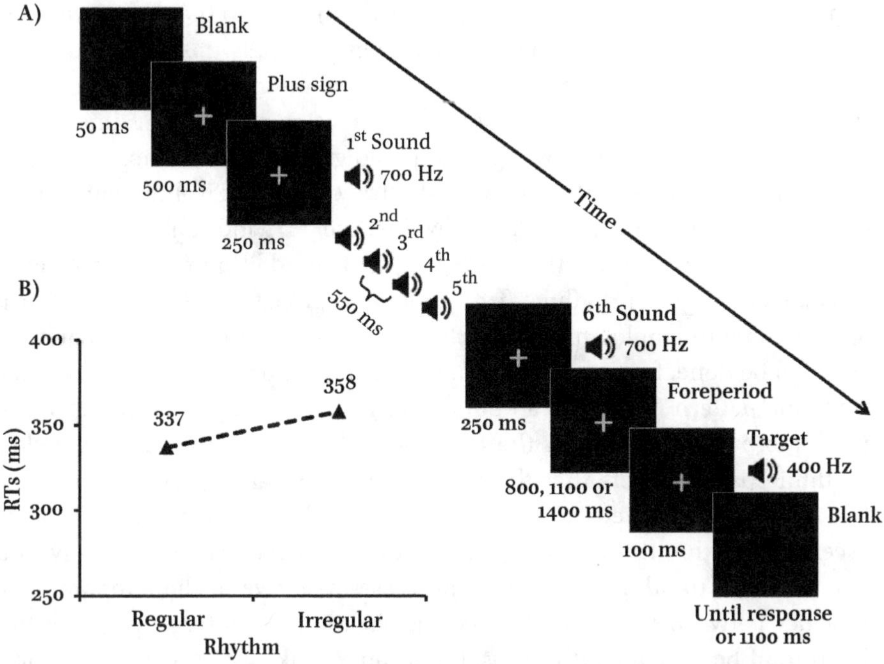


FIGURE 10.3 *(A) Experimental design. The rhythm-based temporal preparation task used by de la Rosa, Sanabria, Capizzi, and Correa (2012). In the original study, the fixation cross (a plus sign, here displayed in gray), which remained on the screen during the entire trial, could be filled with one of three colors (red, green, or blue). Each trial started with the presentation of a 50-ms blank screen. Five hundred milliseconds later, a regular or irregular rhythm was presented at random. To create the regular rhythm, as shown in the figure, all the intervals between the six tones had a fixed duration of 550 ms. Conversely, in the irregular rhythm each interval duration was selected randomly among the following values: 150, 350, 550, 750, or 950 ms. After a variable foreperiod (800, 1100, or 1400 ms), the target tone (a 400-Hz sound lasting for 100 ms) was presented. Participants had to press the "b" key as fast as possible to the onset of the target tone. The next trial began following the response to the target, or after 1100 ms. (B) Regularity effects. Participants' RTs are plotted as a function of Rhythm (regular, irregular). In line with the hypothesis that the presentation of a regular rhythm improves performance, participants were faster in responding to targets appearing after a regular rhythm compared to an irregular rhythm. The procedure and data are adapted from de la Rosa et al. (2012).*

they were explicitly told that this rhythm had no bearing on the task and could, therefore, be ignored.

Figure 10.3B shows participants' RTs plotted as a function of Rhythm (regular, irregular). As depicted, participants were faster to detect the target that appeared after the regular rhythmic pattern compared to the target preceded by an irregular sequence. This demonstrates that it is possible to develop temporal preparation on the basis of regular auditory rhythms. Please note that rhythms represent a powerful way of orienting attention in time also in the visual modality. Some examples of paradigms that have employed visual rhythms can be found in the following studies (Correa and Nobre, 2008; Correa et al., 2014; Doherty, Rao, Mesulam, & Nobre, 2005; Triviño et al., 2011).

As a last note, it is worth mentioning that in addition to non-predictive (regular vs. irregular) rhythms (e.g., de la Rosa et al., 2010; Lange, 2010), it is also possible to employ predictive rhythms to induce temporal preparation. This can be done, for example, by manipulating the pace (fast or slow) of the rhythmic pattern (e.g., Correa et al., 2014; Triviño et al., 2011). That is, a fast rhythmic pace predicts that the target would appear early, whereas a slow rhythmic pace predicts that the target would appear late. A rhythm-based benefit is also observed using predictive rhythms. However, it is difficult to ascertain whether this advantage is due to the exogenous effects of rhythms alone, or also to other endogenous processes sensitive to the temporal contingency between the rhythmic pace and the moment of target onset, or to a mixture of both factors. It is, thus, important to take into account these features of the experimental design when thinking of how to investigate rhythm-based preparation.

## 5 Conclusions and Practical Considerations

In this chapter, we provided a general overview of the typical experimental procedures developed to measure temporal preparation. This conclusive section offers some final advice that may help determine the best procedure to use for addressing a given research question. To this aim, we start by briefly considering the relationship between different temporal preparation phenomena (see Correa, 2010, for a theoretical discussion on this issue). If, for example, the focus of the study is measuring temporal orienting, foreperiod, and sequential effects within the same task, there are no disadvantages to using a temporal orienting paradigm. Correa and colleagues (Triviño et al., 2010; see also Correa et al., 2010b) developed an adapted and quick version of the temporal orienting task that allows the evaluation of the three temporal preparation effects at the same time. By contrast, if one is interested in investigating the ERP correlates of both temporal orienting and sequential effects, it is advisable to also have a neutral condition in which the temporal cue is substituted with a non-informative warning signal. Otherwise, it has been suggested that the ERP correlates of sequential effects may be masked by the controlled temporal orienting processing (Capizzi et al., 2013).

Another example of the reciprocal influence between different temporal preparation effects can be found in Sanabria, Capizzi, and Correa (2011). The authors showed that temporal preparation driven by auditory rhythms was effective at the long foreperiod only when catch-trials (i.e., trials in which a target is not presented) were included. This is due to the fact that when the target always occurs, the certainty that it will appear at the long foreperiod, once the short foreperiod has been passed, is so strong that it overcomes the benefits afforded by the auditory rhythms. The presence of catch-trials instead counteracts the foreperiod effect by decreasing participants' certainty about target occurrence (Näätänen, 1972). Note, also, that the inclusion of catch-trials has been useful to unveil the effects of both temporal orienting (Correa et al., 2004) and sequential effects (Capizzi et al., 2015) at the long foreperiod.

Once the best approach to use has been established, it is critical to adapt the chosen paradigm according to the specific target population. As an illustrative example, Mento and colleagues (Mento & Tarantino, 2015; Mento & Vallesi, 2016) have recently developed a simplified child-friendly version of the temporal orienting paradigm. In this version, children imagine being at the zoo and taking pictures of the animals that appear within the camera lens. They are instructed that the color of the lens will help them to predict when the animal will appear.

Other examples of research that have been fruitfully conducted in special populations can be found in the following studies (e.g., Droit-Volet & Coull, 2016; Johnson et al., 2016; Johnson, Burrowes, & Coull, 2015; Mento & Valenza, 2016; Vallesi & Shallice, 2007, for *children*; Bherer & Belleville, 2004; Chauvin et al., 2016; Vallesi, McIntosh, & Stuss, 2009b; Zanto et al., 2011, for *older adults;* Correa et al., 2010b, for *individuals with high and low trait impulsivity;* Marzecova et al., 2013, for *bilinguals*). Moreover, the reader interested in investigating temporal preparation in clinical settings can refer to these studies (Triviño et al., 2016, 2011, 2010; Vallesi et al., 2007a, for *neurological patients*; Correa et al., 2011, for *patients with fibromyalgia*).

To conclude, the increasing number of studies that have investigated temporal preparation across different tasks and populations demonstrates the growing interest in this area of research. Future developments will, thus, include devising novel experimental designs and, importantly, linking the theory of temporal preparation to practice-oriented research.

## Acknowledgements

M.C. was supported by a post-doctoral fellowship from the European Research Council Starting grant n° 313692 (FP7/2007–2013). A.C. was supported by the Junta de Andalucía (Proyectos de Excelencia: SEJ-3054).

## References

Bausenhart, K.M., B. Rolke, & R. Ulrich (2008). Temporal preparation improves temporal resolution: Evidence from constant foreperiods. *Perception & Psychophysics*, *70*(8), 1504–1514.

Bherer, L., & S. Belleville (2004). Age-related differences in response preparation: The role of time uncertainty. *Journal of Gerontology B: Psychology Science and Social Science*, *59*, 66–74.

Bolger, D., J.T. Coull, & D. Schön (2014). Metrical rhythm implicitly orients attention in time as indexed by improved target detection and left inferior parietal activation. *Journal of Cognitive Neuroscience*, *26*(3), 593–605.

Bolger, D., W. Trost, & D. Schön (2013). Rhythm implicitly affects temporal orienting of attention across modalities. *Acta Psychologica*, *142*(2), 238–244.

Breska, A., & L.Y. Deouell (2014). Automatic bias of temporal expectations following temporally regular input independently of high-level temporal expectation. *Journal of Cognitive Neuroscience*, *26*(7), 1555–1571.

Capizzi, M., A. Correa, & D. Sanabria (2013). Temporal orienting of attention is interfered by concurrent working memory updating. *Neuropsychologia*, *51*(2), 326–339.

Capizzi, M., A. Correa, A. Wojtowicz, & R.D. Rafal (2015). Foreperiod priming in temporal preparation: Testing current models of sequential effects. *Cognition*, *134*, 39–49.

Capizzi, M., D. Sanabria, & A. Correa (2012). Dissociating controlled from automatic processing in temporal preparation. *Cognition*, *123*(2), 293–302.

Chauvin, J.J., C.R. Gillebert, G. Rohenkohl, G.W. Humphreys, & A.C. Nobre (2016). Temporal orienting of attention can be preserved in normal aging. *Psychology and aging*, *31*(5), 442–455.

Correa, A. (2010). Enhancing behavioural performance by visual temporal orienting. In Nobre, A.C. and J.T. Coull (Eds.), *Attention and time* (pp. 357–370). Oxford University Press.

Correa, A., P. Cappucci, A.C. Nobre, & J. Lupiáñez (2010a). The two sides of temporal orienting: Facilitating perceptual selection, disrupting response selection. *Experimental Psychology*, *57*(2), 142–148.

Correa, A., G. Cona, S. Arbula, A. Vallesi, & P. Bisiacchi (2014). Neural dissociation of automatic and controlled temporal preparation by transcranial magnetic stimulation. *Neuropsychologia, 65*, 131–136.

Correa, A., J. Lupiáñez, E. Madrid, & P. Tudela (2006a). Temporal attention enhances early visual processing: A review and new evidence from event-related potentials. *Brain Research, 1076*(1), 116–128.

Correa, A., J. Lupiáñez, B. Milliken, & P. Tudela (2004). Endogenous temporal orienting of attention in detection and discrimination tasks. *Perception & Psychophysics, 66*(2), 264–278.

Correa, A., J. Lupiáñez, & P. Tudela (2006b). The attentional mechanism of temporal orienting: Determinants and attributes. *Experimental Brain Research, 169*(1), 58–68.

Correa, A., E. Miró, M.P. Martínez, A.I. Sánchez, & J. Lupiáñez (2011). Temporal preparation and inhibitory deficit in fibromyalgia syndrome. *Brain and Cognition, 75*(3), 211–216.

Correa, A., & A.C. Nobre (2008). Neural modulation by regularity and passage of time. *Journal of Neurophysiology, 100*(3), 1649–1655.

Correa, A., M. Triviño, C. Pérez-Dueñas, A. Acosta, & J. Lupiañez (2010b). Temporal preparation, response inhibition and impulsivity. *Brain and Cognition, 73*(3), 222–228.

Coull, J.T. (2011). Discrete neuroanatomical substrates for generating and updating temporal expectations. In Dehaene, S. & E. Brannon (eds.), *Time and Number in the Brain: Searching for the Foundations of Mathematical Thought*. Elsevier.

Coull, J.T. (2009). Neural substrates of mounting temporal expectation. *PLoSBiology, 7*(8), e1000166.

Coull, J.T., & A.C. Nobre (1998). Where and when to pay attention: the neural systems for directing attention to spatial locations and to time intervals as revealed by both pet and fMRI. *Journal of Neuroscience, 18*(18), 7426–7435.

Cutanda, D., A. Correa, & D. Sanabria (2015). Auditory temporal preparation induced by rhythmic cues during concurrent auditory working memory tasks. *Journal of Experimental Psychology. Human Perception and Performance, 41*(3), 790–797.

de la Rosa, M.D., D. Sanabria, M. Capizzi, & A. Correa (2012). Temporal preparation driven by rhythms is resistant to working memory interference. *Frontiers in Psychology, 3*, 308.

Di Luca, M., & D. Rhodes (2016). Optimal perceived timing: Integrating sensory information with dynamically updated expectations. *Scientific Reports, 6*, 28563.

Doherty, J.R., A. Rao, M.M. Mesulam, & A.C. Nobre (2005). Synergistic effect of combined temporal and spatial expectations on visual attention. *Journal of Neuroscienc, 25*(36), 8259–8266.

Drazin, D.H. (1961). Effects of foreperiod, foreperiod variability, and probability of stimulus occurrence on simple reaction time. *Journal of Experimental Psychology, 62*, 43–50.

Droit-Volet, S., & J.T. Coull (2016). Distinct developmental trajectories for explicit and implicit timing. *Journal of Experimental Child Psychology, 150*, 141–154.

Elithorn, A., & C. Lawrence (1955). Central inhibition: Some refractory observations. *Quarterly Journal of Experimental Psychology, 11*, 211–220.

Ellis, R.J., & M.R. Jones (2010). Rhythmic context modulates foreperiod effects. *Attention, Perception & Psychophysics, 72*(8), 2274–2288.

Gibbon, J. (1977). Scalar expectancy theory and Weber's law in animal timing. *Pychological Review, 84*, 279–385.

Grondin, S. (2010). Timing and time perception: A review of recent behavioral and neuroscience findings and theoretical directions. *Attention, Perception & Psychophysics, 72*(3), 561–582.

Ivry, R.B., & R.E. Hazeltine (1995). Perception and production of temporal intervals across a range of durations: Evidence for a common timing mechanism. *Journal of Experimental Psychology. Human Perception and Performance, 21*(1), 3–18.

Johnson, K.A., M. Bryan, K. Polonowita, D. Decroupet, & J.T. Coull (2016). Isochronous sequential presentation helps children orient their attention in time. *Frontiers in Psychology, 7*, 1417.

Johnson, K.A., E. Burrowes, & J.T. Coull (2015). Children can implicitly, but not voluntarily, direct attention in time. *PloS One, 10*(4), e0123625.

Jones, M.R., H. Moynihan, N. MacKenzie, & J. Puente (2002). Temporal aspects of stimulus-driven attending in dynamic arrays. *Psychological Science, 13*(4), 313–319.

Klemmer, E.T. (1956). Time uncertainty in simple reaction time. *Journal of Experimental Psychology, 51*(3), 179–184.

Klemmer, E.T. (1957). Simple reaction time as a function of time uncertainty. *Journal of Experimental Psychology, 54*(3), 195–200.

Lange, K. (2010). Can a regular context induce temporal orienting to a target sound? *International Journal of Psychophysiology: Official Journal of the International Organization of Psychophysiology, 78*(3), 231–238.

Large, E.W., & M.R. Jones (1999). The dynamics of attending: How people track time-varying events. *Psychological Review, 106*(1), 119–159.

Los, S.A. (2010). Foreperiod and the sequential effect: Theory and data. In Nobre, A.C. & J.T. Coull (Eds.), *Attention and time* (pp. 289–302). Oxford: Oxford University Press.

Los, S.A. (2013). The role of response inhibition in temporal preparation: Evidence from a go/no-go task. *Cognition, 129*(2), 328–344.

Los, S.A., & C.E. van den Heuvel (2001). Intentional and unintentional contributions to nonspecific preparation during reaction time foreperiods. *Journal of Experimental Psychology. Human Perception and Performance, 27*(2), 370–386.

Marzecova, A., M. Bukowski, A. Correa, M. Boroso, J. Lupiañez, & Z. Wodniecka (2013). Tracing the bilingual advantage in cognitive control: The role of flexibility in temporal preparation and category switching. *Journal of Cognitive Psychology, 25*, 586–604.

Mattes, S., & R. Ulrich (1997). Response force is sensitive to the temporal uncertainty of response stimuli. *Perception & Psychophysics, 59*(7), 1089–1097.

Mento, G., & V. Tarantino (2015). Developmental trajectories of internally and externally driven temporal prediction. *PLoS ONE, 10*(8), 1–18.

Mento, G., V. Tarantino, A. Vallesi, & P.S. Bisiacchi (2015). Spatiotemporal neurodynamics underlying internally and externally driven temporal prediction: A high spatial resolution ERP study. *Journal of Cognitive Neuroscience, 27*(3), 425–439.

Mento, G., & E. Valenza (2016). Spatiotemporal neurodynamics of automatic temporal expectancy in 9-month old infants. *Scientific Reports, 6*, 36525.

Mento, G., & A. Vallesi (2016). Spatiotemporally dissociable neural signatures for generating and updating expectation over time in children: A High Density-ERP study. *Developmental Cognitive Neuroscience, 19*, 98–106.

Miniussi, C., E.L. Wilding, J.T. Coull, & A.C. Nobre (1999). Orienting attention in time: Modulation of brain potentials. *Brain, 122*, 1507–1518.

Näätänen, R. (1972). Time uncertainty and occurrence uncertainty of the stimulus in a simple reaction time task. *Acta Psychologica, 36*, 492–503.

Naccache, L., E. Blandin, & S. Dehaene (2002). Unconscious masked priming depends on temporal attention. *Psychological Science, 13*(5), 416–424.

Niemi, P., & R. Näätänen (1981). Foreperiod and simple reaction time. *Psychological Bulletin, 89*(1), 133–162.

Nobre, A.C. (2001). Orienting attention to instants in time. *Neuropsychologia, 39*, 1317–1328.

Nobre, A.C., A. Correa, & J.T. Coull (2007). The hazards of time. *Current Opinion in Neurobiology, 17*(4), 465–470.

Nobre, A.C., & J.T. Coull (2010). *Attention and time.* Oxford: Oxford University Press.

Posner, M.I., C.R. Snyder, & B.J. Davidson (1980). Attention and the detection of signals. *Journal of Experimental Psychology, 109*(2), 160–174.

Rohenkohl, G., J.T. Coull, & A.C. Nobre (2011). Behavioural dissociation between exogenous and endogenous temporal orienting of attention. *PLoS ONE, 6*(1), 1–5.

Rohenkohl, G., A.M. Cravo, V. Wyart, & A.C. Nobre (2012). Temporal expectation improves the quality of sensory information. *Journal of Neuroscience 32*(24), 8424–8428.

Sanabria, D., M. Capizzi, & A. Correa (2011). Rhythms that speed you up. *Journal of Experimental Psychology. Human Perception and Performance, 37*(1), 236–244.

Sanabria, D., & A. Correa (2013). Electrophysiological evidence of temporal preparation driven by rhythms in audition. *Biological Psychology, 92*(2), 98–105.

Steinborn, M.B., & R. Langner (2012). Arousal modulates temporal preparation under increased time uncertainty: Evidence from higher-order sequential foreperiod effects. *Acta Psychologica, 139*(1), 65–76.

Steinborn, M.B., B. Rolke, D. Bratzke, & R. Ulrich (2008). Sequential effects within a short foreperiod context: Evidence for the conditioning account of temporal preparation. *Acta Psychologica, 129*(2), 297–307.

Triviño, M., M. Arnedo, J. Lupiáñez, J. Chirivella, & A. Correa (2011). Rhythms can overcome temporal orienting deficit after right frontal damage. *Neuropsychologia, 49*(14), 3917–3930.

Triviño, M., A. Correa, M. Arnedo, & J. Lupiáñez (2010). Temporal orienting deficit after prefrontal damage. *Brain, 133*(4), 1173–1185.

Triviño, M., A. Correa, J. Lupiáñez, M.J. Funes, A. Catena, X. He, & G.W. Humphreys (2016). Brain networks of temporal preparation: A multiple regression analysis of neuropsychological data. *NeuroImage, 142*, 489–497.

Vallesi, A. (2010). Neuroanatomical substrates of foreperiod effects. In Nobre, A.C. & J.T. Coull (Eds.), *Attention and time* (pp. 303–316). Oxford: Oxford University Press.

Vallesi, A., S. Arbula, & P. Bernardis (2014). Functional dissociations in temporal preparation: evidence from dual-task performance. *Cognition, 130*(2), 141–151.

Vallesi, A., V.N. Lozano, & A. Correa (2013). Dissociating temporal preparation processes as a function of the inter-trial interval duration. *Cognition, 127*(1), 22–30.

Vallesi, A., A.R. McIntosh, T. Shallice, & D.T. Stuss (2009a). When time shapes behavior: fMRI evidence of brain correlates of temporal monitoring. *Journal of Cognitive Neuroscience, 21*(6), 1116–1126.

Vallesi, A., A.R. McIntosh, & D.T. Stuss (2009b). Temporal preparation in aging: A functional MRI study. *Neuropsychologia, 47(13)*, 2876–2881.

Vallesi, A., A. Mussoni, M. Mondani, R. Budai, M. Skrap, & T. Shallice (2007a). The neural basis of temporal preparation: Insights from brain tumor patients. *Neuropsychologia, 45*(12), 2755–2763.

Vallesi, A., & T. Shallice (2007). Developmental dissociations of preparation over time: Deconstructing the variable foreperiod phenomena. *Journal of Experimental Psychology. Human Perception and Performance, 33*(6), 1377–1388.

Vallesi, A., T. Shallice, & V. Walsh (2007b). Role of the prefrontal cortex in the foreperiod effect: TMS evidence for dual mechanisms in temporal preparation. *Cerebral Cortex, 17*(2), 466–474.

Van Ede, F., M. Niklaus, & A.C. Nobre (2017). Temporal expectations guide dynamic prioritization in visual working memory through attenuated α oscillations. *Journal of Neuroscience, 37*(2), 437–445.

Woodrow, H. (1914). The measurement of attention. *Psychological Monographs, 17*, 158.

Zanto, T.P., P. Pan, H. Liu, J. Bollinger, A.C. Nobre, & A. Gazzaley (2011). Age-related changes in orienting attention in time. *Journal of Neuroscience, 31*(35), 12461–12470.

CHAPTER 11

# Temporal Order and Synchrony Judgments: A Primer for Students

*Maria Kostaki and Argiro Vatakis*

## 1 Introduction

Every day we perceive events that are multisensory, well aligned in time, and unified. This percept is a given fact and we rarely think that the synchrony of those events is not the rule but rather the exception. It is now known that the perceived order or subjective simultaneity of two sensory stimuli might not correspond to their actual physical order or objective synchrony, respectively. This was, however, unknown to astronomers back in the early 19th century who measured stellar transits (i.e., estimation of the position of a star across the reticules of the telescope between successive beats of a clock or metronome) via the 'ear and eye' method. This method was considered quite accurate, however, it was later observed that astronomers' judgments deviated from each other for intervals that could reach in some cases the 800 ms (e.g., Mollon & Perkins, 1996). These deviations eventually led Gustav Fechner and Wilhelm Wundt to establish the fields of psychophysics and experimental psychology, respectively (Aghdaee, Battelli, & Assad, 2014).

The systematic discrepancy between objective and subjective stimulus timing (i.e., order and simultaneity) was partially accounted for by sensory *arrival latencies*. That is, differences in the time needed for the signals to be detected by the sense organs and transmitted to the appropriate processing centers (Sternberg & Knoll, 1973). Light, for example, travels through the air much faster than sound (i.e., approximately 300.000.000 m/s for light and 343 m/s for sound; Spence & Squire, 2003). When detected though, a visual stimulus needs more time to be transduced to the retina as compared to an auditory stimulus to get processed by the hair cells in the inner ear (King & Palmer, 1985). Thus, the sensory arrival latencies are identified in two different levels, the physical and the neuronal level.

At the physical level, one of the parameters affecting stimulus arrival times is the distance of the stimulus' origin from the observer. As the distance of the multisensory event (e.g., audiovisual) increases, the arrival times of the auditory input lag even more than the visual ones. Take, for example, thunders in

 | DOI 10.1163/9789004280205_012

the sky, first you see the illumination and then you hear the rattle. Physical time differences in the travelling times of light and sound signals are expected to be cancelled out by neural transmission differences at a distance of approximately 10 m from the observer, a distance known as the '*horizon of simultaneity*' (Pöppel, Schill, & von Steinbüchel, 1990). For audiovisual stimuli presented within the limits of this 'horizon', sound arrival times precede those of light, whereas beyond this horizon, the light is the signal that takes precedence (Pöppel et al., 1990). Yet, arrival latencies at both the physical and the neuronal levels are not fixed (King, 2005). Nonetheless, the underlying brain mechanisms manage to deal with these differences, thus, allowing the percept of synchrony for a large range of event distances (Keetels & Vroomen, 2012) and stimulus characteristics (King, 2005; Vatakis, 2013).

Stimuli presented within the horizon of simultaneity and close in time are perceived as synchronous even though they are physically asynchronous (Stein & Meredith, 1993). One potential account of how simultaneity is perceived is the mechanism of the temporal window of integration (TWI; e.g., Colonius & Diederich, 2012; King, 2005; Lewkowicz & Ghazanfar, 2009; Spence & Squire, 2003; van Wassenhove et al., 2007; Vatakis 2013; Vroomen & Keetels, 2010). This 'window' represents the temporal range at which the brain tolerates asynchronies in the presented stimuli (e.g., visual or auditory leads/lags) so as to integrate the multisensory event and perceive the inputs as simultaneous (Diedrich & Collonius, 2015; Stevenson & Wallace, 2013). Research has shown that the TWI has a sensory bias with higher tolerance to visual as compared to auditory leads given the naturally occurring arrival delays of the sound. Thus, for simultaneity to be perceived, the TWI is asymmetric characterized by a visual shift, which is referred to as a visual bias (i.e., the visual stimulus has to be presented before the auditory stimulus for synchrony to be perceived; Hirsh & Sherrick 1961; Lewald & Guski, 2003; Munhall, Gribble, Sacco, & Ward, 1996; Slutsky & Recanzone, 2001; Vatakis, Navarra, Soto-Faraco, & Spence, 2008; Zampini, Shore, & Spence, 2003; Zampini, Guest, Shore, & Spence, 2005a). The systems' tolerance to asynchronies, however, is malleable since it has been shown that the system can be recalibrated when adapted to specific asynchronies (e.g., Fujisaki, Shimojo, Kashino, & Nishida, 2004; Navarra, Vatakis, Zampini, Soto-Faraco, Humphreys, & Spence, 2005; Vatakis, Navarra, Soto-Faraco, & Spence, 2007; Vroomen et al., 2004). Thus, a potential shift of the TWI towards audition can be attained by manipulation of depth cues, sensory exposure, or experience (e.g., music experts vs. non-musicians; King, 2005; Petrini et al., 2009; Silva et al., 2014; Spence & Squire, 2003). Moreover, the TWI can also be modulated by the temporal ventriloquism effect (i.e., the phenomenon where the sound ventriloquizes vision into temporal alignment; Morein-Zamir, Soto-Faraco, & Kingstone, 2003; Spence & Squire, 2003).

Spatial proximity of the auditory and visual inputs is another parameter that affects integration of the incoming inputs (Lewald & Guski, 2003). For instance, two events that are close in space, time, and structure are usually perceived as emanating from the same underlying cause, while in the case of large spatial displacements the percept is associated with two different events originating from different sources (Kording et al., 2007). Thus, it is proposed that the brain uses causal inference to make estimations about optimal cue combinations (Kayser & Shams, 2015; Kording et al., 2007; Shams & Beierholm, 2010). In other words, the brain decides on whether the two stimuli originate from the same source and, subsequently, whether to integrate or segregate them (Bayesian causal inference; Kording et al., 2007). Thus, one could infer that the ecological validity of the stimuli presented could enhance the likelihood of multisensory integration and, thus, ones' percept of synchrony (Aschersleben, 1999; but see van Eijk, 2008).

Kohlrausch et al. (2013) have also suggested that the potential differences noted in the perception of synchrony between various audiovisual stimuli (i.e., simple stimuli such as flashes and beeps, or ecologically valid stimuli such as a bouncing ball that have inherent anticipatory and predictive characteristics) may be attributed to the apparent causality of the event and not to the visual event predictability per se. More specifically, the claim is that an ecological valid multisensory stimulus is expected to promote the impression that the visual stimulus causes the auditory stimulus. Thus, this implied causal relationship leads to the expectation that the auditory stimulus cannot precede the visual one. Such expectations are not present in simple stimulation, thus leading to shifts in the judgment of synchrony between different types of stimulation (i.e., less tolerance to auditory leading asynchronies for ecological valid stimuli; Kohlrausch et al., 2013).

Many other lower- and higher-level parameters have been reported to affect both participants' perception of synchrony and sensitivity to asynchrony (see Vatakis, 2013, for a review). Stimulus characteristics such as intensity, duration, type, and content (Eg & Behne, 2015), as well as task characteristics, and attentional and decisional mechanisms have been reported to modulate the TWI and, thus, to affect participants' multisensory synchronous percept (Garcia-Perez & Alcala-Quintana, 2012; Keetels & Vroomen, 2012; Schneider & Bavelier, 2003; Shore, Spence, & Klein, 2001; Spence, Shore, & Klein, 2001; Sternberg & Knoll, 1973; Zampini et al., 2005b). Moreover, adaptation to specific stimulus asynchronies (Fujisaki et al., 2004; Vatakis, Navarra,

Soto-Faraco, & Spence, 2007) and experience with time-dependent activities such as video games have been found to improve the perception of synchrony (Donohue, Woldorff, & Mitroff, 2010). Age, on the other hand, has been found to deteriorate the perception of temporal order suggesting larger temporal windows of integration, while perception of synchrony has been found to remain unaffected (Bedard & Barnett-Cowan, 2016). These parameters tap either on the physical or the neuronal level of arrival latencies affecting the range of the TWI.

## 2 Tasks for the Measurement of Synchrony Perception

Research on the perception of synchrony has made use of various tasks for the measurement of perceptual latencies (differences in processing speed) between different sensory modalities, stimulus characteristics, participant groups, cue types (e.g., prior-entry), or in temporal recalibration (Garcia-Perez & Alcala-Quintana, 2015b). The two most widely used tasks are the: temporal order judgment (TOJ) task and the binary simultaneity judgment (SJ-2) task. In the TOJ task, stimuli of different modalities are presented at various stimulus onset asynchronies (SOAs) and participants have to decide which stimulus was presented first or second (i.e., visual/auditory-first or visual/auditory-second response). Similarly, in the SJ-2 task, participants are stimulated bimodally with various SOAs but their response now is on whether the two sensory inputs were simultaneously presented or not (i.e., synchronous or asynchronous response).

Another commonly used task for measuring the perception of synchrony is the ternary simultaneity judgment task (SJ-3; Kohlrausch, et al., 2013; Kuling, van Eijk, Juola, & Kohlrausch, 2012; Ulrich, 1987; van Eijk, Kohlrausch, Juola, & van de Par, 2008). This task is a combination of the two previously described tasks (i.e., TOJ and SJ-2). That is, the participants are presented with synchronous and asynchronous bimodal stimulation and they have to decide whether the stimuli were presented in synchrony or not and in case of an asynchronous response to report which sensory input was presented first. Thus, the three possible answers to this task are: synchronous/asynchronous, auditory-leading, or visual-leading stimulation.

Other tasks have also been implemented for the perception of synchrony but not commonly used such as reaction time tasks (Cardoso-Leite, Gorea, & Mamassian, 2007; Diedrich & Colonius, 2015; Leone & McCourt, 2015), perceptual fusion tasks (where participants decide whether they perceived a unified event or not; Stevenson & Wallace, 2013), and two-interval forced choice tasks (where participants decide which of two bimodal events was synchronous; Stevenson & Wallace, 2013; Yarrow et al., 2016; see Chapter 13, this volume; or which bimodal event had a shorter or longer duration; Linares & Holcombe, 2014). Measurements of perceptual latencies are not restricted to audiovisual stimulation but extend to other modality combinations such as audiotactile and visuotactile stimulations (e.g., Noel, Wallace, Orchard-Mills, Alais, & Van der Burg, 2015). In the present chapter, however, we are going to focus on audiovisual events and the three most commonly used tasks: TOJ, SJ-2, and SJ-3.

## 3 The Parameters Associated with Synchrony Perception

The raw data obtained from the TOJ, SJ-2, and SJ-3 tasks are most commonly processed so as to obtain the: *point of subjective simultaneity* (PSS; note that PSS differs from the point of *objective* simultaneity due to the latencies described in the introduction section), *just noticeable difference* (JND; standard deviation of the distribution, SD, in SJ tasks), and TWI (see Table 11.1). Although these derived parameters are interpreted in the same way across tasks, it is as yet unclear whether the different tasks actually measure the same exact perceptual processes. The discrepancies noted in these sensitivity parameters when these tasks are compared have raised concerns on whether these measures reflect differences in participants' sensitivity to the event synchrony/asynchrony or reported biases and experimental manipulations (e.g., Garcia-Perez & Alcala-Quintana 2012; 2015a,b; Nicholls, Lew, Loetscher, & Yates, 2011; Spence & Parise, 2010; van Eijk et al., 2008; Vatakis et al., 2008; Vroomen & Keetels, 2010; Yates & Nicholls, 2011). We will explore these discrepancies later in this chapter, but let's first explain what the parameters of these measures represent for each task and how they are calculated.

### 3.1 *TOJ Measures of Sensitivity*

In the TOJ task, the PSS is an indirect measure of the perceived simultaneity of the stimuli presented (Garcia-Perez & Alcala-Quintana, 2015a). It represents the amount of asynchrony that must be present between bimodal inputs at which participants cannot reliably detect their temporal order (i.e., indirect perception of synchrony; see Table 11.1). Thus, participant responses for temporal order at the PSS point are near chance level (for which modality was presented first or second) assuming no inherent response biases. To compute the PSS in a TOJ task, the percentage of the visual-first (or auditory-first) responses

TABLE 11.1 *An overview of the sensitivity measures obtained from the TOJ and SJ tasks.*

<table>
<tr><th>Task<br>Measures</th><th>TOJ [Auditory or Visual first?]</th><th>SJ [Synchronous or Asynchronous?]</th></tr>
<tr><td rowspan="2">PSS</td><td>Indirect measure</td><td>Direct measure</td></tr>
<tr><td colspan="2">of perceived synchrony</td></tr>
<tr><td rowspan="2">JND</td><td colspan="2">The smallest interval needed for reliable perception of</td></tr>
<tr><td>temporal order</td><td>asynchrony</td></tr>
<tr><td rowspan="2">TWI</td><td colspan="2">The range of tolerance for perception</td></tr>
<tr><td>of synchrony (i.e., inability to detect temporal order)</td><td>of synchrony</td></tr>
</table>

for each stimulus condition is plotted as a function of the SOA (i.e., the percentage/proportion of trials across conditions that the participants reported 'visual first' percepts). The data are fitted to an S-shaped psychometric function and the 50% crossover point on this curve is the PSS (i.e., the SOA at which it is assumed that the participants perceive inputs as simultaneous; see Coren et al., 2004, for further details; see Figure 11.1). In general, positive PSS values indicate a visual leading requirement in stimulus presentation for synchrony to be perceived, while negative PSS values indicate a visual lagging requirement.

The JND represents the smallest interval at which the participants can reliably decide which sensory input of the two presented was first (see Table 11.1). The steepness of the curve at the 50% point reflects participants' sensitivity to temporal asynchronies. This measure can also be expressed as the JND value and computed as half the difference between the 25% and the 75% point on the same curve (see Figure 11.1). Typically, a steep slope results in small JNDs and, thus, in high participant sensitivity in the detection of asynchronies in the stimuli presented (i.e., high temporal resolution).

Finally, the TWI represents the range of tolerance in audiovisual asynchronies within which the perceptual system integrates the sensory inputs and, thus, reliable detection of order is not possible. The range of the TWI is computed as the [PSS ± JND]. The left TWI (i.e., PSS-JND) represents participants' insensitivity to detect order when the auditory stimulation is leading, while the right TWI (i.e., PSS + JND) represents participants' insensitivity to a leading visual stimulation.

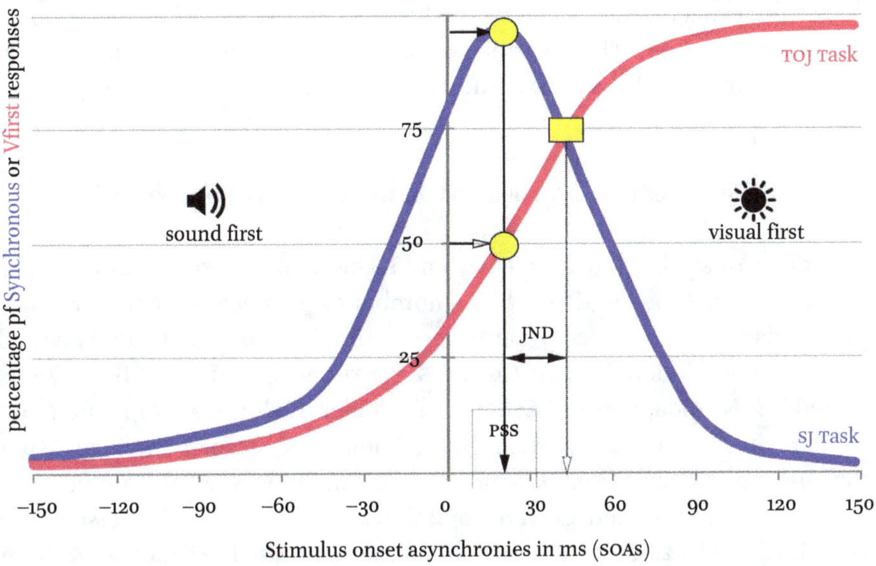


FIGURE 11.1 *Percentage of hypothetical 'synchronous' responses from an* SJ *task as a function of the* SOA *(blue bell-shaped curve) and hypothetical 'visual first' responses from a* TOJ *task plotted as a function of the* SOA *(magenta sigmoid curve). Negative* SOAs *indicate an auditory lead in the audiovisual pair presented and positive* SOAs *indicate a visual lead. The* PSS *values in this example are positive (i.e., visual leads are required) for both tasks.*

### 3.2 *SJ Measures of Sensitivity*

In the SJ-2 task, the PSS represents the interval at which the participants perceive the incoming inputs as simultaneous, the JND represents the smallest interval at which the participants can reliably detect the asynchrony in the audiovisual event presented, and the TWI represents the range of tolerance within which participants perceive synchrony (see Table 11.1). To compute these measures, the percentage/proportion of the 'synchronous' responses (i.e., trials across conditions in which the participants reported that the stimuli presented were in synchrony) is plotted as a function of the SOA and a bell-shaped Gaussian psychometric curve is obtained (see Figure 11.1). The peak of this curve corresponds to the PSS value and the JND can be computed as the mean SOA for the 75% point of the same curve (see Stone et al., 2001; Myung, 2003, for a detailed explanation). The range of the TWI is computed as in the TOJ task. Typically, the left and right TWIs derived for an SJ task are not symmetrical. The right TWI tends to be longer (i.e., participants report simultaneous responses for larger intervals when the visual stimulus is leading) due to the expected visual bias or the bias to perceive events as synchronous (Vatakis et al., 2008). The measures of sensitivity obtained from the SJ-3 task are a combination of the analysis described in the SJ-2 and TOJ data.

## 4 Differences in the Measures Obtained from TOJ and SJ Tasks

A number of studies have used TOJ and SJ tasks to investigate how the perception of synchrony is affected by a number of parameters such as the: type of stimulation (e.g., speech or non-speech stimuli, vestibular, auditory, visual, or tactile stimulation; Barnett-Cowan & Harris, 2009, 2011; Eg & Behne, 2015; Fujisaki & Nishida, 2009; Li & Cai, 2014; Maier, Di Luca, & Noppeney, 2011; Sanders et al., 2011; Vatakis et al., 2008; Vroomen & Stekelenburg, 2010), participant group tested (e.g., patients with schizophrenia, video gamers, older adults; Bedard & Barnett-Cowan, 2016; Capa, Duval, Blaison, & Giersch, 2014; Donohue et al., 2010), attentional manipulations used (Schneider & Bavelier, 2003), potential confounds and biases (see Keetels & Vroomen, 2012, for a review), adaptation and recalibration effects (e.g., Fujisaki et al., 2004; Vroomen et al., 2004), and perceptual training (Cecere, Gross, & Thut, 2016; Stevenson et al., 2013). These studies revealed marked differences between the parameters obtained from the two tasks. For instance, the mean PSS values across TOJ studies were mainly shifted towards audition (i.e., auditory-leading), whereas in SJ studies the mean PSS values were generally visually shifted (i.e., visual-leading; see van Eijk et al., 2008, for an extended literature review; but also see Leone & McCourt, 2015; Linares & Holcombe, 2014). A central question in the study of synchrony perception is, therefore, whether or not the TOJ and SJ tasks utilized refer to the same or different perceptual processes (e.g., Binder, 2015; Garcia-Perez & Alcala-Quintana, 2015a; Keetels & Vroomen, 2010; Love et al., 2013; Spence & Parise, 2010; van Eijk et al., 2008; Vatakis et al., 2008).

The debate about whether the two tasks tap into common underlying processes stems back in the '70s. On the basis of the independent-channels models described by Sternberg and Knoll (1973), it has been argued that a central timing mechanism receives the signals from the two stimuli presented, which arrive with randomly distributed arrival latencies, and applies a ternary decision rule to the arrival-time difference between the two signals in order to determine order or synchrony judgments. Thus, it was assumed that the two tasks were based on the same internal events (perceptual latency model; see Allan, 1975) and, thus, the perception of successiveness/asynchrony was a necessary and sufficient condition for the perception of temporal order. According to Hirsh (1959), however, the perception of asynchrony is a necessary but not sufficient condition for the perception of order (two stage models; Jaskowski, 1991; Poppel, 1988). For instance, in a TOJ task, apart from the detection of synchrony/asynchrony (first stage), the participant must also identify the order of the presented stimuli (second stage). Thus, the TOJ task involves more processes than those required for the SJ task. To-date, however, no agreement has been reached on whether these differences are due to perceptual or decisional processes or both. A growing number of studies are currently investigating this issue, both in perceptual and neuronal level, adopting identical experimental set-ups and participant groups for both the TOJ and SJ tasks (e.g., Binder, 2016; Garcia-Perez & Alcala-Quintana, 2012; 2015; Linares & Holcombe, 2014; Love, Petrini, Cheng, & Pollick, 2013; Machulla, Di Luca, & Ernst, 2016; Matthiews et al., 2016; Stevenson & Wallace, 2013; van Eijk et al., 2008; Vatakis et al., 2008).

### 4.1 *Divergent Perceptual Processes*

Some researchers have proposed a potential dissociation between the mechanisms involved during the execution of the TOJ and SJ tasks. Zampini et al. (2003), for instance, have argued that the TOJ task may reflect processes related to temporal discrimination, while SJ tasks may be related more to temporal binding mechanisms. Vatakis et al. (2008) also argued that the two tasks might not measure the same aspects of temporal perception. In their study, they measured the participants' sensitivity in the SJ and TOJ tasks using temporal recalibration in a simple pair of audiovisual stimuli with an audiovisual speech stimulus as the adaptor. The results showed positive PSS values for both tasks but these values were not correlated across tasks on an individual basis. If the PSS values across tasks were correlated then one could support that they measure the same underlying processes. Yet this was not the case. Similarly, no correlation was found for the JND values obtained from the two tasks. Vatakis and colleagues attributed the differences in the PSS values to the nature of the SJ task that could potentially bias participants' responses toward a simultaneous rather an asynchronous response, given that matched events tend to also be matched in time, and, thus, should be in synchrony.

Moreover, van Eijk et al. (2008) argued for the adoption of different terminology as a function of task so as to avoid potential interpretations of obtained data as representing the same underlying perceptual processes. Specifically, Van Eijk and colleagues tested the effect of experimental method used and stimulus type presented on the audiovisual temporal percept across the same participants (not in a temporal adaptation, Vatakis et al., 2008; or prior entry paradigm, Yates et al., 2011). They used simple (i.e., light flashes and sound clicks) and complex stimuli (i.e., bouncing balls and impact sounds) at various SOAs and asked participants to provide a TOJ, an SJ-2, and an SJ-3 response.

They measured the PSS, the slope of the corresponding curves, and the TWI for the SJ tasks. Their results showed visual leading (positive) PSS values for both the SJ and TOJ task (with large inter-participant deviations). PSS differences from the TOJ and SJ task were only obtained for the complex stimuli presented, while no differences were found for the simple stimuli used. Moreover, no PSS correlations between the TOJ and SJ-2 or SJ-3 tasks were obtained for both types of stimuli, while highly correlated PSS values were obtained between the SJ-2 and SJ-3 tasks. Additionally, for the bouncing ball stimulus, the slope values and the width of synchrony range between TOJ and SJ-2 (but not SJ-3) tasks were significantly correlated, while for the simple stimulus pair no correlations between TOJ and SJ-2 or SJ-3 were obtained. Thus, van Eijk and colleagues proposed that the PSS values are dependent on the experimental method utilized and possibly related to different underlying perceptions. They also argued that the inconsistent PSS values observed in the TOJ task could be accounted for by differences in participants' response strategy, making the TOJ task a less reliable task for synchrony perception.

Similarly, Love et al. (2013) suggested that the SJ and TOJ tasks do not represent the same underlying processes of the perception of synchrony. Their study extended the results of van Eijk et al. (2008) by using five different stimulus types at different levels of complexity (i.e., beep-flash, beep-flash-constant-visual, beep-flash-drumming, point-light-drumming, and face-voice stimuli). The aim of their study was to investigate whether the previously observed PSS differences (van Eijk et al., 2008) were consistent across different stimulus types. To eliminate any potential confounds due to stimulus duration,[1] they mostly created stimuli with equal overall duration. They measured the PSS and the TWI. The PSS obtained from the TOJ task was auditory shifted (i.e., auditory-leading), while, for the SJ task, the PSS values were visual shifted (i.e., visual-leading). Regarding the TWIs, narrower windows were obtained for the TOJ task using simple stimuli as compared to the SJ task, while for complex speech stimuli, the TWI was longer for the TOJ task as compared to the SJ task. For the rest of the complex stimuli, no significant differences were obtained in the TWI across tasks. Similar to van Eijk et al., Love et al. found no correlation for the PSS or TWI between the two tasks. Participants were more accurate to detect asynchrony in the SJ task when the auditory modality was leading as compared to visual leads for all stimulus types except for the speech pairs.

1 The increasing SOAs between two signals increases linearly the stimulus duration potentially providing extra cues when judging asynchrony, while this is not valuable when judging stimulus order. Although, it would be best to ensure equal stimulus duration for both the TOJ and SJ tasks.

Additionally, regarding the potential effect of stimulus duration, they found no evidence of SJ differences (see also Stevenson & Wallace, 2013), while performance in the TOJ task was very poor resulting in the exclusion of over 65% of the participants. In general, Love et al.'s results are in line with van Eijk et al.'s findings, thus supporting that a single task is not sufficient to explore the many different factors underlying the percept of synchrony.

As can be seen from the above-mentioned studies, differences in the PSS values across stimulus types and across studies are not consistent. This may be due to the different stimulation used, the experimental set-ups, or analyses implemented. Together, however, these differences argue for the potential that the two tasks used for the measure of synchrony might not tap into the same underlying processes.

### 4.2 *Partially Overlapping Mechanisms and Potential Sources of Bias*

Fewer studies have also argued for the potential that the SJ and TOJ tasks share at least in part some common mechanisms of temporal perception. Maier et al. (2011), for example, supported this for the perception of synchrony for audiovisual speech stimuli. In their study, they utilized speech and non-speech stimuli and fitted their data using both parametric (Gaussian and cumulative Gaussian psychometric functions) and non-parametric methods. Their results showed that the TWI for the TOJ task was wider than that obtained from the SJ task for the speech but not for the non-speech stimuli. This finding was attributed to an attentional shift of the participants' focus to the onsets of the audiovisual signal during the TOJ task, making participants ignore the temporal information of the rest of the stimuli, whereas the SJ task required judgments based on the combined auditory and visual signals. The PSS and JND parameters obtained in the TOJ and SJ task were found to be significantly correlated (in the non-parametric fitting). Thus, although one could argue for some common processing, one could also claim that the analysis method can potentially affect the results and their interpretation.

Linares and Holcombe's (2014) TOJ and SJ comparison focused on the role of biases as a potential explanation for the differences noted between the two tasks (Garcia-Perez & Alcala-Quintana, 2012; Schneider & Bavelier, 2003; Shore, Spence, & Klein, 2001; Yarrow, Jahn, Durant, & Arnold, 2011; Vatakis et al., 2008). Using simple stimuli they found positive (i.e., visual leading) PSS values both for the TOJ and SJ tasks (at individual level, larger deviations were obtained in the TOJ task with both positive and negative PSS values). Moreover, no PSS correlations were found between the two tasks. Thus, Linares and Holcombe attributed the different perceptual latencies (i.e., PSSs)' across tasks at partially distinct sets of biases. For instance, the PSS obtained from the TOJ task could have been affected by potential prior entry effects (where differential attentional allocation can result in perceptual latency differences), response biases over one response as compared to another (particularly for SOAs near the PSS), and task difficulty biases (i.e., difficulty to order stimuli in time). Conversely, the PSS obtained in an SJ task could have been affected by potential prior entry effects but also due to the asymmetric criterion bias (due to perceptual latencies).

Stevenson and Wallace (2013) also argued that it is possible for the two tasks to share some common underlying processes for the "*ascription of temporal identity at a stimulus level*". Specifically, they investigated the effect of task and stimulus type on the TWI at different statistical criterion levels to check whether and how the criterion level may affect TWI outcomes. They used simple and complex (speech and non speech) stimuli and asked participants to perform an SJ and a TOJ task. Their results showed that the TWI was dependent both on the task and stimulus type. The SJ task yielded wider TWIs as compared to the TOJ task showing that additional processing steps were required after the low-level analysis of the temporal relationship of the stimulus pair. Similarly, the speech stimuli yielded wider TWIs as compared to non-speech stimuli (for criterion level at 50%). Moreover, the right side of the TWI (both at 50 and 70% criterion levels) was longer for the SJ than the TOJ task, while for speech stimuli it was symmetrical as compared to the asymmetrical TWI for non-speech stimuli with criterion levels impacting this symmetry. Contrary to what Linares and Holcombe reported for individual performance levels, Stevenson and Wallace found strong within participant TWI correlations across tasks. On the basis of this finding, they proposed that while task and stimulus types may affect differentially the PSS and TWI values, the tasks strongly correlate in their elicitation of the TWI.

Similarly, Garcia-Perez and Alcala-Quintana (2015) argued that differences in performance across the SJ and TOJ tasks could be due to task-dependent decisional and response processes that operate on *the "timing processes that are identical under both tasks"*. In previous studies, the reported performance was based on a curve fitting relative to the task at hand and systematic discrepancies between TOJ and SJ tasks were obtained. According to Garcia-Perez and Alcala-Quintana (2012; 2015b), however, this kind of analysis could not distinguish whether the differences obtained across stimuli, tasks, or experimental manipulations were due to timing, decisional, or response processes. To address this problem, Garcia-Perez and Alcala-Quintana (2012) developed a computational model of timing judgments to address the individual contribution of each component (i.e., timing, decisional, and response processes). Their model was based on independent-channels models and, thus, on the principle that sensory inputs from different modalities are processed independently from the different sensory pathways to create a perceived onset time (Garcia-Perez & Alcala-Quintana, 2015a). Then, a decision rule is applied on the perceived onset of the sensory inputs and a timing judgment is made (e.g., auditory first, visual first, or simultaneous response). Finally, this judgment elicits the response processes and the participant provides his/her response according to the task at hand. The model's sensory parameters affecting observed performance were isolated and kept invariant across tasks, whereas decisional parameters could vary across tasks, thus showing the potential differences between the tasks used to measure the perception of synchrony (Garcia-Perez & Alcala-Quintana, 2012; for more details see Chapter 12, this volume).

The Garcia-Perez and Alcala-Quintana model has been extended to explain previously reported differences of the TOJ and SJ tasks (Garcia-Perez & Alcala-Quintana, 2015a,b). For instance, Garcia-Perez and Alcala-Quintana (2015b) attributed the PSS values obtained in the TOJ and SJ tasks utilized in Matthews and Welch's (2015) study to a left visual field advantage and different resolution at the decision mechanism (that is, different perceived onset asynchronies were required to perceive asynchrony between the hemifields), but not to differential low-level temporal characteristics (i.e., visual acuity). In general, for stimuli and conditions similar across tasks, the differences obtained in temporal judgments were due to decisional and response processes between the two tasks rather than different timing processes (Garcia-Perez & Alcala-Quintana, 2015a). Thus, this is among the first approaches to investigate how SJs and TOJs differ in the decisional space (Garcia-Perez & Alcala-Quintana, 2012, 2015a; Matsuzaki et al., 2014; Regener, Love, Petrini, & Pollick, 2015).

Similarly, Matthews et al. (2016) supported that the differences between the TOJ and SJ tasks stem from decisional and response related factors. Specifically, in their study they showed that decisional factors govern both the relative speed (i.e., reaction time) and accuracy of the relative timing judgments. Using a SJ and a TOJ task within a visual RSVP task, Matthews and colleagues found that reaction times increased with uncertainty near the task-specific decision boundaries. Thus, for SJ they found faster reaction times to synchronous as compared to asynchronous inputs, while the opposite pattern was obtained for TOJ. Overall, they found smaller reaction time (RT) patterns (RT to synchronized stimuli/RT to ± threshold asynchrony) for the SJ than for the TOJ task (although not consistent across participants) suggesting that decisional and not stimulus driven differences affected the RTs obtained in the two tasks. Further testing of this RT pattern to other types of stimuli (not only visual) still remains to be done, as well as testing its reliability for timing judgments.

Machulla et al. (2016) also supported that TOJ and SJ tasks share, at least partly, common processes and, thus, they are not independent. They tested PSS and JND values across the SJ and TOJ tasks using simple audiovisual, audiotactile, and visuotactile stimulation and they found that the PSSs obtained between the two tasks were correlated. Moreover, the PSS values from the TOJ task were negative across the different pairs of stimuli (i.e., audiovisual, audiotactile, and visuotactile), while the PSS values from the SJ task were positive only for the visuotactile pair. It is important to note that the fitting data procedure was based on a non-parametric method (to allow for asymmetries around the PSS).

Finally, an interesting proposal was recently put forward by Parise and Ernst (2016), who argued that a general mechanism could potentially explain perceptual processes such as those governing causality, synchrony, and order. Specifically, they developed a model by borrowing the structure of a neural mechanism that detects motion and motion direction in the visual system (known as the Hassenstein-Reichardt detector or elementary motion detector) and modified it so as to explain aspects of multisensory perception. In brief, this mechanism contains detectors/subunits (termed as multisensory correlation detectors) that receive sensory information from different senses of spatially aligned receptive fields. The detectors' inputs are subjected to low-pass temporal filtering (i.e., temporal shifts are applied) and the outputs are either multiplied or subtracted to detect causality (correlation) and temporal order, respectively. Through this structure, one can explain the spatiotemporal characteristics of multisensory integration in a single general mechanism accounting for both neuronal and behavioral level outcomes (Parise & Ernst, 2016).

Recently, therefore, more and more researchers are supporting that the TOJ and SJ tasks may not be independent but instead they may share common mechanisms in the low level of the timing judgment or even one general mechanism. See Table 11.2 for an overview of the studies mentioned above. To further contribute to this discussion, the next section briefly covers what happens at the neuronal level when utilizing these two tasks.

### 4.3 *The Underlying Neuronal Processes of the SJ and TOJ Tasks*

It is important now to examine what happens in the brain when someone performs a synchrony/temporal judgment task. From a neuronal level perspective, Binder (2015) explored within participant differences in neural activation between TOJ and SJ tasks for a simple audiovisual stimulation using event-related fMRI. One of the main findings of this study was that the active areas elicited by both tasks overlapped with regions usually associated with spatial selective attention. Thus, timing judgments of audiovisual sensory inputs activate regions that are used during tasks based on spatial information. Another interesting finding of this study, is that the TOJ task activated additional regions in the left-hemisphere sites (in the prefrontal, parietal, and temporo-occipital regions) as compared to the SJ task. This additional activity made Binder to suggest that the TOJ task needs further processing potentially due to the formation of stimuli representations as "*separate and temporally ordered sensory events.*" This suggestion is consistent with the assumptions of the two-stage models (Jaskowski, 1991; Poppel, 1988), where an additional level is required to form the internal representation of order and reach a decision (Binder, 2016).

Using more complex stimuli and a mixed block/event-related fMRI design, Love et al. (submitted), supported that the SJ and TOJ task have "*divergent neural mechanisms*" despite the common brain activity elicited for both tasks. In line with Binder's (2015) results, Love et al. found that during the TOJ task, but not the SJ, several regions in the left hemifield were activated (middle occipital, middle frontal, precuneus, and superior medial frontal cortex), while the left middle occipital cortex (MOC) areas were deactivated. These findings suggest differential neural mechanisms for the two temporal tasks challenging the notion that the two tasks are based on the same cognitive architecture using the same sensory information (i.e., perceptual latency between the sensory inputs; Love et al., submitted).

Miyazaki et al. (2016) also supported that from the neuronal perspective, the two tasks are based on different mechanisms. In their study, they utilized unimodal tactile stimuli to test the neural activity between the TOJ and SJ tasks and they found specific brain activation patterns for each of the two tasks. More specifically, during the TOJ task more areas were activated as compared to those during the SJ task (i.e., left ventral and bilateral dorsal premotor cortices and left posterior parietal cortex for the TOJ task and posterior insular cortex for the SJ task) providing support for the hypothesis that not only the two tasks engage their own specific processes but also that the TOJ task involves more processes than SJ and that SJ processes are included in those for TOJ (Miyazaki et al., 2016).

To sum up, despite the growing number of studies on the comparison of the two main tasks for synchrony perception, the results and, thus, the conclusions are still inconsistent across studies. The different types of stimuli used along with the different SOAs, the number of participants, the various techniques followed to fit the data, and the different parameters derived from each study (PSS, JND, or TWI), do not allow for a direct comparison of the existing SJ and TOJ studies. Moreover, research on the mechanisms of the perception of synchrony and order have not as yet expressed a clear position in relation to the initially proposed hypotheses of perceptual latency and two-stage models (but see Binder, 2016; Garcia-Perez & Alcala-Quintana, 2012).

TABLE 11.2 *An overview of the differences between the studies that have directly compared the TOJ, SJ, and SJ-3 tasks. (A: Auditory, V: Visual, AV: Audiovisual).*

<table>
<tr><th>Authors</th><th>Task</th><th>Stimulus type</th><th colspan="3">Parameters for participants' sensitivity</th><th>TOJ Fitting</th><th>SJ-2 Fitting</th><th>SJ-3 fitting</th><th>Correlations</th></tr>
<tr><td rowspan="4">Love et al. (2013)</td><td rowspan="4">TOJ, SJ</td><td rowspan="4">5 types of stimuli (simple AV, complex, speech)</td><td>TWI</td><td colspan="2">PSS</td><td rowspan="4">Gausian cummulative distribution (V first)</td><td rowspan="4">Gausian probability density function for synchronous responces</td><td rowspan="4">n/a</td><td rowspan="4">No correlation of PSS or TWI across tasks</td></tr>
<tr><td></td><td>TOJ</td><td>SJ</td></tr>
<tr><td>TOJ < SJ for simple stimuli</td><td rowspan="2">A leading values (mainly)</td><td rowspan="2">V leading values</td></tr>
<tr><td>TOJ > SJ for speech</td></tr>
<tr><td rowspan="6">van Eijk et al. (2008)</td><td rowspan="6">TOJ, SJ, SJ-3</td><td rowspan="6">simple and complex non-speech AV stimuli</td><td>TWI for SJs only</td><td>PSS</td><td></td><td rowspan="6">Gausian cummulative distribution (V-first)</td><td rowspan="6">Two Gausian functions for synchronous responses</td><td rowspan="6">A-first and V-first fitted separately</td><td rowspan="6">No correlations of PSS between TOJ and SJ-2 or SJ-3 tasks</td></tr>
<tr><td></td><td>TOJ</td><td>SJ | SJ3</td></tr>
<tr><td rowspan="2">No effect of stimulus type</td><td rowspan="2">No effect of stimulus type</td><td>SJ-2 PSS> SJ-3 PSS</td></tr>
<tr><td>Larger positive PSS in ball</td></tr>
<tr><td>Effect of task in right TWI: SJ-2>SJ-3</td><td rowspan="2">A & V leading values</td><td rowspan="2">V leading values</td></tr>
<tr><td>Effect of stimulus type in left TWI: bouncing ball < flash</td></tr>
</table>

| | | | | | | | | | |
|---|---|---|---|---|---|---|---|---|---|
| Stevenson & Wallace (2013) | TOJ, SJ | simple and complex (speech and non-speech) AV stimuli | TWI (at 50 & 70%) | | | Gausian cummulative distribution (V-first) | Two Gausian functions for synchronous responses & for non zero PSS refitting of data | n/a | Correlated TWIS across tasks |
| | | | | | | | | | |
| | | | Speech TWI > simple & complex non-speech TWI | | | | | | |
| | | | SJ TWI > TOJ TWI | | | | | | |
| | | | Left & Right TWI: speech > simple & complex non-speech TWI | | | | | | |
| Linares & Holcombe (2014) | TOJ, SJ | simple AV stimuli | | PSS (ΔL; perceptual latency) | | Cummulative normal curves (V-first) | Two Gausian functions for synchronous responses & for non zero PSS *refitting of data* | n/a | No correlation of PSS found between TOJ - SJ tasks |
| | | | | TOJ | SJ | | | | |
| | | | | A & V leading values | V leading values | | | | |
| | | | | Overall positive values | | | | | |

TABLE 11.2 *An overview of the differences between the studies that have directly compared the TOJ, SJ, and SJ-3 tasks. (A: Auditory, V: Visual, AV: Audiovisual).* (cont.)

| Authors | Task | Stimulus type | Parameters for participants' sensitivity | | | TOJ Fitting | SJ-2 Fitting | SJ-3 fitting | Correlations |
|---|---|---|---|---|---|---|---|---|---|
| Matthews et al. (2016) | TOJ, SJ in RSVP task | simple V stimuli | RT ratio (RT at oms/RT at ±133ms)<br><br>SJ RT ratio < TOJ RT ratio | | | Psychometric function (1 / (1 + exp(-K (X-Xo)) (LetterFirst) | Signal detection theory (d' for each target asynchrony) | n/a | n/a |
| Leone et al. (2015) | TOJ, SJ | simple AV stimuli | | PSS<br>TOJ<br>V leading values | <br>SJ<br>V leading values | Cummulative distribution for V-first (least-squares method) | Gausian probability density function for synchronous responses | n/a | n/a |

| Vatakis et al. (2008) | TOJ, SJ | simple AV stimuli | | PSS | | Gausian cummulative distribution for V-first | Gausian probability density function for synchronous responses | n/a | No correlations of PSS between TOJ and SJ-2 tasks |
|---|---|---|---|---|---|---|---|---|---|
| | | | | TOJ | SJ | | | | |
| | | | | V leading values | V leading values | | | | |
| Maier et al. (2011) | TOJ, SJ | speech AV stimuli | | PSS (parametric fitting) | | Parametric and non-parametric fitting | Parametric and non-parametric fitting | n/a | Correlated PSS and JND across tasks only with non-parametric fitting |
| | | | | TOJ | SJ | | | | |
| | | | | V leading values | V leading values | | | | |
| Machulla et al. (2016) | TOJ, SJ | simple AV, AT, VT stimuli | | PSS | | Non-parametric fitting | Non-parametric fitting | | Correlated PSS across tasks |
| | | | | TOJ | SJ | | | | |
| | | | | A leading values | A leading for AV, AT | | | | |
| | | | | | V leading for VT | | | | |

## 5 Which Task Should One Use to Measure Synchrony Perception?

The SJ and TOJ tasks have been treated as equivalent and used interchangeably when studying synchrony (Keetels & Vroomen, 2012). In view of the recent findings, however, the question of which task to use in ones' study arises. In many studies, it has been argued that the TOJ data are more variable in terms of PSS as compared to the SJ data (van Eijk et al., 2008), which could be due to the inherent response biases (Garcia-Perez & Alcala-Quintana, 2012; Schneider & Bavelier, 2003; Vatakis et al., 2008) that cannot be distinguished from perceptual effects (Garcia-Perez & Alcala-Quintana, 2013). Garcia-Perez and Alcala-Quintana's (2012) findings support that the response bias is larger in TOJ experiments (resulting in shallower psychometric function at the 50% point; Garcia-Perez & Alcala-Quintana, 2013) because participants are required to guess which stimulus was presented first even when they perceived multiple stimulations as a single event. In this respect, performance measures are contaminated and possibly cannot be directly compared to those obtained from SJ tasks (Garcia-Perez & Alcala-Quintana, 2012). Yarrow et al. (2011) questioned the suitability of SJ tasks supporting that they are not appropriate for understanding the underlying mechanisms of "*apparent timing distortions*". As they support, it is not clear for the SJ task whether the PSS reflects differences in perceptual latencies and/or shifts in the criterion used to decide between synchronous and asynchronous sensory inputs.

Vatakis et al. (2008) have also expressed concerns about inherent response bias in the SJ task given that participants may be biased toward binding the incoming sensory inputs and treating them as a unified percept. This bias according to Vatakis et al. cannot affect the performance in the TOJ task where participants need to judge the order of presentation of the stimulus pair. These arguments are based on their findings of worst JND values in the SJ as compared to the TOJ task. Similar JND effects were also obtained by Barnett-Cowan and Harris (2009), who tested the temporal sensitivity to vestibular stimulation in relation to auditory, visual, and tactile stimulation.

Moreover, the comparison between the measures obtained from the SJ-2 and SJ-3 tasks revealed that although the auditory leading boundary in both SJ tasks was shorter than the visual, the PSS values and visual-first boundary for the SJ-2 task were larger as compared to those obtained from the SJ-3 task (van Eijk et al., 2008). These results led van Eijk et al. to suggest that for the perception of synchrony the SJ-3 task is potentially a better choice. The suitability of the ternary SJ task (i.e., SJ-3) for temporal sensitivity measures has also been proposed by other researchers (e.g., Garcia-Perez & Alcala-Quintana, 2013; Schneider & Bavelier, 2003; Spence & Parise, 2010; Ulrich, 1987; Zampini, Shore,

& Spence, 2008) that suggested that the TOJ task should be avoided in research focusing on PSS values and SJ-3 tasks should be preferred instead.

Garcia-Perez and Alcala-Quintana (2015b) are also in line with this suggestion by taking it a step further to the decision space. Given that participants in an SJ-3 task may perceive asynchrony but are unable to identify stimulus order, Garcia-Perez and Alcala-Quintana proposed a potential extra division of the existing decisional space (synchronous, auditory first, visual first) in order to cater for this forth type of judgment (asynchronous but cannot report the order). The authors suggest that such a division could potentially also explain why the TWI in a TOJ task is wider than that for an SJ task, however, these arguments need to be further investigated.

A task recommendation that is well accepted by the whole community is yet to be defined. This is mainly due to the unresolved issues of what are the underlying mechanisms governing each task and what the actual data obtained from each task really means in terms of the perception of synchrony. Thus, more research is needed comparing the three tasks and for the time being one's choice of a task should be dependent on the specific question asked in a given study.

## 6 Individual Differences

Generally, mean PSS values are mainly positive for SJ tasks and negative for TOJ tasks. On an individual level though, both auditory- and visual-leading PSSs have been reported for both tasks (Linares & Holcombe, 2014; Stone et al., 2001; van Eijk et al., 2008). For instance, while some participants perceive simultaneity when the visual stimulus is leading, some others perceive simultaneity when the sound stimulus is leading. As Stone et al. supported PSS values are significantly different between most individuals as well as in each individual and the estimated population mean PSS value. These consistent individual differences should not be disregarded but instead, should get more attention so as to better understand the underlying causes of such differences (Spence & Squire, 2003).

Moreover, there is a controversy on whether the measures of sensitivity on an individual level correlate or not across tasks. For instance, many researchers (Fujisaki & Nishida, 2009; Linares & Holcombe, 2014; Love et al., 2013; van Eijk et al., 2008; Vatakis et al., 2008; Vroomen & Stekelenburg, 2010) reported that within participants the TWI or PSS values were not correlated across tasks. On the other hand, Stevenson and Wallace (2013) found strong TWI correlations on an individual level. Linares and Holcombe (2014) suggested that to evaluate the obtained differences in the estimation of the PSS between the two tasks, one should consider both the consistency of PSS across tasks and the consistency of the same values across different moments in time in an individual level (see also Odegaard & Shams, 2016). Stone et al. (2001) who tested PSS values in an individual level using a SJ task with a pair of simple stimuli at various SOAs, found that individual PSS values varied between negative and positive values (i.e., −21 to +150 ms). Moreover, testing the PSS values from a SJ task across different stimulus distances, Stone et al. found highly correlated PSS values showing that PSS values within observers are rather stable and independent of stimulus distance. More recent findings on highly positive correlated PSS values across SJ and TOJ tasks, suggest a consistency in sensitivity on an individual level with participants with large JNDs in a SJ task having also large JNDs in a TOJ task (Machulla et al., 2016). Thus, it is as yet not clear how binding and the perception of synchrony are modulated across tasks and stimulus properties in an individual level but there is growing evidence that interindividual differences may be systematic (Eg & Behne, 2015; Linares & Holcombe, 2014; Stone et al., 2001; van Eijk et al., 2008).

## 7 Criteria for Excluding Participants

When collecting data for the study of the perception of synchrony you may sometimes need to exclude some data, thus in this section we will take a brief look at the criteria used for removing inappropriate data. Excluding participant' data in a TOJ task is more widespread that in a SJ task for similar stimulation and experimental settings (e.g., Love et al., 2013; Matthews et al., 2016). This may imply that the TOJ task is a more difficult task to perform than the SJ. Different criteria are used across studies as it relates to potential data exclusion. One of these criteria is whether the participant's data are fitted to the curves used (Love et al., 2013). The criterion for the goodness of fit between data and fitted function is usually the $r^2$ value. Values below 0.5 are taken as indices for exclusion while values above this level lead to the maintenance of the data for further analysis. There are also occasions where the participants' performance in the task (inability to perform above chance levels; Matthews et al., 2016; van Eijk et al., 2008) and/or PSS values exceed the tested asynchronies range and, thus, are excluded from further analyses (Matthews et al., 2016; Vatakis & Spence, 2008; Zampini et al., 2003). In other studies, each participant's PSS values are compared to the group PSS values with those falling two standard deviations away from the mean being excluded (Vatakis et al., 2008).

Love et al. (2013) had large exclusion rates in the TOJ task of their study (over 63% for complex stimuli with manipulations of duration) and this suggests that exclusion rates should not be disregarded in literature. They may indicate either a potential outcome measure or that the functions used to fit the TOJ data are inappropriate (Love et al., 2013). In any case, it is important to keep in mind that participants' data exclusion needs to be done carefully following specific criteria that will allow for further comparisons and/or replications.

## 8 Differences in Fitting Data Procedures

The most recent research in synchrony perception mostly fits Gaussian and cumulative Gaussian psychometric functions to SJ and TOJ data. Even the fitting procedures, however, are not consistent across studies and this may cause difficulties in comparing results across studies. The curve used to fit the data can affect the estimation of the PSS either precluding its estimation or biasing the PSS *"away from the mean of the distribution and towards the median"* (Linares & Holcombe, 2014; Maier et al., 2011). Nevertheless, the obtained psychometric function is usually modeled either by a cumulative Gaussian (Leone & McCourt, 2015; Linares & Holcombe, 2014; Love et al., 2013; Stevenson & Wallace, 2013; van Eijk et al., 2008) or a logistic function posing yet another issues in the comparison of the reported data across studies. Similarly, to analyze data obtained from a SJ task, a bell-shaped psychometric function is used to fit the "synchronous" response curve (see Stone et al., 2001). Generally though, this fitting procedure is conducted on an individual basis and mean values across participants are computed for each parameter. Despite the function used to fit the data, the parameters obtained from this task are derived using the formulas described earlier in this chapter. Thus, it is clear that even within a study, where the stimuli and asynchronies across tasks are the same, the parameters involved in the different fitted functions in the TOJ and SJ tasks can lead to differences in a study's outcomes (Garcia-Perez & Alcala-Quintana, 2012).

A shortcoming of the Gaussian functions trying to fit the data curve is its symmetry while the data that we want to fit are asymmetric (i.e., participants report "synchronous" for larger visual leading asynchronies; Alcala-Quintana & Garcia-Perez, 2013). To capture asymmetries in data from SJ tasks, some studies have used two cumulative Gaussians functions to fit the synchrony response curve allowing for different slopes in the two halves (e.g., Hillock, Powers, & Wallace, 2011; Stevenson & Wallace, 2013; van Eijk et al., 2008). Thus, one function was fitted to the synchronous responses when sound was leading (left TWI) and the other to the synchronous responses when sound was lagging (right TWI). The maximum synchronous response proportion was calculated for both halves and the intersection of the two curves may be used to estimate each participant's PSS (Stevenson & Wallace, 2013). Although, researchers also calculate the goodness of fit between obtained and predicted data using either ANOVA comparing predicted data points and actual measured data points corresponding to the SOAs presented (Stevenson & Wallace, 2013) or the $R^2$ parameter (Love et al., 2013; Vatakis et al., 2008), the symmetry problem remains.

Researchers have also used non-parametric functions to fit their data (both SJ and TOJ; see Machulla et al., 2016; Maier et al., 2011) allowing asymmetries between the two halves. Comparing the parametric and non-parametric fitting, Maier et al. have found differences in the correlations obtained between the parameters of the two tasks (i.e., parametric functions did not reveal significant correlations, while non-parametric functions revealed some significant correlations) showing that the data analysis method affects the obtained results. Signal detection procedures have also been used to describe each participant's temporal precision (d') across SOAs (Matthews et al., 2016) and, subsequently, these values are fitted to the selected functions as would have happened with proportions of "visual first" or "synchronous" responses.

It should be clear, therefore, that there is no consistent analysis used across studies to describe the observed TOJ or SJ performance. The main problem though, regarding the fitting of arbitrary functions to data is that these arbitrary functions (although they describe adequately the tendency of the data) cannot explain the differences obtained in sensitivity measures between the tasks (Alcala-Quintana & Garcia-Perez, 2012; Garcia-Perez & Alcala-Quintana, 2013) and do not address the potential link of the data to the sensory and decisional parameters that may affect judgments (Garcia-Perez & Alcala-Quintana, 2012). Garcia-Perez and Alcala-Quintana (2012, 2015a,b), therefore, propose the use of the model they constructed on the basis of the independent channels models of timing judgments in order to better describe participants' performance and gain insight at how the different levels of perception (sensory, decisional, response) affect sensitivity to synchrony/asynchrony and temporal order perception. It is still, however, early days for the community to suggest other models or to adopt one specific model.

## 9 Conclusion

In this chapter, we aimed to describe the TOJ and SJ tasks as well as the differences they yield in the estimation of perceptual latency. As it has been described, research has shown that many factors affect the PSS and the TWI values such as the task, the stimulus type, the analysis method, inherent biases, decisional factors, as well as the individual. Progress has also been made in comparing the different tasks and trying to disentangle the underlying processes governing an order and simultaneity response. Future research will help us to further clarify the uncertainties associated with the tasks used to measure synchrony perception, thus allowing us for a better view of how the brain manages to maintain the percept of synchrony where really there is none.

## References


Aghdaee, S.M., L. Battelli, & J.A. Assad (2014). Relative timing: From behaviour to neurons. *Philosophical Transactions of the Royal Society B, 369*, 20120472.

Allan, L.G. (1975). The relationship between judgments of successiveness and judgments of order. *Perception & Psychophysics, 18*, 29–36.

Aschersleben, G. (1999). On the asymmetry of the temporal contiguity window: Commentary on Lewkowicz. In G. Aschersleben, T. Bachmann, & J. Müsseler (Eds.), *Cognitive contributions to the perception of spatial and temporal events* (pp. 421–424). Amsterdam: Elsevier.

Barnett-Cowan, M., & L.R. Harris (2009). Perceived timing of vestibular stimulation relative to touch, light and sound. *Experimental Brain Research, 198*, 221–231.

Barnett-Cowan, M., & L.R. Harris (2011). Temporal processing of active and passive head movement. *Experimental Brain Research, 214*, 27–35.

Bedard, G., & M. Barnett-Cowan (2016). Impaired timing of audiovisual events in the elderly. *Experimental Brain Research, 234*, 331–340.

Binder, M. (2015). Neural correlates of audiovisual temporal processing- comparison of temporal order and simultaneity judgments. *Neuroscience, 300*, 432–447.

Capa, R.L., C.Z. Duval, D. Blaison, & A. Giersch (2014) Patients with schizophrenia selectively impaired in temporal order judgments. *Schizophrenia Research, 156*, 51–55.

Cardoso-Leite, P., A. Gorea, & P. Mamassian (2007). Temporal order judgment and simple reaction times: Evidence for a common processing system. *Journal of Vision, 7*, 11.

Cecere, R., J. Gross, & G. Thut (2016). Behavioural evidence for separate mechanisms of audiovisual temporal binding as a function of leading sensory modality. *European Journal of Neuroscience, 43*, 1561–1568.

Colonius, H., & A. Diederich (2012). Focused attention vs. crossmodal signals paradigm: Deriving predictions from the time-window-of-integration model. *Frontiers in Integrative Neuroscience, 6*, 62.

Coren, S., L.M. Ward, & J.T. Enns (2004). *Sensation and perception*, 6th eds. New York: Wiley.

Diederich, A., & H. Colonius, (2015). The time window of multisensory integration: Relating reaction times and judgments of temporal order. *Psychological Review, 122*, 232–241.

Donohue, S.E., M.G. Woldorff, & S.R. Mitroff (2010). Video game players show more precise multisensory temporal processing abilities. *Attention, Perception, & Psychophysics, 72*, 1120–1129.

García-Pérez, M.A., & R. Alcalá-Quintana (2012). On the discrepant results in synchrony judgment and temporal-order judgment tasks: A quantitative model. *Psychonomic Bulletin & Review, 19*, 820–846.

García-Pérez, M.A., & R. Alcalá-Quintana (2013). Shifts of the psychometric function: Distinguishing bias from perceptual effects. *The Quarterly Journal of Experimental Psychology, 66*, 319–337.

García-Pérez, M.A., & R. Alcalá-Quintana (2015a). Converging evidence that common timing processes underlie temporal-order and simultaneity judgments: A model-based analysis. *Attention, Perception, & Psychophysics, 77*, 1750–1766.

García-Pérez, M.A., & R. Alcalá-Quintana (2015b). The left visual field attentional advantage: No evidence of different speeds of processing across visual hemifields. *Consciousness and Cognition, 37*, 16–26.

Fujisaki, W., & S. Nishida (2009). Audio-tactile superiority over visuo-tactile and audiovisual combinations in the temporal resolution of synchrony perception. *Experimental Brain Research, 198*, 245–259.

Fujisaki, W., S. Shimojo, M. Kashino, & S. Nishida (2004). Recalibration of audiovisual simultaneity. *Nature Neuroscience, 7*, 773–778.

Hillock, A.R., A.R. Powers, & M.T. Wallace (2011). Binding of sights and sounds: Age-related changes in multisensory temporal processing. *Neuropsychologia, 49*, 461–467.

Hirsh, I.J. (1959). Auditory perception of temporal order. *Journal of the Acoustical Society of America, 31*, 759–767.

Hirsh, I.J., & Sherrick, C.E. Jr. (1961). Perceived order in different sense modalities. *Journal of Experimental Psychology, 62*, 423–432.

Jaskowski, P. (1991). Two-stage model for order discrimination. *Perception & Psychophysics, 50*, 76–82.

Kayser, C., & L. Shams (2015). Multisensory causal inference in the brain. *PLoS Biology, 13*(2), e1002075.

Keetels, M., & J. Vroomen (2012). Perception of synchrony between the senses. In Murray, M.M. & M.T. Wallace (Eds). *Frontiers in the neural bases of multisensory processes* (pp. 147–177). London: Taylor and Francis.

King, A.J. (2005). Multisensory integration: Strategies for synchronization. *Current Biology, 15*, R339–R341.

King, A.J., & A.R. Palmer (1985). Integration of visual and auditory information in bimodal neurons in the guinea-pig superior colliculus. *Experimental Brain Research, 60*, 492–500.

Kohlrausch, A., R. Eijk, J.F. Juola, I. Brandt, & S. Par (2013). Apparent causality affects perceived simultaneity. *Attention, Perception, & Psychophysics, 75*, 1366–1373.

Körding, K.P., U. Beierholm, W.J. Ma, S. Quartz, J.B. Tenenbaum, & L. Shams (2007). Causal inference in multisensory perception. *PLoS ONE 2*(9), e943.

Kuling, I.A., R.L. Eijk, J.F. Juola, & A. Kohlrausch (2012). Effects of stimulus duration on audio-visual synchrony perception. *Experimental Brain Research, 221*, 403–412.

Lewald, J., & R. Guski (2003). Cross-modal perceptual integration of spatially and temporally disparate auditory and visual stimuli. *Cognitive Brain Research, 16*, 468–478.

Lewkowicz, D.J., & A.A. Ghazanfar (2009). The emergence of multisensory systems through perceptual narrowing. *Trends in Cognitive Sciences, 13*, 470–478.

Leone, L.M., & M.E. McCourt (2015). Dissociation of perception and action in audiovisual multisensory integration. *European Journal of Neuroscience, 42*, 2915–2922.

Li, S.X., & Y.C. Cai (2014). The effect of numerical magnitude on the perceptual processing speed of a digit. *Journal of Vision, 14*, 18.

Linares, D., & A.O. Holcombe (2014). Differences in perceptual latency estimated from judgments of temporal order, simultaneity and duration are inconsistent. *i-Perception, 5*, 559–571.

Love, S.A., K. Petrini, A. Cheng, & F.E. Pollick (2013). A psychophysical investigation of differences between synchrony and temporal order judgments. *PLoS One, 8*, e54798.

Love, S.A., K. Petrini, C.R. Pernet, M. Latinus, & F.E. Pollick (submitted). Overlapping but divergent neural correlates underpinning synchrony and temporal order judgments. *NeuroImage*.

Machulla, T-K., M Di Luca., & M.O. Ernst (2016). The consistency of crossmodal synchrony perception across the visual, auditory, and tactile senses. *Journal of Experimental Psychology: Human Perception and Performance*, 1–46.

Maier, J.X., M. Di Luca, & U. Noppeney (2011). Audiovisual asynchrony detection in human speech. *Journal of Experimental Psychology: Human Perception and Performance, 37*, 245–256.

Matsuzaki, K.S., H. Kadota, T. Aoyama, S. Takeuchi, H. Sekiguchi, T. Kochiyama, & M. Miyazaki (2014). Distinction between neural correlates of audiovisual temporal order and simultaneity judgments. *International Journal of Psychophysiology, 94*, 193.

Matthews, N., & L. Welch (2015). Left visual field attentional advantage in judging simultaneity and temporal order. *Journal of Vision, 15*(2), 1–13.

Matthews, N., L. Welch, R. Achtman, R. Fenton, & B. FitzGerald (2016). Simultaneity and temporal order judgments exhibit distinct reaction times and training effects. *PLoS ONE, 11*(1), e0145926, 1–17.

Miyazaki, M., H. Kadota, K.S. Matsuzaki, S. Takeuchi, H. Sekiguchi, T. Aoyama, & T. Kochiyama (2016). Dissociating the neural correlates of tactile temporal order and simultaneity judgements. *Scientific Reports, 6*, 1–10.

Mollon, J., & A.J. Perkins (1996). Errors of judgement at Greenwich in 1796. *Nature, 380*, 101–102.

Morein-Zamir, S., S. Soto-Faraco, & A.F. Kingstone (2003). Auditory capture of vision: Examining temporal ventriloquism. *Cognitive Brain Research, 17*, 154–163, doi: 10.1016/S0926-6410(03)00089-2.

Munhall, K.G., P. Gribble, L. Sacco, & M. Ward (1996). Temporal constraints on the McGurk effect. *Perception & Psychophysics, 58*, 351–362.

Myung, I.J. (2003). Tutorial on maximum likelihood estimation. *Journal of Mathematical Psychology, 47*, 90–100.

Navarra, J., A. Vatakis, M. Zampini, S. Soto-Faraco, W. Humphreys, & C. Spence (2005). Exposure to asynchronous audiovisual speech increases the temporal window for audiovisual integration of non-speech stimuli. *Cognitive Brain Research, 25*, 499–507.

Nicholls, M.E., M. Lew, T. Loetscher, & M.J. Yates (2011). The importance of response type to the relationship between temporal order and numerical magnitude. *Attention, Perception & Psychophysics, 73*, 1604–1613.

Noel, J.-P., M.T. Wallace, E. Orchard-Mills, D. Alais, & E. Van der Burg (2015). True and perceived synchrony are preferentially associated with particular sensory pairings. *Scientific Reports, 5*, 17467.

Odegaard, B., & L. Shams (2016). The brain's tendency to bind audiovisual signals is stable but not general. *Psychological Science, 27*(4), 583–591.

Parise, C.V., & M.O. Ernst (2016). Correlation detection as a general mechanism for multisensory integration. *Nature Communications, 7*, 11543.

Petrini, K., S. Dahl, D. Rocchesso, C.H. Waadeland, F. Avanzini, A. Puce, & F.E. Pollick (2009). Multisensory integration of drumming actions: Musical expertise affects perceived audiovisual asynchrony. *Experimental Brain Research, 198*(2–3), 339–352.

Pöppel, E. (1988). Mindworks: Time and conscious experience. Boston, MA: Harcourt.

Pöppel, E., K. Schill, & N. Steinbüchel (1990). Sensory integration within temporally neutral system states: A hypothesis. *Naturwissenschaften, 77*, 89–91.

Regener, P., S. Love, K. Petrini, & F. Pollick (2015). Audiovisual temporal perception differences in autism spectrum disorder revealed by a model-based analysis. *Perception, 44*, 190–191.

Sanders, M.C., N.Y. Chang, M.M. Hiss, R.M. Uchanski, & T.E. Hullar (2011). Temporal binding of auditory and rotational stimuli. *Experimental Brain Research, 210*, 539–547.

Schneider, K.A., & D. Bavelier (2003). Components of visual prior entry. *Cognitive Psychology, 47*, 333–366.

Shams, L., & U.R. Beierholm (2010). Causal inference in perception. *Trends in Cognitive Sciences, 14*, 425–432.

Shore, D.I., C. Spence, & R.M. Klein (2001). Visual prior entry. *Psychological Science, 12*, 205–212.

Silva, C.C., C. Mendonça, S. Mouta, R. Silva, J.C. Campos, & J. Santos (2014). Correction: Depth cues and perceived audiovisual synchrony of biological motion. *PLOS One, 9*(1).

Slutsky, D.A., & G.H. Recanzone (2001). Temporal and spatial dependency of the ventriloquism effect. *NeuroReport, 12*, 7–10.

Spence, C., & C. Parise (2010). Prior-entry: A review. *Consciousness and Cognition, 19*, 364–379.

Spence, C., D.I. Shore, & R.M. Klein (2001). Multisensory prior entry. *Journal of Experimental Psychology: General, 130*, 799–832.

Spence, C., & S. Squire (2003). Multisensory integration: Maintaining the perception of synchrony. *Current Biology, 13*, R519–R521.

Stein, B.E., & A.M. Meredith (1993). *The merging of the senses.* Cambridge, MA: MIT Press.

Sternberg, S., & R.L. Knoll (1973). The perception of temporal order: Fundamental issues and a general model. In Kornblum, S. (Ed.), *Attention and Performance IV* (pp. 629–685). New York: Academic Press.

Stevenson, R.A., & M.T. Wallace (2013). Multisensory temporal integration: Task and stimulus dependencies. *Experimental Brain Research, 227*, 249–261.

Stevenson, R.A., M.M. Wilson, A.R. Powers, & M.T. Wallace (2013). The effects of visual training on multisensory temporal processing. *Experimental Brain Research, 225*, 479–489.

Stone, J.V., N.M. Hunkin, J. Porrill, R. Wood, V. Keeler, M. Beanland, M. Port, & N.R. Porter (2001). When is now? Perception of simultaneity. *Proceedings Biological Sciences-The Royal Society, 268*, 31–38.

Ulrich, R. (1987). Threshold models of temporal-order judgments evaluated by a ternary response task. *Perception & Psychophysics, 42*, 224–239.

Eijk, R.L.J., A. Kohlrausch, J.F. Juola, & S. Par (2008). Audiovisual synchrony and temporal order judgments: Effects of experimental method and stimulus type. *Attention, Perception, & Psychophysics, 70*, 955–968.

Wassenhove, V., Grant, K.W., & D. Poeppel (2007). Temporal window of integration in auditory-visual speech perception. *Neuropsychologia, 45*, 598–607.

Vatakis, A. (2013). The role of stimulus properties and cognitive processes in the quality of the multisensory perception of synchrony. In L. Albertazzi (Ed.). *Handbook of Experimental Phenomenology: Visual Perception of Shape, Space and Appearance* (pp. 243–263). UK: John Wiley and Sons.

Vatakis, A., J. Navarra, S. Soto-Faraco, & C. Spence (2008). Audiovisual temporal adaptation of speech: Temporal order versus simultaneity judgments. *Experimental Brain Research, 185*, 521–529.

Vatakis, A., J. Navarra, S. Soto-Faraco, & C. Spence (2007). Temporal recalibration during asynchronous audiovisual speech perception. *Experimental Brain Research, 181*, 173–181.

Vatakis, A., & C. Spence (2008). Evaluating the influence of the 'unity assumption' on the temporal perception of realistic audiovisual stimuli. *Acta Psychologica, 127,* 12–23.

Vroomen, J., & M. Keetels (2010). Perception of intersensory synchrony: A tutorial review. *Attention, Perception, & Psychophysics, 72,* 871–884.

Vroomen, J., M. Keetels, B. Gelder, & P. Bertelson (2004). Recalibration of temporal order perception by exposure to audio-visual asynchrony. *Cognitive Brain Research, 22,* 32–35.

Vroomen, J., & J.J. Stekelenburg (2010). Perception of intersensory synchrony in audiovisual speech: Not that special. *Cognition, 118,* 75–83.

Yarrow, K., N. Jahn, S. Durant, & D.H. Arnold (2011). Shifts of criteria or neural timing? The assumptions underlying timing perception studies. *Consciousness and Cognition, 20,* 1518–1531.

Yarrow, K., S.E. Martin, S. Di Costa, J.A. Solomon, & D.H. Arnold (2016). A roving dual-presentation simultaneity-judgment task to estimate the point of subjective simultaneity. *Frontiers in Psychology, 7,* 416.

Yates, M.J., & M.E.R. Nicholls (2011). Somatosensory prior entry assessed with temporal order judgments and simultaneity judgments. *Attention, Perception, & Psychophysics, 73,* 1586–1603.

Zampini, M., D.I. Shore, & C. Spence (2003). Audio temporal order judgments. *Experimental Brain Research, 152,* 198–210.

Zampini, M., D.I. Shore, & C. Spence (2005b). Audiovisual prior entry. *Neuroscience Letters, 381,* 217–222.

Zampini, M., S. Guest, D.I. Shore, & C. Spence (2005a). Audio-visual simultaneity judgments. *Perception & Psychophysics, 67,* 531–544.

CHAPTER 12

# Perceived Temporal Order and Simultaneity: Beyond Psychometric Functions

*Miguel A. García-Pérez and Rocío Alcalá-Quintana*

## 1 Introduction

Perception via the traditional senses of vision, audition, touch, gustation, or olfaction implies mechanisms (the sense organs) and neural structures (the sensory pathways) that transduce, transmit, and process physical energy (in vision, audition, and touch) or molecules (in gustation and olfaction). The same holds for the non-traditional senses of nociception, thermoception, equilibrioception, and others. In contrast, time does not emanate from a physical source and we do not have a sense organ for time, yet we have a vivid experience of it. Perception of time (chronoception) for brief events manifests in two remarkable abilities arguably subserved by separate processes. One is the ability to discriminate whether or not two punctate (instantaneous) events occurred simultaneously; the other is the ability to discriminate whether or not two brief events lasted the same duration. These punctate or brief events are delivered by presenting stimuli that can be perceived with our senses. Those stimuli are the occasion for some elusive machinery in the brain to extract the signals that render our perception of the time of occurrence of punctate events and our perception of temporal durations.

The duration of a stimulus is defined as the time elapsed between its onset and its offset. Then, perception of the duration of a stimulus presentation requires a second-stage process based on the output of first-stage processes determining the perceived onset and offset of the stimulus. This chapter focuses only on the first-stage processes and, specifically, on the methods used to assess their functioning and the utility of such methods to characterize timing processes. First-stage processes imply capture and transduction at the corresponding sense organ, followed by transmission of the sensory signals up the applicable pathway onto a central mechanism in the brain. Transduction, transmission, and processing of stimulus signals incur temporal delays that differ across sensory modalities but such delays also vary across stimulus types within the same modality and across repeated presentations of the exact same stimulus. When two punctate events occur simultaneously, the arrival times of

DOI 10.1163/9789004280205_013

their signals at the central mechanism reveal differences in speed of processing. Investigating such differences across and within sensory modalities would be simple if arrival times were accessible, that is, if the location of the target center in the brain were identified and the arrival times of sensory signals at that center were recorded electrophysiologically. Because this is currently impossible, indirect behavioral data must be used instead.

The behavioral (psychophysical) methods more widely used for this purpose are the *binary simultaneity judgment* (SJ2) task and the also binary *temporal-order judgment* (TOJ) task, which consist of presenting two stimuli (A and B) with a *temporal offset, temporal delay*, or *stimulus onset asynchrony* (SOA) that varies across trials. Both tasks are variants of the *single-presentation method*, whose defining characteristic is that each trial delivers a single stimulus magnitude (an SOA here) and requests a categorical response from the observer. In an SJ2 trial, observers report whether or not the stimuli A and B were subjectively presented simultaneously; in a TOJ trial, instead, observers report which of the two stimuli appeared to be presented first, with no option to report subjective simultaneity even if that were what the observer perceived. The *ternary simultaneity judgment* (SJ3) task blends SJ2 and TOJ tasks by allowing observers to report the three judgments: A first, B first, or A and B simultaneous (Ulrich, 1987). A further response option has occasionally been allowed for observers to report that presentation was subjectively non-simultaneous but order was impossible to identify (e.g., Weiß & Scharlau, 2011), making up a *4-ary simultaneity judgment* (SJ4) task. Discussion of the SJ4 task will be deferred to a later section of this chapter.

A second set of psychophysical methods has also been used that belongs in the category of *dual-presentation* or *multiple-presentation* methods. In these cases, two or more SOAs are sequentially presented in a trial and the observer is asked to indicate which of them satisfies some condition. For instance, in the so-called *two-alternative forced-choice* (2AFC) task, each trial presents two pairs of stimuli (i.e., a pair of SOAs) and the observer reports in which pair the presentation was more (or less) synchronous (Allan & Kristofferson, 1974; Fouriezos et al., 2007; Grant et al., 2004; Pastore & Farrington, 1996; van de Par & Kohlrausch, 2000; Yarrow et al., 2016); in the *match-to-sample* or *ABX* task, each trial presents a sample pair (i.e., a sample SOA) followed by two other pairs (two more SOAs) and the observer reports in which of the latter two pairs the (a)synchrony was more similar to (or more different from) that of the sample (Hillenbrand, 1984; Liberman et al., 1961; McGrath & Summerfield, 1985; van Eijk et al., 2009). Dual- and multiple-presentation methods are used less often than single-presentation methods and they will not be covered in detail.

Whichever task is used to collect data, observers' responses are tallied to compute the proportion of trials in which each judgment was reported at each of a set of SOAs. The most common method of analysis of data consists of fitting *psychometric functions*, continuous curves that match the path of the data. Figure 12.1 shows sample psychometric functions for SJ2, SJ3, and TOJ tasks without the data that might have given rise to them. When binary responses are involved (i.e., in SJ2 and TOJ tasks), only the psychometric function for one of the responses is needed (Figs. 12.1a and 12.1c); when ternary responses are involved (i.e., in SJ3 tasks), only the psychometric functions for two of the responses are needed although all three are often plotted (Fig. 12.1b).

The form of the function fitted to the data varies greatly in the literature. For TOJ data, the most common option is to fit a cumulative Gaussian or logistic function, which can be referred to an observer model with simple sensory and decisional components. For SJ2 data, the most common option is to fit a scaled Gaussian, which merely provides a description of the data because it cannot be referred to any observer model. Because SJ2 data are generally asymmetric (as illustrated in Fig. 12.1a) its rising and declining parts are sometimes fitted separately using monotonic functions, which describes the data better but again cannot be referred to an observer model. For SJ3 data, a mixture of these strategies is used across the implied psychometric functions (e.g., van Eijk et al., 2008), a strategy with which the probability that *some* response is given often falls short of (or exceeds) unity. The fitted functions are then used to summarize performance via measures such as the *point of subjective simultaneity* (PSS), the *synchrony range* (SR), the *synchrony boundaries* (SBs) or the *just noticeable difference* (JND). The PSS is defined as the SOA at which the psychometric function for "simultaneous" responses peaks in SJ2 or SJ3 tasks, or as the SOA at which the psychometric function evaluates to 0.5 in TOJ tasks (dotted vertical lines in Fig. 12.1). The SR is defined as the range of SOAs within which "simultaneous" responses are more prevalent than any other response in SJ2

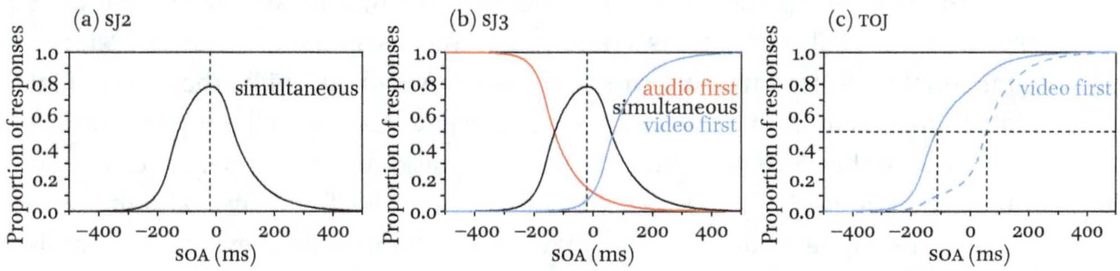


FIGURE 12.1 *Sample psychometric functions for SJ2 data (a), SJ3 data (b), and two sets of TOJ data (c). SOAs are assumed to be delivered via audiovisual stimulus pairs, with negative (positive) SOAs indicating that auditory onset preceded (lagged) visual onset.*

or SJ3 tasks and the SBs are the SOAs at the endpoints of the SR. The JND, also known as the *difference limen* (DL), is half the range of SOAs over which the psychometric function in TOJ tasks increases from, say, 0.25 to 0.75.

The three tasks can be used to estimate all of these measures, but many within-subjects studies involving two or more of the tasks have shown that the estimates are inconsistent (Bedard & Barnett-Cowan, 2016; Binder, 2015; Capa et al., 2014; Donohue et al., 2010; Fujisaki & Nishida, 2009; Li & Cai, 2014; Linares & Holcombe, 2014; Love et al., 2013; Sanders et al., 2011; Schneider & Bavelier, 2003; van Eijk et al., 2008; Vatakis et al., 2008). These inconsistencies are difficult to understand considering that only the question that observers respond to at the end of each trial differs across tasks and, hence, that the underlying timing processes should be invariant. Analogous inconsistencies have been found in within-subjects studies in which performance on single- and dual-presentation tasks was compared (Stevenson & Wallace, 2013).

One reason for these inconsistencies is that conventional psychometric functions only capture observed performance, which is the final outcome of the interplay of sensory (timing), decisional, and response processes involved in any psychophysical task. Then, even on the reasonable assumption that the timing component of performance is identical in SJ2, SJ3, and TOJ tasks, differences in decisional and response components would produce differences in observed performance across them. These differences are somewhat misleading because they do not reflect actual differences in timing processes. Thus, separating out the components of observed performance is needed for a proper characterization of timing processes that is not tainted by the effects of other processes. Accomplishing this separation requires psychometric functions whose mathematical form is derived from a suitable observer model that explicitly represents all of the intervening components. As discussed earlier, cumulative Gaussian or logistic psychometric functions fitted to TOJ data can be referred to an observer model whereby "A first" responses are given when a decision variable exceeds a criterion and "B first" responses are given otherwise. Sensible as this observer model may seem, it makes an inconvenient prediction for SJ2 or SJ3 tasks, namely, that the observer will never give "simultaneous" responses. In other words, if observers had the ability that this model attributes them in the TOJ task (i.e., if they could *always* tell temporal order), why would they lose that ability in SJ2 or SJ3 tasks and report perceived simultaneity so often? Or, in reverse, if observers' reports of perceived simultaneity in SJ2 and SJ3 tasks are genuine, by what mechanism do they give temporal-order responses in analogous occasions under the TOJ task? The TOJ task precludes reporting simultaneity, which does not imply that such judgments are never made. Then, TOJ data must necessarily be affected by how observers go about (mis)reporting temporal order when they perceive simultaneity.

The goal of this chapter is to describe an observer model that explicitly represents the timing, decisional, and response processes involved in single-presentation timing tasks, thus offering an integrated and coherent account of performance across tasks. The model elaborates on ideas put together by Sternberg and Knoll (1973) in their *independent channels* models, which were further developed by Schneider and Bavelier (2003). The resultant psychometric functions include parameters that characterize the underlying timing, decisional, and response processes. The model has been empirically validated extensively and the focus of this chapter is instead on its underpinnings and on practical aspects of its use for testing hypothesis or making inferences about how timing processes differ across manipulations (e.g., cuing) or across groups of observers (e.g., patients vs. normal controls).

## 2 The Observer Model

The observer model makes explicit assumptions about each of the processes whose participation is needed to respond on each trial. These processes include a *sensory component* that provides the evidence on which timing judgments are based, a *decisional component* that makes a judgment based on the sensory evidence, and a *response component* by which the judgment is expressed. The next three subsections describe each of these components and their referents, also characterizing them formally. The fourth subsection illustrates how the interplay of components shapes the psychometric functions in SJ2, SJ3, and TOJ tasks.

### 2.1 *The Sensory Component*

A punctate temporal event is signaled physically by, for example, the onset of a stimulus. Each of the two stimuli used to deliver an SOA must be peripherally processed by the corresponding sense organ and their neural signals must be transmitted up the corresponding sensory pathway onto a central mechanism. These operations incur delays that vary across sensory modalities and across stimuli within the same modality, resulting in differences in *arrival time* at the central mechanism and providing the evidence for timing judgments. Such delays are not fixed so that variability occurs across repeated presentations of the same stimulus.

The concept of arrival time is broader than what the previous description suggests, which only referred to physiological components. This may suffice in a description for simple stimuli such as a flash of light or a sound beep, for which the arrival time of signals at a central mechanism is perhaps also the time at which the presence of the stimulus is identified by the observer, which is what actually yields the arrival time referred to in the model. With vestibular stimuli such as yaw rotation, signals are surely reaching the brain as soon as the movement starts but the observer will not identify the motion until some aspect of it (amplitude, speed, acceleration, etc.) attains a necessary magnitude determined by the observer's sensitivity. Similarly, for time-varying stimuli such as single-syllable utterances in audiovisual speech, auditory and visual signals reach the brain continually from the nominal onset of the stimulus but the referents for auditory and visual arrival times in the model are the auditory identification of certain sounds and the visual identification of their articulations, respectively.

Across repeated presentations, arrival times will have a distribution that is impossible to determine empirically. Nevertheless, some distributions may be hypothesized that satisfy the physical constraint of causality: The arrival time of a stimulus signal cannot precede the onset of the stimulus itself. If stimulus onset is regarded as the origin of time, suitable distributions for arrival times must have all their probability mass on the positive real line. This constraint rules out the normal distributions included in most observer models in psychophysics. Figure 12.2 shows three plausible candidates, namely, a shifted exponential distribution given by

$$g(t)=\lambda\exp\left[-\lambda(t-\tau)\right],\quad t\geq\tau, \tag{1}$$

a shifted gamma distribution given by

$$g(t)=\frac{\lambda^{\alpha}}{\Gamma(\alpha)}(t-\tau)^{\alpha-1}\exp\left[-\lambda(t-\tau)\right],\quad t\geq\tau, \tag{2}$$

where $\Gamma$ is the gamma function, and a log-normal distribution given by

$$g(t)=\frac{1}{\sqrt{2\pi\sigma x}}\exp\left[\frac{-(\log(t)-\mu)^{2}}{2\sigma^{2}}\right],\quad t\geq 0. \tag{3}$$

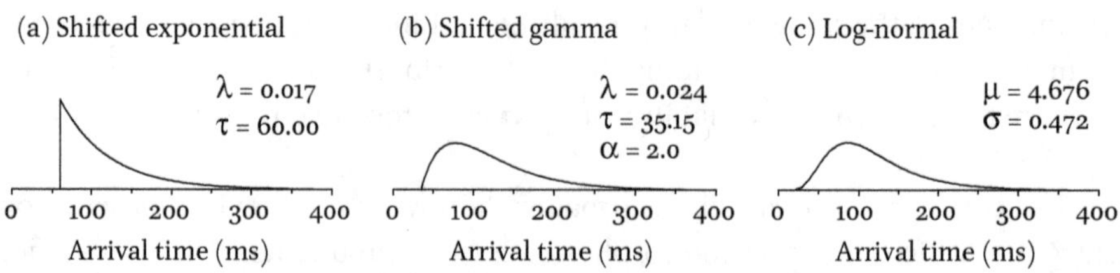


FIGURE 12.2 *Sample distributions of arrival time. (a) Shifted exponential distribution given by Eq. 1. (b) Shifted gamma distribution given by Eq. 2. (c) Log-normal distribution given by Eq. 3. Parameter values are printed in the insets.*

Parameter values are printed in Fig. 12.2 for each distribution, which were chosen so that arrival times have the same mean and variance in all cases. The form of the distribution of arrival times cannot be determined empirically but any particular choice is largely immaterial if it meets the above constraints. The shifted exponential distribution in Eq. 1 is simple and easily tractable mathematically, and it has often been used to model arrival latencies and peripheral processing times (e.g., Colonius & Diederich, 2011; Heath, 1984). In addition, this distribution has proven empirically adequate to account for observed performance in timing tasks (see García-Pérez & Alcalá-Quintana, 2012a, 2012b, 2015a, 2015b, 2015c). Thus, and without loss of generality, these are the distributions that will be used here.

Because SOAs are typically delivered via two different stimuli (and, generally, from different sensory modalities), the parameters of the corresponding distributions of arrival times will differ. In empirical studies, one of the stimuli is regarded as the reference so that SOA is defined as the relative delay with which the other (test) stimulus is presented. Thus, negative (positive) SOAs indicate that the onset of the test precedes (lags) the onset of the reference. Two distributions must thus be considered. The distribution of arrival times for the reference stimulus is subject to the constraint illustrated in Fig. 12.2, because the origin of time is set at its physical onset; in contrast, the (generally different) distribution of arrival times for the test stimulus will be shifted to the right or to the left according to the applicable SOA, potentially encroaching into the negative region of a time line whose origin is not at its own onset but at the onset of the reference. Formally, the arrival times (or *perceived onsets*, or *perceived latencies*) $T_r$ and $T_t$ of reference and test stimuli are random variables with densities $g_r$ and $g_t$ given by

$$g_i(t) = \lambda_i \exp\left[-\lambda_i\left(t-\left(\Delta t_i+\tau_i\right)\right)\right], \quad t \geq \Delta t_i + \tau_i, \quad i \in \{\mathrm{r}, \mathrm{t}\} \tag{4}$$

where $\Delta t_i$ is the onset time of stimulus $i$ and $\lambda_i$ and $\tau_i$ are the parameters of each distribution, which surely vary across observers and maybe also across experimental conditions. By the convention that sets the origin of time at the onset of the reference, $\Delta t_r = 0$ by definition and $\Delta t \equiv \Delta t_t$ is the SOA with which the pair is presented. The top row in Fig. 12.3 shows sample distributions at three different SOAs.

As discussed later, parameters $\lambda_i$ and $\tau_i$ affect the shape of the observed psychometric function and provide information about speed of processing and variability of arrival times: The mean and standard deviation of arrival times for stimulus $i$ are $1/\lambda_i + \tau_i$ and $1/\lambda_i$, respectively. These parameters thus inform of the sensory limits for perception of temporal order. If the standard deviations of arrival times are small, temporal order can be correctly perceived at

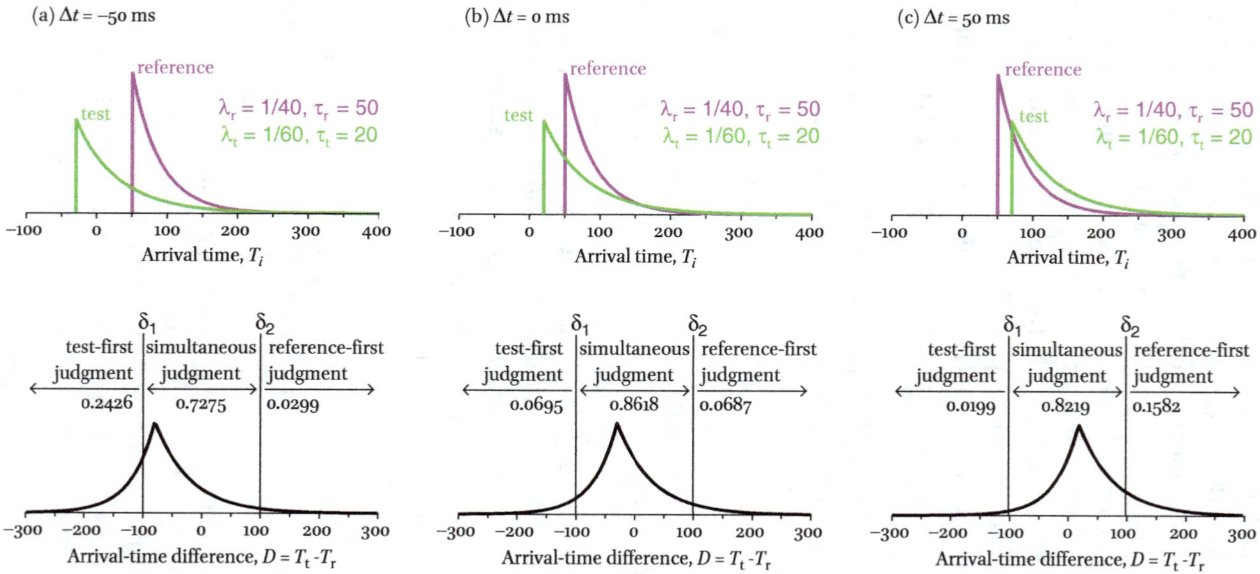


FIGURE 12.3 *Effects of SOA on the distribution of arrival times for reference and test stimuli (top row) and on the distribution of arrival-time differences (bottom row). (a) SOA of −50 ms. (b) SOA of 0 ms. (c) SOA of 50 ms. In the top row, the distribution for the reference stimulus is invariant with SOA because the origin of time is at its onset; the distribution for the test stimulus instead shifts with SOA. In the bottom row, the distribution of arrival-time differences shifts to the right as SOA increases. Vertical lines at $D = \delta_1$ and $D = \delta_2$ demarcate regions associated with each type of judgment (labels at the top of each region); the probability of each judgment (numeral under each label) is the area under the distribution within the applicable region.*

smaller SOAs than when standard deviations are large. Furthermore, large differences in the standard deviation of arrival times across stimuli produce large differences in the accuracy with which temporal order can be perceived at positive vs. negative SOAs (i.e., when the test stimulus precedes or lags the reference stimulus). Finally, large differences in average arrival time across stimuli produce discrepancies between physical and perceptual synchrony.

### 2.2 *The Decisional Component*

On any given trial, the perceived onset of each stimulus is a random value drawn from the corresponding distribution. These perceived onsets provide the evidence for a timing judgment, which the observer makes by application of a decision rule. Sternberg and Knoll (1973) and Schneider and Bavelier (2003) discussed several decision rules tailored to the response format of particular variants of the task. To detach the decisional component from the response component imposed by the task, we will consider a decision rule by which observers only make judgments at this stage, irrespective of the type of response later requested by the task. Observers' spontaneous reports to the effect that sometimes they guessed a response in TOJ trials because they could not tell which stimulus was presented first provides evidence that judgments precede responses and that the three types considered explicitly only in SJ3 tasks are made in all tasks. To better understand why judgments and responses must be separated, consider an experiment in which observers are given the SJ2, SJ3, or TOJ response options at random at the end of each trial. Because response options are revealed only after the stimuli were extinguished, a judgment must have been made before it can be expressed as a response. Such random mixture of trials from several tasks does not seem to have ever been used in timing perception, but Schneider and Komlos (2008) used it in a different context to make a similar point.

The decision variable is the *arrival-time difference* (or *perceived-onset difference*, or *latency difference*) $D = T_t - T_r$, which has the asymmetric Laplace distribution

$$f(d;\,\Delta t) = \begin{cases} \dfrac{\lambda_r \lambda_t}{\lambda_r + \lambda_t} \exp\left[\lambda_r \left(d - \Delta t - \tau\right)\right] & \text{if } d \le \Delta t + \tau, \\[2ex] \dfrac{\lambda_r \lambda_t}{\lambda_r + \lambda_t} \exp\left[-\lambda_t \left(d - \Delta t - \tau\right)\right] & \text{if } d > \Delta t + \tau \end{cases} \tag{5}$$

where $\tau = \tau_t - \tau_r$. Combination of $\tau_t$ and $\tau_r$ into an aggregate parameter $\tau$ is a trivial consequence of tasks that involve differencing: The precise arrival time of each stimulus is immaterial for the judgment and only the difference matters. The unfortunate consequence is that neither $\tau_t$ nor $\tau_r$ can be estimated separately, thus precluding the estimation of average arrival times for each individual stimulus although their variability (i.e., $1/\lambda_i$) can still be estimated. In addition, arrival-time differences are well characterized, with mean $1/\lambda_t - 1/\lambda_r + \tau + \Delta t$ and variance $1/\lambda_t^2 + 1/\lambda_r^2$. The bottom row in Fig. 12.3 shows the distribution of $D$ for each case in the top row, which only shifts location with SOA.

Before an amendment that will be introduced later, the decision rule partitions the domain of $D$ into three regions with boundaries at $\delta_1$ and $\delta_2$ (see the bottom row of Fig. 12.3). Then, "test-first" (TF) judgments arise when $D$ is large and negative ($D < \delta_1$), "reference-first" (RF) judgments arise when $D$ is large and positive ($D > \delta_2$), and "simultaneous" (S) judgments arise when $D$ is small ($\delta_1 \le D \le \delta_2$). The probability of each judgment varies with SOA ($\Delta t$), as this shifts the distribution of arrival-time differences (see the bottom row of Fig. 12.3). Formally, the probabilities $p_{\mathrm{TF}}$, $p_{\mathrm{S}}$, and $p_{\mathrm{RF}}$ of each judgment vary with $\Delta t$ as

$$p_{\mathrm{TF}}(\Delta t) = F(\delta_1; \Delta t), \tag{6a}$$

$$p_{\mathrm{S}}(\Delta t) = F(\delta_2; \Delta t) - F(\delta_1; \Delta t), \tag{6b}$$

$$p_{\mathrm{RF}}(\Delta t) = 1 - F(\delta_2; \Delta t), \tag{6c}$$

where

$$\begin{aligned} F(d;\ \Delta t) &= \int_{-\infty}^{d} f(z;\ \Delta t)\,\mathrm{d}z \\ &= \begin{cases} \dfrac{\lambda_r}{\lambda_r + \lambda_t} \exp\left[\lambda_r \left(d - \Delta t - \tau\right)\right] & \text{if } d \le \Delta t + \tau \\ 1 - \dfrac{\lambda_t}{\lambda_r + \lambda_t} \exp\left[-\lambda_t \left(d - \Delta t - \tau\right)\right] & \text{if } d > \Delta t + \tau \end{cases}. \end{aligned} \tag{7}$$

In principle, $\delta_1$ and $\delta_2$ could be placed anywhere. An asymmetric placement such that $\delta_1 \neq -\delta_2$ reflects a *decisional bias* whereby the absolute magnitude of the arrival-time difference required to make a TF judgment is not the same as that required to make an RF judgment. This may represent a natural bias of the observer but it may also be caused by experimental manipulations that favor one type of judgment over the other. In contrast, when $\delta_1 = -\delta_2$ (as in the bottom row of Fig. 12.3) the decision rule is unbiased. The width $\delta_2 - \delta_1$ reflects the operating resolution of the observer: The narrower this region, the smaller the arrival-time difference that allows the observer to judge temporal order (whether or not such judgment is physically correct; note in the bottom panel of Fig. 12.3a that the probability of an RF judgment is 0.03 even though the test stimulus was presented 50 ms before the reference).

As discussed later, the width and placement of the central region in decision space has consequences on the shape of the psychometric function under all tasks. Hence, estimating the decisional parameters $\delta_1$ and $\delta_2$ allows separating out their influence for an assessment of timing processes. Schneider and Bavelier (2003) noted that multiple location parameters are confounded in SJ2 and TOJ tasks, something that generalizes to all variants of the single-presentation method (García-Pérez & Alcalá-Quintana, 2013; Yarrow et al., 2011). Inspection of Eqs. 6 and 7 reveals that the location parameters affected here by this confound are $\tau$, $\delta_1$, and $\delta_2$, and that only the aggregates $\delta_1 - \tau$ and $\delta_2 - \tau$ can be estimated. A compromise (though suboptimal) solution is to enforce the assumption of no decisional bias, that is, to assume $\delta_1 = -\delta$ and $\delta_2 = \delta$, leaving $\delta$ as the only free parameter representing the half-width of a symmetric central region in decision space. The implications as well as other ways around this confound will be discussed later in this chapter.

### 2.3 *The Response Component*

In each trial, the observer must report the judgment arising from the timing and decisional components just discussed, using the response format that the task allows. In addition, response errors may occur by which observers misreport their judgment. Evidence of response errors is often found empirically in the form of, e.g., exceptional TF responses at large negative SOAs where RF responses have been given in most other trials. These errors may be caused by carelessness, by an unnatural arrangement of the response interface, or by insufficient practice to use it properly. The probability of a response error is generally small, but errors seem to affect some responses more often than others (for empirical examples, see García-Pérez & Alcalá-Quintana, 2012a, 2012b, 2015a, 2015b, 2015c). The response component thus comprises the mapping of judgments onto one of the responses allowed by the task, with a potential for misreporting such judgments due to errors. This component produces the final data that delineate the observed psychometric function.

Mapping judgments onto responses is straightforward in the SJ3 task because there is a distinct response option for each possible judgment. In the absence of response errors, the psychometric functions in the SJ3 task are directly given by the functions $p_{TF}$, $p_S$, and $p_{RF}$ in Eqs. 6 above. Response errors imply that each type of judgment has some probability of being misreported and, in such event, that the two forms that the misreport may take also have different probabilities. Let $\varepsilon_X^{(SJ3)}$ denote the probability of misreporting judgment X in the SJ3 task and let $\kappa_{X-Y}$ denote the probability of misreporting it as response Y. Then, the final psychometric functions for TF, S, and RF responses in SJ3 tasks are

$$\Psi_{\mathrm{TF}}^{(\mathrm{SJ3})}(\Delta t)=\left(1-\varepsilon_{\mathrm{TF}}^{(\mathrm{SJ3})}\right)p_{\mathrm{TF}}(\Delta t)+\varepsilon_{\mathrm{S}}^{(\mathrm{SJ3})}\kappa_{\mathrm{S-TF}}\,p_{\mathrm{S}}(\Delta t)+\varepsilon_{\mathrm{RF}}^{(\mathrm{SJ3})}\kappa_{\mathrm{RF-TF}}p_{\mathrm{RF}}(\Delta t), \tag{8a}$$

$$\Psi_{\mathrm{S}}^{(\mathrm{SJ3})}(\Delta t)=\varepsilon_{\mathrm{TF}}^{(\mathrm{SJ3})}\kappa_{\mathrm{TF-S}}\,p_{\mathrm{TF}}(\Delta t)+\left(1-\varepsilon_{\mathrm{S}}^{(\mathrm{SJ3})}\right)p_{\mathrm{S}}(\Delta t)+\varepsilon_{\mathrm{RF}}^{(\mathrm{SJ3})}\left(1-\kappa_{\mathrm{RF-TF}}\right)p_{\mathrm{RF}}(\Delta t), \tag{8b}$$

$$\Psi_{\mathrm{RF}}^{(\mathrm{SJ3})}(\Delta t)=\varepsilon_{\mathrm{TF}}^{(\mathrm{SJ3})}\left(1-\kappa_{\mathrm{TF-S}}\right)p_{\mathrm{TF}}(\Delta t)+\varepsilon_{\mathrm{S}}^{(\mathrm{SJ3})}\left(1-\kappa_{\mathrm{S-TF}}\right)p_{\mathrm{S}}(\Delta t)+\left(1-\varepsilon_{\mathrm{RF}}^{(\mathrm{SJ3})}\right)p_{\mathrm{RF}}(\Delta t). \tag{8c}$$

These expressions are easily unpacked with the help of the tree diagram in figure 12.4b of García-Pérez and Alcalá-Quintana (2012a), which is not reproduced here. Note also that Eqs. 8 reduce to Eqs. 6 if all εs are zero (i.e., in the absence of response errors). The psychometric functions in Fig. 12.1b for the SJ3 task arise from these equations (without response errors) when $1/\lambda_t = 40$, $1/\lambda_r = 90$, $\tau = 60$, and $\delta = 100$.

Mapping judgments onto responses is also straightforward in SJ2 tasks because TF and RF judgments are unambiguously aggregated into a category of non-simultaneous judgments. Under our notation for error parameters, the psychometric function for S responses in SJ2 tasks is

$$\Psi_{\mathrm{S}}^{(\mathrm{SJ2})}(\Delta t)=\varepsilon_{\mathrm{TF}}^{(\mathrm{SJ2})}p_{\mathrm{TF}}(\Delta t)+\left(1-\varepsilon_{\mathrm{S}}^{(\mathrm{SJ2})}\right)p_{\mathrm{S}}(\Delta t)+\varepsilon_{\mathrm{RF}}^{(\mathrm{SJ2})}p_{\mathrm{RF}}(\Delta t) \tag{9}$$

(see the tree diagram in figure 12.4a of García-Pérez & Alcalá-Quintana, 2012a), which reduces to Eq. 6b when all εs are zero. The psychometric function in Fig. 12.1a arises from this equation (without response errors) also when $1/\lambda_t = 40$, $1/\lambda_r = 90$, $\tau = 60$, and $\delta = 100$, which explains why the psychometric function for S responses is identical in Figs. 12.1a and 12.1b.

Finally, TOJ tasks force observers to give TF or RF responses upon S judgments. The mapping thus requires an extra *response bias* parameter ξ reflecting the probability with which the observer gives RF responses in such cases. Incorporating response errors as before (see the tree diagram in figure 12.4c of García-Pérez & Alcalá-Quintana, 2012a), the psychometric function for RF responses in TOJ tasks is

$$\Psi_{\mathrm{RF}}^{(\mathrm{TOJ})}(\Delta t)=\varepsilon_{\mathrm{TF}}^{(\mathrm{TOJ})}p_{\mathrm{TF}}(\Delta t)+\xi p_{\mathrm{S}}(\Delta t)+\left(1-\varepsilon_{\mathrm{RF}}^{(\mathrm{TOJ})}\right)p_{\mathrm{RF}}(\Delta t), \tag{10}$$

which reduces to Eq. 6c plus a fraction of Eq. 6b when all εs are zero. Also in the absence of errors, when $\xi = 0$ (i.e., when the observer has a strong response bias in the direction of never giving RF responses upon S judgments) Eq. 10 reduces to Eq. 6c and, thus, the psychometric function for RF responses in the TOJ task is identical to that for RF responses in the SJ3 task. Alternatively, when $\xi = 1$ (i.e., when the observer has a strong bias in the direction of always giving RF responses upon S judgments) Eq. 10 reduces to the sum of Eqs. 6b and 6c and, thus, the psychometric function for RF responses in the TOJ task is identical to the sum of the psychometric functions for RF and S responses in the SJ3 task. The psychometric functions shown in Fig. 12.1c for the TOJ task arise from Eq. 10 also without response errors and with parameter values as before ($1/\lambda_t = 40$, $1/\lambda_r = 90$, $\tau = 60$, and $\delta = 100$); additionally, $\xi = 0.85$ for the continuous curve whereas $\xi = 0.15$ for the dashed curve. For cases in which $\xi = 0.5$, see Fig. 12.4 below.

Error parameters also have an effect on the final shape of the psychometric functions under all tasks, as does the response bias parameter $\xi$ in TOJ tasks. These effects will be described in the next section. Estimating the response bias parameter $\xi$ that operates in TOJ tasks is crucial for an understanding of some nagging empirical results; in contrast, estimating error parameters (the εs and κs in Eqs. 8–10 above) is theoretically uninteresting but useful to remove contamination affecting estimates of the timing and decisional parameters described above.

### 2.4 *Overall Effects on the Shape of the Psychometric Function*

In an empirical study, the observed psychometric function reflects the joint influence of all the components just discussed. They affect the overall shape, location, and slope of the psychometric function so that changes in any of these characteristics across experimental conditions or across groups of observers cannot be arbitrarily attributed to any of the components. Recourse to model-based psychometric functions for the analysis of data is needed to separate out those influences. We mentioned earlier that this is not achievable in full when the data are collected with single-presentation methods because the task confounds some of the parameters of interest. Ways around this problem will be discussed later in this chapter, but it is useful at this point to consider several scenarios that illustrate the way in which all parameters contribute to shaping the observed psychometric functions. We will leave response errors aside in this presentation, which generally affect only the asymptotes of the psychometric functions (for a discussion and illustration of them in the context of SJ3 tasks, see García-Pérez & Alcalá-Quintana, 2012b).

Figure 12.4 shows the psychometric functions that arise under two different forms for the relative distributions of arrival times of test and reference stimuli (top row) and two different widths for the symmetric central region in

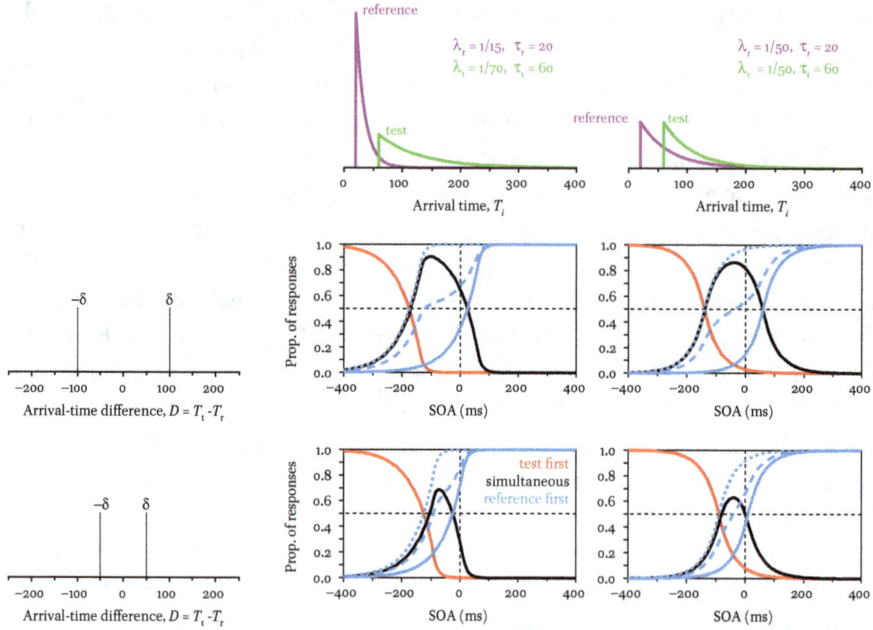


FIGURE 12.4 *Psychometric functions in SJ2, SJ3, and TOJ tasks in four different scenarios resulting from the combination of two cases for the distributions of arrival times (top row) and two cases for the width of the (symmetric) central region in decision space (left column). In each of the other four panels, solid red, black, and blue curves are the psychometric functions for TF, S, and RF responses, respectively, in the SJ3 task (see legend in the bottom-left panel). The solid black curve is also the psychometric function for S responses that would be observed in an SJ2 task. The solid blue curve is also the psychometric function for RF responses that would be observed in a TOJ task if $\xi = 0$; the dashed and dotted blue curves are those that would be observed in the TOJ task if $\xi = 0.5$ or $\xi = 1$, respectively.*

decision space (left column). Consider first the top-right panel in the 2×2 array of psychometric functions, for the case in which the arrival-time distributions for reference and test stimuli differ in offset (parameters $\tau_r$ and $\tau_t$) but not in spread (parameters $\lambda_r$ and $\lambda_t$). The psychometric function for S responses in SJ2 or SJ3 tasks (black curve) is symmetric because $\lambda_r = \lambda_t$, and peaks at $\tau_r - \tau_t$. In the panel underneath, when the central region in decision space is narrower, the psychometric function for S responses keeps these characteristics but it is narrower and shorter as a result of higher resolution to judge temporal order. In both cases, the psychometric function for RF responses in the TOJ task can span a broad range of shapes and locations due to the response-bias parameter $\xi$ (blue curves), which decouples observed performance in SJ and TOJ tasks even when timing and decisional parameters are identical.

By comparison, $\lambda_r \neq \lambda_t$ (panels on the left side of the 2×2 array) renders asymmetric functions that are analogously affected by the width of the central region and by the mismatch between $\tau_r$ and $\tau_t$. However, note that the peak of the psychometric function for S responses in SJ2 and SJ3 tasks (black curves) does not occur at $\tau_r - \tau_t$ in these conditions.

Given their multiple determinants, PSSs or JNDs are not dependable for a characterization of timing processes. The data ask instead for an alternative analysis that can separate out these influences. Fitting model-based psychometric functions in which these influences are captured by distinct parameters thus allows proper inferences about how experimental manipulations or group membership affect timing, decisional, or response processes. The next section summarizes the empirical evidence supporting this model of performance in timing tasks.

## 3 Empirical Evidence Supporting the Model

A theoretical model allows extracting latent aspects that are not directly accessible via PSSs, SRs, JNDs, or other indices based on observed performance. However, these benefits can only be gained if the model offers a satisfactory account of empirical data, not only in terms of fitting data adequately but also, and more importantly, when model predictions are supported by the data. If this is the case, estimated model parameters can then be subsequently analyzed in search for an interpretation of experimental outcomes in terms of timing, decisional, and response processes.

Testing the adequacy of a model requires assessing its success at fitting empirical data. Although a model can never be proven correct, one expects an adequate model to pass goodness-of-fit tests the stated percentage of times (García-Pérez, 2017): At, say, the 5% significance level, an adequate model will be rejected in about 5% of the occasions; a meaningfully larger number of rejections indicates inadequacy. The model gains further support under stringent goodness-of-fit tests involving predictions to the effect that some parameters must remain invariant across experimental manipulations.

This model has been subjected to the latter type of test extensively, in the two forms described next. Firstly, recall that timing parameters (i.e., $\lambda_r$, $\lambda_t$, $\tau_r$, and $\tau_t$) reflect delays in capture and transmission of sensory information that should not be affected by which question the observer responds to at the end of each trial (i.e., whether data are collected with SJ2, SJ3, or TOJ tasks). These parameters should thus be invariant across tasks, whereas response and decisional parameters would reasonably vary across tasks. If the model is adequate, common values for these timing parameters should provide a satisfactory account of SJ and TOJ data collected in within-subjects studies under otherwise identical experimental conditions. This type of data have been reported in a number of independent studies (e.g., Capa et al., 2014; Fujisaki & Nishida, 2009; Li & Cai, 2014; Linares & Holcombe, 2014; Matthews & Welch, 2015; Sanders et al., 2011; Schneider & Bavelier, 2003; van Eijk et al., 2008). An analysis of the 455 data sets from those studies supported the expectation of common timing parameters across tasks: The model including common timing parameters for all tasks was rejected in 24 (5.27%) of the occasions at the 5% significance level (see García-Pérez & Alcalá-Quintana, 2012b, 2015a, 2015b).

Secondly, because the model includes parameters that separately characterize the distribution of arrival latencies for test and reference stimuli, manipulations that affect the sensory processing of one stimulus but not the other should result in data that can be accounted for with common parameter values for the non-manipulated stimulus along with parameter values that vary for the other stimulus across the conditions in which it is manipulated. The analysis of SJ2 data on asynchronous audiovisual speech in an experiment in which only the visual stimulus was manipulated in four different ways (Magnotti, Ma, & Beauchamp, 2013) supported this prediction: Fitted under the stated constraint, the model was rejected at the 5% significance level for only 1 of 16 observers (6.25%; see García-Pérez & Alcalá-Quintana, 2015c).

In all of the analyses just mentioned, model psychometric functions followed very closely the path of empirical data to capture characteristics that conventional psychometric functions (i.e., arbitrary logistic or Gaussian functions) could not accommodate. These include asymmetries in SJ2 and TOJ data, relatively broad plateaus in SJ2 data, and intermediate regions of reduced slope in TOJ data. This ability to account for subtle features of the data provides additional qualitative support for the model and indicates that observed performance reflects the interplay of timing, decisional, and response processes. As discussed in the next section, summary performance measures such as PSSs, SRs, or JNDs are insufficient (and misleading) to identify relevant differences in timing processes across groups or experimental conditions.

## 4 PSSs, JNDs, and SRs vs. Interpretation of Model Parameters

The following discussion will leave the TOJ task aside. As shown in Fig. 12.4, the irrelevant response-bias parameter $\xi$ strongly affects the shape of the observed psychometric function so as to make PSSs and JNDs uninformative and uninterpretable. To keep things simple, the discussion will focus on the SJ2 task although consideration of the SJ3 task yields analogous outcomes. Also, we will consider the SR instead of the JND.

The top row in Fig. 12.5 shows two scenarios regarding timing processes, with latency distributions that are in both cases identical for test and reference stimuli except for a shorter average latency for the reference stimulus. Compared to the left side, the scenario on the right side depicts latency distributions that are narrower and closer to stimulus onset, as one might expect in an experimental condition in which (or in a group of observers for whom) latencies are shorter and subject to less variability. Thus, on the left side, arrival latencies for reference and test stimuli have means $1/\lambda_r + \tau_r = 100$ and $1/\lambda_t + \tau_t = 140$, respectively, whereas their standard deviations are identically valued at $1/\lambda_r = 1/\lambda_t = 50$; on the right side, arrival latencies have instead means $1/\lambda_r + \tau_r = 45$ and $1/\lambda_t + \tau_t = 85$ and standard deviations $1/\lambda_r = 1/\lambda_t = 25$. One would generally like to know whether differences between groups or experimental conditions occur at the level of timing processes (i.e., their speed and variability), and these parameters convey such information.

These two scenarios regarding timing processes can be combined with the two different scenarios regarding decisional processes that are illustrated on the left of Fig. 12.5. At the top, the central region for simultaneity judgments is broader than it is at the bottom (i.e., $\delta = 100$ at the top, compared to $\delta = 50$ at the bottom). A broader central region implies that observers need larger differences in arrival time to identify temporal order (i.e., they have less ability to tell small differences in arrival time). This is also another characteristic that may differentiate groups or experimental conditions, and one that a researcher would like to know about.

Because these aspects are captured by model parameters (i.e., $\delta$, the $\lambda$s, and the $\tau$s), fitting model-based psychometric functions to estimate them provides all the necessary information for inferences about the timing and decisional components of observed performance, and also about how they vary across groups or experimental conditions (for detailed examples, see García-Pérez & Alcalá-Quintana, 2015b, 2015c). Fitting the model and obtaining parameter estimates is straightforward with the software described in Alcalá-Quintana and García-Pérez (2013).

This method of analysis is in sharp contrast with the routine calculation of PSSs and SRs (or JNDs), whose values are printed in each panel in the 2×2 array of psychometric functions at the bottom right of Fig. 12.5. Note that the PSS is entirely immune to these differences in timing and decisional processes, sitting at an SOA of −40 ms in all cases. In turn, the SR is also almost identical across differences in timing processes (center and right columns) and it is only slightly affected by differences in decisional processes (center and bottom rows).

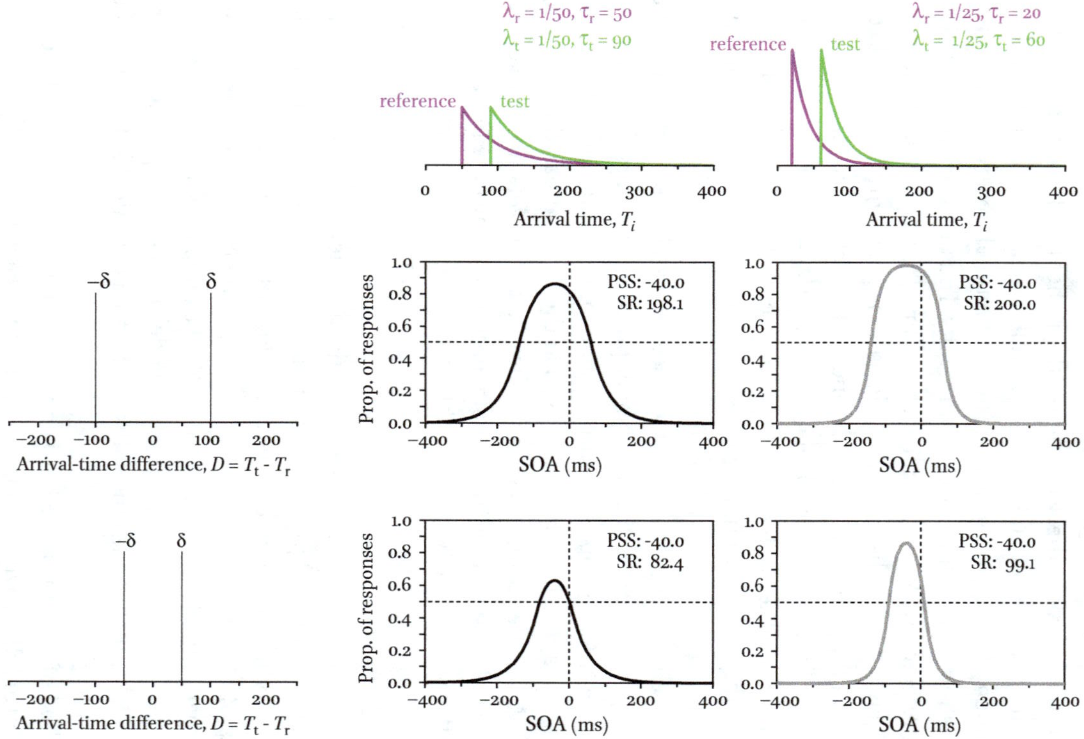


FIGURE 12.5 *PSSs and SRs (or JNDs) are insufficient to assess performance differences across groups or experimental conditions. The 2×2 array of psychometric functions at the bottom right arise from the combination of two scenarios regarding timing processes (top row) and two scenarios regarding decisional processes (left column). Each panel shows the resultant psychometric function in an SJ2 task along with the value of the PSS and the SR in each case. Layout as in Fig. 12.4.*

PSSs and SRs cannot portray differences in the underlying processes. Note that fitting arbitrary psychometric functions in these conditions will also not do justice to the data. The scaled Gaussian that is often fitted to SJ2 data cannot approximate the shapes described by the psychometric functions in Fig. 12.5, and their parameters cannot be referred to underlying processes either.

## 5 Shortcomings, Variants, and Extensions

This section discusses two variants of timing tasks that are useful for addressing some of the issues that arise in the use of single-presentation methods, namely, the likely presence of a fourth type of judgment and the problems arising from an inescapable confound.

### 5.1 *The SJ4 Task*

As presented thus far, the model assumes that only three judgments are possible: stimulus A subjectively first, stimulus B subjectively first, or A and B subjectively simultaneous. This assumption embodies the theoretical position that perception of non-simultaneity is necessary and sufficient for perception of temporal order: If observers perceive asynchrony, they also identify temporal order (Allan & Kristofferson, 1974; Baron, 1969). An alternative view (Hirsh & Sherrick, 1961) is that perception of non-simultaneity is a necessary but insufficient condition for perception of temporal order: Perception of asynchrony may still not allow observers to identify temporal order. This latter stance assumes a fourth type of judgment whose existence has been disputed for decades. There is, however, direct and indirect evidence of its presence.

Direct evidence comes from the only (to our knowledge) empirical study in which an SJ4 task was used to allow observers to report this fourth judgment (Weiß & Scharlau, 2011). Although this judgment was never reported by some of their observers, aggregated data in the two experimental conditions of their experiment 1 (see their figure 5) revealed that these judgments are maximally prevalent at SOAs around either end of the SR, with a notch around the point where simultaneous responses are maximally prevalent. These data suggest that the fourth judgment is associated with latency differences that are sufficiently large to judge non-simultaneity but not enough to identify temporal order. In other words, instead of the three regions in which decision space was partitioned in the model, extra regions flanking the central region for S judgments seem necessary. Figure 12.6 shows an extension of the model along these lines and illustrates how the probability of each judgment varies with SOA, namely,

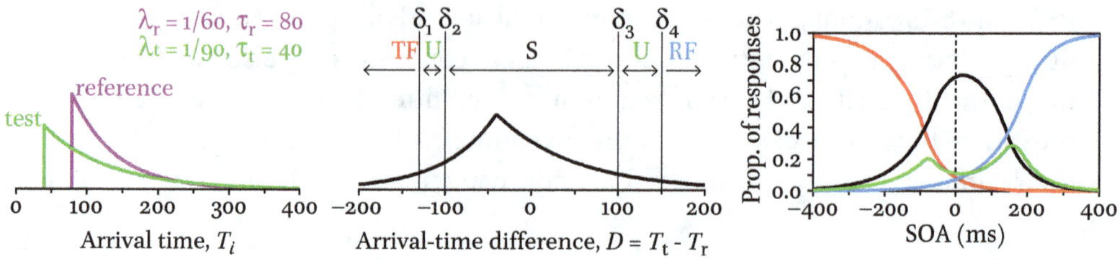


FIGURE 12.6 *Model extension to cover the SJ4 task. The decision space (center panel) includes two regions for "uncertain order" (U) judgments flanking the region for S judgments, besides the outer regions for TF and RF judgments. Note that the extra regions may have different widths. The right panel shows the resultant psychometric function for each judgment category (colored as their labels are in the center panel) under the sample arrival-time distributions for test and reference in the left panel.*

$$p_{\text{TF}}(\Delta t) = F(\delta_1; \Delta t) \tag{11a}$$

$$p_{\text{S}}(\Delta t) = F(\delta_3; \Delta t) - F(\delta_2; \Delta t) \tag{11b}$$

$$p_{\text{RF}}(\Delta t) = 1 - F(\delta_4; \Delta t) \tag{11c}$$

$$p_{\text{U}}(\Delta t) = F(\delta_4; \Delta t) - F(\delta_3; \Delta t) + F(\delta_2; \Delta t) - F(\delta_1; \Delta t) \tag{11d}$$

where U stands for "uncertain about order, though not simultaneous". Error parameters can also be introduced in this model.

Indirect evidence on the U judgment also exists that comes from the joint analysis of SJ2 and TOJ data in within-subjects studies. García-Pérez and Alcalá-Quintana (2015a, 2015b) reported substantial evidence to the effect that the central region in decision space is broader in TOJ tasks than it is in SJ2 tasks. This result can easily be interpreted under the partition illustrated in Fig. 12.6. In SJ2 tasks, only the boundaries $\delta_2$ and $\delta_3$ are operative because arrival-time differences within $[\delta_2, \delta_3]$ render S judgments and all others render non-simultaneous judgments (whether or not temporal order was additionally identified). In contrast, only $\delta_1$ and $\delta_4$ are operative in TOJ tasks because arrival-time differences lower than $\delta_1$ render TF judgments, arrival-time differences greater than $\delta_4$ render RF judgments, and all others force observers to misreport their inability to judge temporal order (whether they perceived non-simultaneity or simultaneity in such cases). Because $\delta_4 - \delta_1 \geq \delta_3 - \delta_2$, the empirical observation of a broader central region in TOJ tasks is consistent with the existence of the U judgment.

To look further into this issue, we conducted a study using the SJ4 task with identical visual stimuli (Gabor patches) presented on the left and on the right of a central fixation point on a monitor running at 60 Hz so that SOAs varied in steps of 16.7 ms. The patch on the right was regarded as the reference and data were collected in three consecutive 192-trial sessions with SOAs determined by an adaptive procedure (García-Pérez, 2014). On each trial, left and right stimuli were abruptly presented with the prescribed SOA and they were both removed simultaneously 750 ms after the onset of the stimulus presented second. Observers then used a keyboard to enter the response that described their judgment. Figure 12.7 shows data and fitted psychometric functions (incorporating error parameters) for each of 19 observers, plus a summary panel.[1]

Most notably, fitted curves follow the path of the data very closely in all cases. At the 5% significance level, the $G^2$ statistic rejected the model for observer #3 only, but even in that case the agreement between data and fitted curves is visibly good. In general, U responses occurred at the small SOAs expected under the model, with an imbalance at negative vs. positive SOAs that is consistent with the allowance for decision ranges of different width at negative vs. positive arrival-time differences (see Fig. 12.6). Some observers gave U responses relatively frequently (e.g., #1 and #5). Other observers gave U responses sparingly (e.g., #9 and #14) or not at all (e.g., #2 and #7). Interestingly, the latter observers' data show bumps of "improper" temporal-order responses (i.e., RF responses at negative SOAs and TF responses at positive SOAs) where U responses would be expected (see red and blue data points and curves at small positive and negative SOAs for observers #2, #6, #7, #8, #9, #10, #12, #13, #14, #15, #18, and #19). It is unclear whether these bumps reflect misreports (i.e., temporal-order responses upon U judgments) or authentic reversals of subjective temporal order, but their presence here is consistent with previous evidence to the same effect, as discussed next.

Firstly, the SJ3 data of van Eijk et al. (2008) showed bumps of improper temporal-order responses that are also found in earlier data sets and which prompted Ulrich (1987) to dismiss independent-channels models. Yet, an analysis under our SJ3 model showed that misreports account for this empirical feature within the realm of independent-channels models (García-Pérez & Alcalá-Quintana, 2012b). The present SJ4 data suggest that U judgments—which must be misreported as TF or RF responses in SJ3 tasks—also contribute to these bumps.

---

1 The data, the MATLAB code used to estimate model parameters and assess goodness of fit, and other related Online Material are available in the book's GitHub repository.

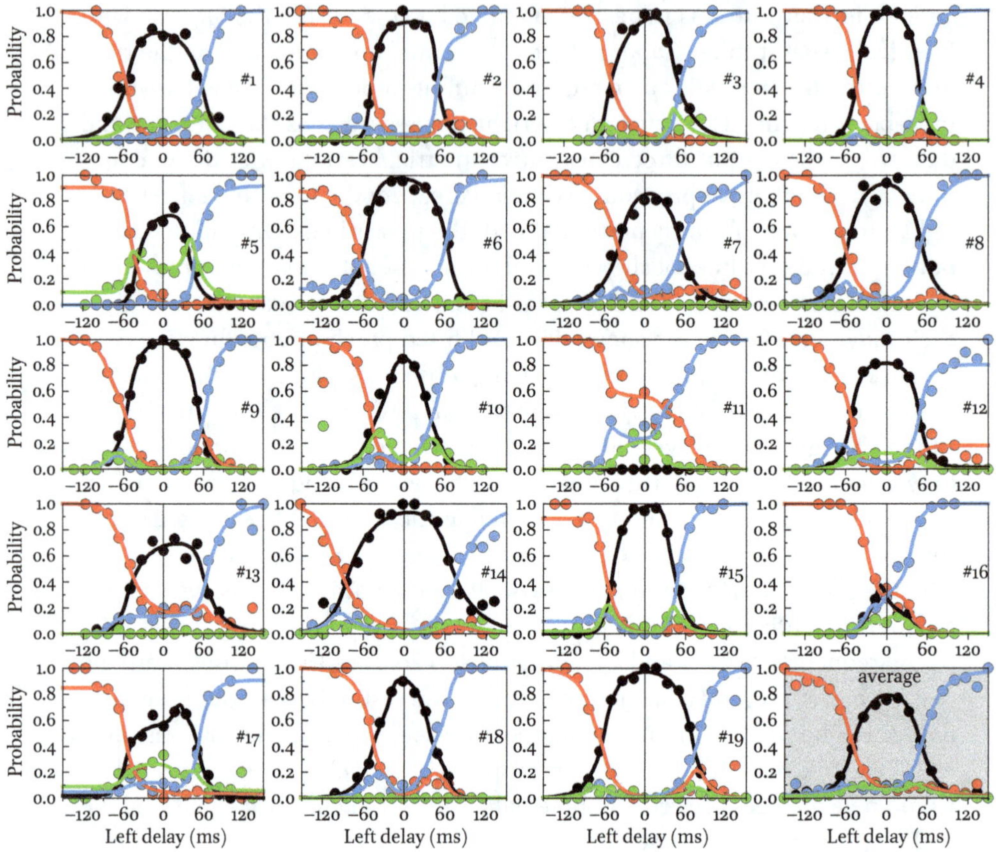


FIGURE 12.7 *SJ4 data and fitted psychometric functions. The grayed panel at the bottom right shows aggregated data and averaged psychometric functions across observers, and note that U responses in that panel display a pattern analogous to that reported by Weiß and Scharlau (2011) also for their aggregated data. Color code as in Fig. 12.6.*

Secondly, improper temporal-order responses at the SOAs where U responses are expected (or given) may indicate authentic reversals of perceived temporal order, not guesses upon U judgments. The cause of such reversals is elusive but they have been reported in TOJ tasks involving tactile stimuli with arms crossed, resulting in N-shaped psychometric functions that are at odds with the sigmoidal shapes obtained without arm crossing and with the comparatively narrow and peaked shape of SJ2 data with or without arm crossing (see Cadieux, Barnett-Cowan, & Shore, 2010; Fujisaki & Nishida, 2009; Yamamoto & Kitazawa, 2001;; see also Heed & Azañón, 2014).

We assessed whether N-shaped TOJ data could arise from the presence of a region for U judgments in decision space that, by a mechanism still to be unraveled, yields reversals of perceived temporal order. For this purpose, we derived a theoretical TOJ curve for each observer in Fig. 12.7 by making reasonable (but speculative) assumptions about how they would have responded in a TOJ task. Specifically, RF and TF responses in the SJ4 task would directly transfer into the same responses in the TOJ task, S responses in the SJ4 task would be evenly split into TF and RF responses in the TOJ task, and U responses in the SJ4 task would be split according to the imbalance of TF and RF responses at each SOA in the SJ4 task. If $\Psi_{TF}$, $\Psi_S$, $\Psi_{RF}$, and $\Psi_U$ are the psychometric functions for TF, S, RF, and U responses in Fig. 12.7, the psychometric function for RF responses in the TOJ task would thus be $\Psi_{RF}^{(TOJ)}(\Delta t) = \Psi_{RF}(\Delta t) + \Psi_S(\Delta t)/2 + \Psi_U(\Delta t) \times \Psi_{RF}(\Delta t)/[\Psi_{RF}(\Delta t) + \Psi_{TF}(\Delta t)]$. The results (not shown) revealed a diversity of patterns similar to that reported by Cadieux et al. (2010, figure 2), including N-shaped functions. It is still unclear why reversals of perceived temporal order occur, why they seem absent in some observers, or why their prevalence varies across experimental conditions, but their relation to intermediate regions for U judgments in decision space makes the SJ4 task a useful tool to investigate these issues.

It is interesting to note that observer #11 gave very few U responses and did not give any S response. One might surmise that this arises from an exquisite ability to perceive temporal order and a lack of S and U regions in decision space (i.e., $\delta_1 \approx \delta_2 \approx \delta_3 \approx \delta_4$). However, the path of TF and RF responses speaks against this surmise and suggests instead that observer #11 approached the SJ4 task essentially as a TOJ task in which S and U judgments are instead misreported as TF or RF responses.

During debriefing, some observers reported effortless identification that the left stimulus had been presented first, even if only by a very short time. These spontaneous reports are consistent with the outcome (see the Online Material) that the interval $[\delta_1, \delta_2]$ was generally estimated to be narrower than the interval $[\delta_3, \delta_4]$, as illustrated in Fig. 12.6. Because the sequential display of two visual stimuli at nearby spatial positions induces beta apparent movement (Larsen, Farrel, & Bundesen, 1983), this outcome suggests a predisposition to perceive rightward beta movement, with (potential) leftward beta movement giving instead an impression of non-simultaneous presentations without clear identification that the right stimulus was presented first. This is speculative at present, but the SJ4 task also proves useful tool to investigate this issue.

It is important to stress that the SJ4 model was fitted with allowance for different arrival-time distributions of test and reference stimuli (i.e., $\lambda_t \neq \lambda_t$

and $\tau \neq 0$). Because of the confound mentioned earlier and further discussed in the next section, this fitting approach requires the constraint that $\delta_2 = -\delta_3$. Estimates of $\lambda_r$ and $\lambda_t$ nevertheless turned out to be very similar on a subject-by-subject basis (see the Online Material), which seems reasonable in retrospect given that (1) reference and test stimuli were identical Gabor patches differing only in their location in the visual field and (2) location in the visual field does not affect the processing speed of visual stimuli (García-Pérez & Alcalá-Quintana, 2015b). An alternative approach to fitting these data under the constraints that $\lambda_r = \lambda_t$ and $\tau_t = \tau_r$ is discussed in the next subsection.

### 5.2 *Decisional Bias and the Dual-presentation Task*

We mentioned that single-presentation methods always confound several parameters in any model of psychophysical performance. Unfortunately, they always confound sensory parameters with decisional parameters, thus precluding the characterization of timing processes and the identification of the cause of observed differences across conditions. Then, the confound affects studies aimed at assessing *prior entry* (a hypothetical sensory acceleration that hastens the processing of attended stimuli) or *temporal recalibration* (an adjustment of subjective synchrony due to prolonged exposure to asynchronous stimulation). To test these hypotheses, one needs to tell whether observed differences across conditions are due to timing or to decisional processes, something that turns out to be impossible with single-presentation methods. Figure 12.8 illustrates this confound for an SJ3 task in a hypothetical experiment on prior entry with audiovisual stimuli. The column labeled "natural processing" reflects a condition in which neither stimulus is favored so that visual and auditory arrival times have their natural distributions; the column labeled "sensory (visual) acceleration" shows what visual arrival times might be like if processing of the visual signal is speeded up: Arrival times advance, as determined by $\tau_v = 20$ compared to $\tau_v = 60$ in the former case.

The panels underneath show the psychometric functions that would be obtained in each condition. Whether the central region in decision space is centered (middle row) or displaced (bottom row), visual acceleration displaces the psychometric functions to the left. However, the experimental manipulation presumably inducing visual acceleration might instead induce only a decisional bias. Then, compared to the psychometric functions at the top-left of the 2×2 array, decisional bias will also produce a leftward shift of the psychometric functions (bottom-left panel). Then, sensory acceleration without decisional bias (top-right panel, grayed) produces the same observable effect as decisional bias without sensory acceleration (bottom-left panel, grayed). Given that single-presentation methods confound sensory and decisional processes

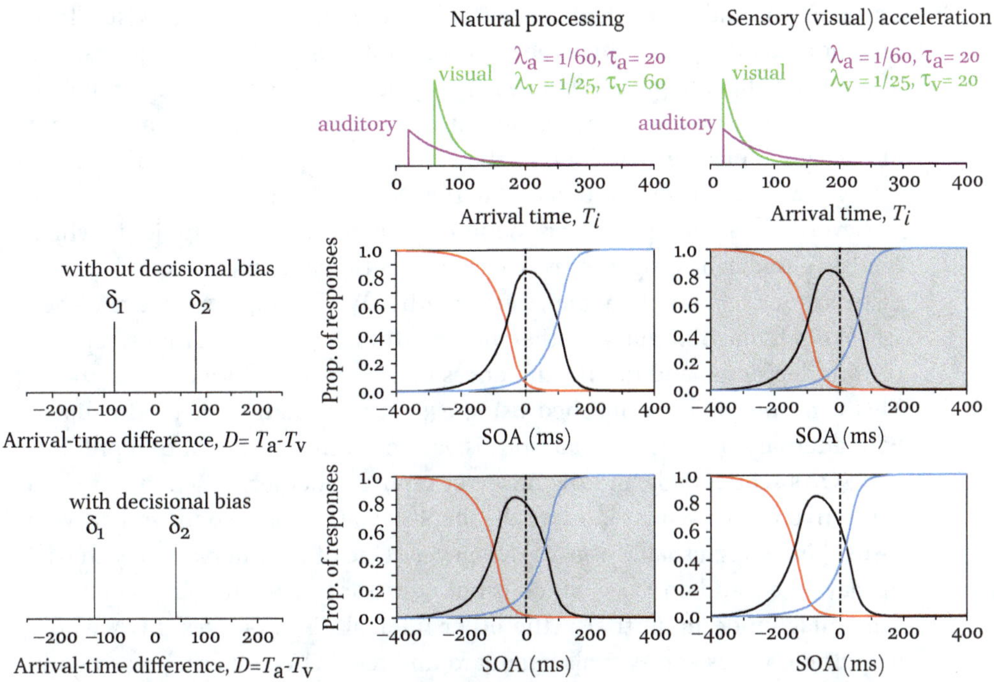


FIGURE 12.8 *Single-presentation methods cannot distinguish decisional bias from sensory acceleration. The grayed panels in the 2×2 array depict identical psychometric functions that can result either from pure sensory acceleration (top-right panel) or from pure decisional bias (bottom-left panel). Layout as in Fig. 12.4.*

and do not allow telling apart their respective influences, the prior-entry (or temporal-recalibration) hypotheses cannot be tested with them.

Methodological confounds that preclude testing for prior entry with single-presentation methods were pointed out by Spence and Parise (2010) but ways around this problem have not been devised thus far. We will show here that a dual-presentation method and a ternary response format allow separating out sensory and decisional contributions to observed performance (see also García-Pérez & Alcalá-Quintana, 2013).

Before discussing this issue, we should note that single-presentation methods are still useful when SOAs are delivered with stimuli that can be reasonably assumed to render identical arrival-time distributions. In these conditions, $\lambda_r = \lambda_t = \lambda$ and $\tau_t = \tau_r$ so that $\tau = 0$. Because the confound affects parameters $\tau$, $\delta_1$, and $\delta_2$, the constraint $\tau = 0$ resolves it and allows estimating decisional bias. García-Pérez and Alcalá-Quintana (2015a, 2015b) illustrated these particular circumstances in several empirical cases in which test and reference stimuli were identical visual patterns presented at different spatial locations on a monitor, and once it had been established that position in the visual field does not alter the processing speed of a visual stimulus. These conditions hold also for the SJ4 data presented in the previous subsection, something that allows an analysis without the constraint $\delta_2 = -\delta_3$ and permits an assessment of decisional bias. Presentation of the results of this alternative approach to fitting the above SJ4 data is deferred to the Online Material.

Nevertheless, this approach is inappropriate in prior entry studies in which the issue under investigation is whether attentional manipulations alter the distribution of arrival times and, thus, whether $\lambda_r \neq \lambda_t$ and/or $\tau_t \neq \tau_r$ even when test and reference stimuli are otherwise identical. In these conditions, separating out decisional and timing influences on observed performance is crucial. This can only be accomplished using a dual-presentation method coupled with a ternary response format. In this variant of the task, each trial presents two SOAs sequentially and the observer reports whether presentations were subjectively more synchronous in the first interval, in the second, or they were instead indistinguishable as to (a)synchrony. One of the intervals in each trial displays the standard SOA, which is not necessarily synchronous but has the same magnitude in all trials;[2] the other interval displays a test SOA whose magnitude varies across trials. Standard and test SOAs are also presented in both possible orders across trials. The observer's responses are then tallied to compute the proportion of trials in which each judgment (test more synchronous, standard more synchronous, or standard and test indistinguishable as to synchrony) was reported under each order of presentation (test SOA presented first or test SOA presented second) at each test SOA.

The model for this dual-presentation task is a straightforward extension of the model described earlier for single-presentation tasks, although derivation of model psychometric functions is somewhat more elaborate. A manuscript in preparation will present such extension in detail, including an empirical test of the validity of the model. Here we will only describe the model succinctly with an eye to illustrating how this task solves the inescapable problems of single-presentation methods and their inappropriateness for testing the prior-entry or temporal-recalibration hypotheses. In a nutshell, the model assumes that observers gather arrival-time differences $D_1$ and $D_2$ in each interval on each trial, whose individual distributions are given by Eq. 5 above for the

2 Sets of trials involving several standard SOAs can be interwoven in a session but the subsequent analysis is conducted separately for each standard SOA, just as if each of them had been used in a separate session (e.g., Allan & Kristofferson, 1974; Pastore & Farrington, 1996; Yarrow et al., 2016).

SOA used in each interval. Because observers are asked to report the interval in which presentation was subjectively more synchronous, the decision variable is the difference of *perceived offsets* (unsigned perceived asynchronies) $D = |D_2| - |D_1|$, whose probability distribution exists in closed form. The decision space also includes three regions so that the observer judges the first (alternatively, second) interval to be more synchronous when $D > \delta_2$ (alternatively, $D < \delta_1$) and judges both intervals to be indistinguishably (a)synchronous when $\delta_1 \leq D \leq \delta_2$. Leaving aside the extension that incorporates parameters for response errors, model psychometric functions for this task are shown in Fig. 12.9 under the same scenarios used in Fig. 12.8 to illustrate the inappropriateness of single-presentation methods.

It is immediately obvious by visual inspection of the 2×2 array of panels at the bottom right of Fig. 12.9 that lack of decisional bias results in psychometric functions that superimpose for both orders of presentation of test and standard SOA in each trial (middle row) whereas decisional bias renders psychometric functions that differ across presentation orders (bottom row). In either case, sensory acceleration (right column) produces psychometric functions that are rigidly shifted laterally compared to their location in the absence of sensory acceleration (middle column), regardless of whether or not the baseline condition involves decisional bias. In other words, sensory acceleration and decisional bias produce distinct effects on the observed psychometric functions, and these effects are captured by model parameters.

Then, the dual-presentation task with a ternary response format allows separating out timing and decisional processes while entirely removing the response bias that might also contaminate the data if observers were forced to report one of the intervals as more synchronous when they actually judge them to be equally (a)synchronous. This task thus allows estimating all the relevant parameters ($\lambda_r$, $\lambda_t$, $\tau$, $\delta_1$, and $\delta_2$) that are needed to express observed performance in terms of timing and decisional processes. Hence, prior-entry and temporal-recalibration hypotheses can be tested with this task. We should stress that timing and decisional components cannot be separated out if a binary response format is used instead (i.e., if the option to report indistinguishability is not given) or if responses are aggregated across orders of presentation of standard and test SOA.

## 6 Conclusion

Research on timing requires collecting data on observers' performance. Such data inform of the speed of sensory processing for each of the stimuli used to deliver SOAs under the selected experimental conditions, but observed performance is also modulated by decisional and response processes. An analysis that separates the contribution of the three components of performance is needed for a proper assessment of timing processes. We have shown that a model-based analysis is useful for this purpose and that classical measures of observed performance (PSSs, JNDs, or SRs) mix up these contributions misleadingly.

We have also shown that the single-presentation tasks most often used in empirical studies (SJ2, SJ3, and TOJ tasks) confound timing and decisional components, lending them unsuitable for studies in which differences in

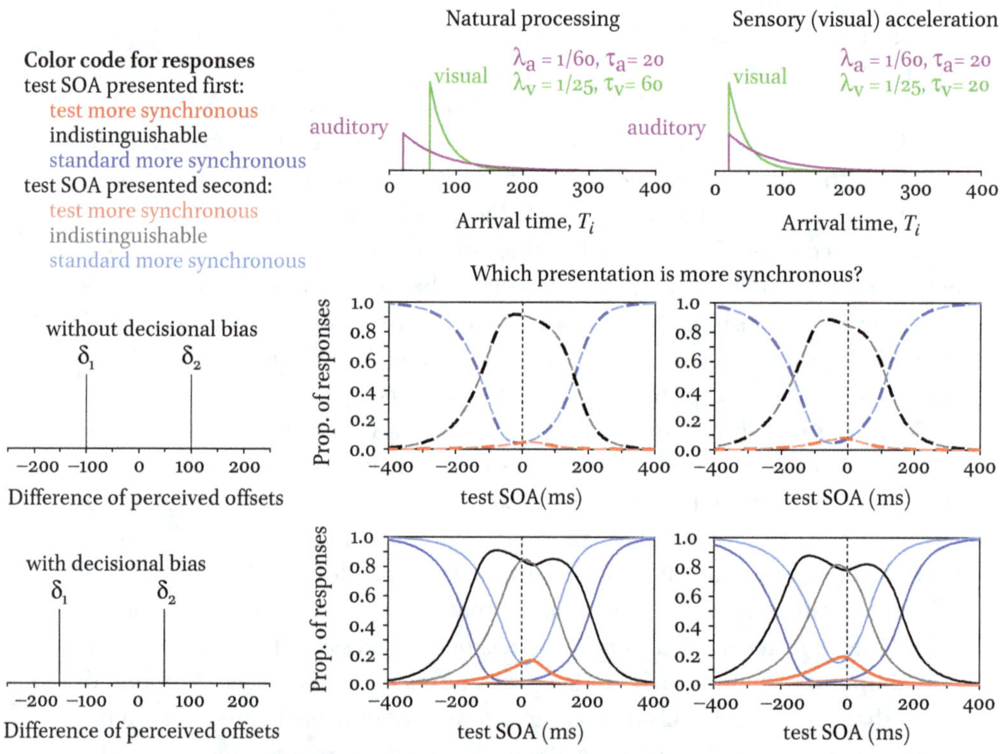


FIGURE 12.9 *Distinguishing decisional bias from sensory acceleration with a dual-presentation task and a ternary response format. Without loss of generality, the standard SOA is assumed to be a synchronous presentation of visual and auditory stimuli whose individual arrival-time distributions are indicated in the top row in the baseline (normal) condition and under an experimental manipulation that speeds up processing of the visual stimulus. The left column shows the decision space with boundaries that incorporate decisional bias (bottom row) or lack thereof (middle row). The 2×2 array of panels at the bottom right shows that sensory acceleration and decisional bias have distinguishable effects on the observed psychometric functions.*

timing processes across conditions are under scrutiny (e.g., prior entry) or for studies in which differences in decisional processes also across conditions are under scrutiny (e.g., temporal recalibration).

The above does not mean that single-presentation tasks are useless. In a number of situations, research interests focus on model parameters not affected by the confound (e.g., the variability of arrival latencies), experimental conditions constrain performance in a way that the confound is bypassed (e.g., test and reference stimuli are identical in all respects), or interest lies in qualitative aspects (e.g., whether the presence of U judgments warrants the use of an SJ4 task). In such cases, single-presentation tasks are useful, but model-based analyses are still needed to extract all the information that the data can provide. The software included in the Online Material accompanying this chapter (for model-based analyses of SJ4 data) supplements the software in Alcalá-Quintana and García-Pérez (2013; for model-based analysis of SJ2, SJ3, and TOJ data) to facilitate this task.

## Acknowledgements

This work was supported by grants PSI2012-32903 and PSI2015-67162-P from Ministerio de Economía y Competitividad (Spain). Parts of the computations were carried out on EOLO, the MECD- and MICINN-funded HPC for climate change at Moncloa Campus of International Excellence, Universidad Complutense.

## References

Alcalá-Quintana, R., & M.A. García-Pérez (2013). Fitting model-based psychometric functions to simultaneity and temporal-order judgment data: MATLAB and R routines. *Behavior Research Methods*, *45*, 972–998.

Allan, L.G., & A.B. Kristofferson (1974). Successiveness discrimination: Two models. *Perception & Psychophysics*, *15*, 37–46.

Baron, J. (1969). Temporal roc curves and the psychological moment. *Psychonomic Science*, *15*, 299–300.

Bedard, G., & M. Barnett-Cowan (2016). Impaired timing of audiovisual events in the elderly. *Experimental Brain Research*, *234*, 331–340.

Binder, M. (2015). Neural correlates of audiovisual temporal processing: Comparison of temporal order and simultaneity judgments. *Neuroscience*, *300*, 432–447.

Cadieux, M.L., M. Barnett-Cowan, & D.I. Shore (2010). Crossing the hands is more confusing for females than males. *Experimental Brain Research*, *204*, 431–446.

Capa, R.L., C.Z. Duval, D. Blaison, & A. Giersch (2014). Patients with schizophrenia selectively impaired in temporal order judgments. *Schizophrenia Research*, *156*, 51–55.

Colonius, H., & A. Diederich (2011). Computing an optimal time window of audiovisual integration in focused attention tasks: Illustrated by studies on effect of age and prior knowledge. *Experimental Brain Research*, *212*, 327–337.

Donohue, S.E., M.G. Woldorff, & S.R. Mitroff (2010). Video game players show more precise multisensory temporal processing abilities. *Attention, Perception, & Psychophysics*, *72*, 1120–1129.

Fouriezos, G., G. Capstick, F. Monette, C. Bellemare, M. Parkinson, & A. Dumoulin (2007). Judgments of synchrony between auditory and moving visual stimuli. *Canadian Journal of Experimental Psychology*, *61*, 277–292.

Fujisaki, W., & S. Nishida (2009). Audio-tactile superiority over visuo-tactile and audiovisual combinations in the temporal resolution of synchrony perception. *Experimental Brain Research*, *198*, 245–259.

García-Pérez, M.A. (2014). Adaptive psychophysical methods for nonmonotonic psychometric functions. *Attention, Perception, & Psychophysics*, *76*, 621–641.

García-Pérez, M.A. (2017). Thou shalt not bear false witness against null hypothesis significance testing. *Educational and Psychological Measurement*, *77*(4), 631–662.

García-Pérez, M.A., & R. Alcalá-Quintana (2012a). On the discrepant results in synchrony judgment and temporal-order judgment tasks: A quantitative model. *Psychonomic Bulletin & Review*, *19*, 820–846.

García-Pérez, M.A., & R. Alcalá-Quintana (2012b). Response errors explain the failure of independent-channels models of perception of temporal order. *Frontiers in Psychology*, *3*, 94.

García-Pérez, M.A., & R. Alcalá-Quintana (2013). Shifts of the psychometric function: Distinguishing bias from perceptual effects. *Quarterly Journal of Experimental Psychology*, *66*, 319–337.

García-Pérez, M.A., & R. Alcalá-Quintana (2015a). Converging evidence that common timing processes underlie temporal-order and simultaneity judgments: A model-based analysis. *Attention, Perception, & Psychophysics*, *77*, 1750–1766.

García-Pérez, M.A., & R. Alcalá-Quintana (2015b). The left visual field attentional advantage: No evidence of different speeds of processing across visual hemifields. *Consciousness and Cognition*, *37*, 16–26.

García-Pérez, M.A., & R. Alcalá-Quintana (2015c). Visual and auditory components in the perception of asynchronous audiovisual speech. *i-Perception*, *6*(6), 1–20.

Grant, K.W., V. van Wassenhove, & D. Poeppel (2004). Detection of auditory (cross-spectral) and auditory-visual (cross-modal) synchrony. *Speech Communication*, *44*, 43–53.

Heath, R.A. (1984). Response time and temporal order judgement in vision. *Australian Journal of Psychology*, *36*, 21–34.

Heed, T., & E. Azañón (2014). Using time to investigate space: A review of tactile temporal order judgments as a window onto spatial processing in touch. *Frontiers in Psychology*, 5:76.

Hillenbrand, J. (1984). Perception of sine-wave analogs of voice onset time stimuli. *Journal of the Acoustical Society of America*, *75*, 231–240.

Hirsh, I.J., & C.E. Sherrick (1961). Perceived order in different sense modalities. *Journal of Experimental Psychology*, *62*, 423–432.

Larsen, A., J.E. Farrell, & C. Bundesen (1983). Short- and long-range processes in visual apparent movement. *Psychological Research*, *45*, 11–18.

Li, S.-X., & Y.-C. Cai (2014). The effect of numerical magnitude on the perceptual processing speed of a digit. *Journal of Vision*, *14*(12), 1–9.

Liberman, A.M., K.S. Harris, J.A. Kinney, & H. Lane (1961). The discrimination of relative onset-time of the components of certain speech and nonspeech patterns. *Journal of Experimental Psychology*, *61*, 379–388.

Linares, D., & A.O. Holcombe (2014). Differences in perceptual latency estimated from judgments of temporal order, simultaneity and duration are inconsistent. *i-Perception*, *5*, 559–571.

Love, S.A., K. Petrini, A. Cheng, & F.E. Pollick (2013). A psychophysical investigation of differences between synchrony and temporal order judgments. *PLoS ONE*, *8*(1), e54798.

Magnotti, J.F., W.J. Ma, & M.S. Beauchamp (2013). Causal inference of asynchronous audiovisual speech. *Frontiers in Psychology*, *4*:798.

Matthews, N., & L. Welch (2015). Left visual field attentional advantage in judging simultaneity and temporal order. *Journal of Vision*, *15*(2), 1–13.

McGrath, M., & Q. Summerfield (1985). Intermodal timing relations and audio-visual speech recognition by normal-hearing adults. *Journal of the Acoustical Society of America*, *77*, 678–685.

Pastore, R.E., & S.M. Farrington (1996). Measuring the difference limen for identification of order of onset for complex auditory stimuli. *Perception & Psychophysics*, *58*, 510–526.

Sanders, M.C., N.-Y.N. Chang, M.M. Hiss, R.M. Uchanski, & T.E. Hullar (2011). Temporal binding of auditory and rotational stimuli. *Experimental Brain Research*, *210*, 539–547.

Schneider, K.A., & D. Bavelier (2003). Components of visual prior entry. *Cognitive Psychology*, *47*, 333–336.

Schneider, K.A., & M. Komlos (2008). Attention biases decisions but does not alter appearance. *Journal of Vision*, *8*(15), 1–10.

Spence, C., & C. Parise (2010). Prior-entry: A review. *Consciousness and Cognition*, *19*, 364–379.

Sternberg, S., & R.L. Knoll (1973). The perception of temporal order: Fundamental issues and a general model. In Kornblum, S. (Ed.), *Attention and Performance IV* (pp. 629–685). New York: Academic Press.

Stevenson, R.A., & M.T. Wallace (2013). Multisensory temporal integration: Task and stimulus dependencies. *Experimental Brain Research, 227*, 249–261.

Ulrich, R. (1987). Threshold models of temporal-order judgments evaluated by a ternary response task. *Perception & Psychophysics, 42*, 224–239.

van Eijk, R.L.J., A. Kohlrausch, J.F. Juola, & S. van de Par (2008). Audiovisual synchrony and temporal order judgments: Effects of experimental method and stimulus type. *Perception & Psychophysics, 70*, 955–968.

van Eijk, R.L.J., A. Kohlrausch, J.F. Juola, & S. van de Par (2009). Temporal interval discrimination thresholds depend on perceived synchrony for audio-visual stimulus pairs. *Journal of Experimental Psychology: Human Perception and Performance, 35*, 1254–1263.

van de Par, S., & A. Kohlrausch (2000). Sensitivity to auditory-visual asynchrony and to jitter in auditory-visual timing. In Rogowitz, B.E. & T.N. Pappas (Eds.), *Proceedings of SPIE: Human vision and electronic imaging V* (Vol. 3959, pp. 234–242). Bellingham, WA: SPIE Press.

Vatakis, A., J. Navarra, S. Soto-Faraco, & C. Spence (2008). Audiovisual temporal adaptation of speech: Temporal order versus simultaneity judgments. *Experimental Brain Research, 185*, 521–529.

Weiß, K., & I. Scharlau (2011). Simultaneity and temporal order perception: Different sides of the same coin? Evidence from a visual prior-entry study. *Quarterly Journal of Experimental Psychology, 64*, 394–416.

Yamamoto, S., & S. Kitazawa (2001). Reversal of subjective temporal order due to arm crossing. *Nature Neuroscience, 4*, 759–765.

Yarrow, K., N. Jahn, S. Durant, & D.H. Arnold (2011). Shifts of criteria or neural timing? The assumptions underlying timing perception studies. *Consciousness and Cognition, 20*, 1518–1531.

Yarrow, K., S.E. Martin, S. Di Costa, J.A. Solomon, & D.H. Arnold (2016). A roving dual-presentation simultaneity-judgment task to estimate the point of subjective simultaneity. *Frontiers in Psychology, 7*:416.

CHAPTER 13

# Collecting and Interpreting Judgments about Perceived Simultaneity: A Model-Fitting Tutorial

*Kielan Yarrow*

## 1 Introduction

In this chapter, I consider the simultaneity judgement (SJ) as a measure of the relative time perceived between two events, outlining the basic experimental design, the kind of data it generates, and how these data can be interpreted via the parameters of fitted models. After a brief overview of data collection methods, I outline the steps involved in both generating model predictions for plausible observer models and determining a single set of best-fitting model parameters, which maximise the likelihood that the model produced the data. I do so in a way intended to make sense to the competent programmer with limited formal mathematical expertise, making reference to accompanying Matlab code (see book's GitHub repository). Although I will focus on fitting simple detection-theoretic models, I also consider alternative approaches to treating SJ data, and briefly review the ways in which more complex models can be conceived and tested. I subsequently extend my discussion to consider a ternary choice, where participants can indicate either simultaneity or one of two possible orders, and also a novel task requiring a choice about which of two intervals contains the *most* simultaneous stimulus pair.

## 2 Chapter Architecture

In the following sections, I will discuss various abstract concepts that are often made concrete within a set of accompanying Matlab code (see book's GitHub repository). To facilitate understanding, I have adopted a cross referencing scheme. At various points in the chapter, I include footnotes to code references, which are shown within triangular brackets, e.g., <SimultaneityNoisyCriteria 112>. These indicate that the concept that is being discussed has been implemented in a Matlab function with that name. The number is the line number at which the relevant code begins. Given that I am not a mathematician, and that this chapter is intended to be comprehensible to non-mathematicians,

I have generally avoided including formal equations in the text, except where they seem particularly helpful to assist understanding (or with implementing the ideas that are being discussed).

## 3 Judging Relative Time

As timing researchers, we are often interested in how observers perceive the timing of events. For example, we might wish to assess how the timing between a brief sound and a flash of light is experienced. There are many ways in which we might operationalise this assessment, but a classic approach is to provide multiple trials containing different relative timings between events (hereafter referred to as *stimulus onset asynchronies* or SOAS) and have our observer make a simple judgement on each trial. The temporal order judgement (TOJ; e.g., "which came first") was popular for many years (e.g., Sternberg & Knoll, 1973) but recently the simultaneity or SJ (e.g., "were they simultaneous?") has become increasingly popular. This may reflect the comparative ease with which observers perform these two tasks: Participants tend to say that the TOJ task is harder than the SJ task (Love, Petrini, Cheng, & Pollick, 2013) and also make more errors than one would predict in the TOJ based only on estimates of sensory precision derived from other timing tasks (García-Pérez & Alcalá-Quintana, 2012a; Yarrow et al., 2016). Here, I will primarily address the SJ task in considerable detail, but also briefly introduce some variant tasks towards the end of the chapter.

## 4 The Simultaneity Judgement Experiment

The basic SJ design is simple: Present observers with pairs of stimuli on each trial, specifying the SOA between them. Across the experiment, present many different SOAS in a random order, and see how often participants judge each SOA to be simultaneous. However, we will need to define the range of SOAS that will be used, and how often each is presented. A classic approach is to use the method of constant stimuli, in which each possible SOA is presented an equal number of times across the experiment. In this case, we must still select the set of SOAS to test. Given a finite number of trials, there is an inevitable trade-off between the resolution implied by the step sizes we use and the range of SOAS we wish to cover. In the SJ task, it is important to adequately sample both of the transition points from perceived succession to perceived simultaneity. For example, in an audiovisual (AV) task, we need to capture the boundary where observers change from perceiving A then V (i.e., perceived succession) to perceiving synchrony, and also the boundary where they change from perceiving synchrony to perceiving V then A (i.e., back to perceived succession once again, but now in the opposite direction). For many participants, this implies sampling a rather wide range of SOAs. One potential problem is that in order to capture observers who report synchrony over a wide range, we end up sampling many times at extreme SOAs, which may appear trivially non-synchronous to experienced observers.

For this reason, some researchers prefer to use adaptive methods to select the SOA on each trial. For the SJ, these approaches generally attempt to place most trials near to the transition boundaries (from succession to synchrony and back again) while still adequately sampling the regions lying both between them and at the extremes. For example, Yarrow et al. (2013) used an approach loosely based on Rosenberger and Grill (1997) where the distribution from which trials are selected starts off being uniform, but is modified after each decision based on how the participant responds. The aim is to end up with a bimodal distribution that peaks over both transition boundaries. A similar goal can also be achieved in various ways via modified and/or interleaved staircases (e.g., Arnold, Petrie, Gallagher, & Yarrow, 2015; García-Pérez, 2014).

No approach is perfect. The method of constant stimuli can be wasteful, and is likely to establish a Bayesian prior that might bias perception towards the centre of the tested range (Miyazaki, Yamamoto, Uchida, & Kitazawa, 2006).[1] Some adaptive approaches imply a statistical dependency between successive stimuli, which is not really desirable, and the distribution of SOAs is likely to be uneven, and also to vary between conditions, particularly where they induce different biases. Different researchers will weigh these concerns differently. Hence, the only advice on which I suspect all researchers in this area would agree is that some pilot work with the population of interest is crucial before finalising the method for selecting SOAs across trials.

A further issue that should be considered closely before formal data collection begins is the accuracy and precision with which the desired SOAs are being generated by the lab hardware and software. Achieving precise stimulus timing is generally not trivial despite the assumed capacities of modern computers. This chapter is not the place to make a detailed comparison of different rigs and their technical limitations. Instead, I offer some brief advice: Check the timing of your stimuli over a fairly large number of trials using an oscilloscope or some similar method, and *never* assume that your computer is simply doing what you think you told it to do.

1 This will also be true of adaptive methods, but here the centre of the test range is more likely to conform to a participant's natural bias.

## 5 Interpreting SJ Data with Observer Models

Having run an experiment as outlined above, on each trial you will typically have: 1) an SOA and 2) a decision (i.e., synchronous or not). Such trial-by-trial data from SJs are commonly summarised as the proportion of times each SOA was judged synchronous.[2] You will then be confronted by a set of data similar to those plotted in Figure 13.1. These data form the basis of a psychometric function, plotting performance (proportion judged synchronous) against the tested SOA. Often, it is helpful to further summarise the data, for example to produce one or more dependent variables for inferential statistical analyses. How should this be achieved? Although I will briefly consider some alternatives in Section 9 of this manuscript, the approach I focus on in the majority of this chapter is the use of parametric observer models to summarise SJ data.

We could summarise data with any mathematical function that looks about right, and indeed this is the approach that has often been taken in characterizing SJ data, with the function of choice being a (truncated and/or vertically rescaled) Gaussian (Stone et al., 2002; Vroomen & Keetels, 2010). However, this arbitrariness comes at a cost. Firstly, if the function has not been derived from any meaningful observer model, there is little reason to believe that it will adequately summarise data in a wide range of situations. Secondly, and relatedly, the parameters of the model will have only a superficial descriptive meaning. By contrast, the parameters of an observer model have meanings that are clearly defined, being tied to the latent processes that have been hypothesised to generate the observations. Furthermore, they can be compared with the same parameters derived when similar observer models are defined and fitted to other tasks (such as the TOJ).

What do I mean by an "observer model"? In short, I mean a model in which a series of well-defined (but often quite abstract) processing steps have been hypothesised to intervene between perception and response. Here, I will be working with fairly simple observer models. They are based on the assumption that each of the two sensory signals to be compared must pass along a neural

2 This kind of summary is less useful where each SOA is only sampled once, as sometimes occurs in situations where SOAs have a random component. An example would be experiments comparing the time of an action to an event (where the event is presented around the time that the action is expected to occur, but this cannot be known for certain in advance; e.g., Yarrow, Sverdrup-Stueland, Roseboom, & Arnold, 2013). Another example would be data generated using an adaptive procedure without a fixed step size. Note that although such data will be more difficult to graph, the model-fitting procedures outlined in this chapter will still work perfectly well.

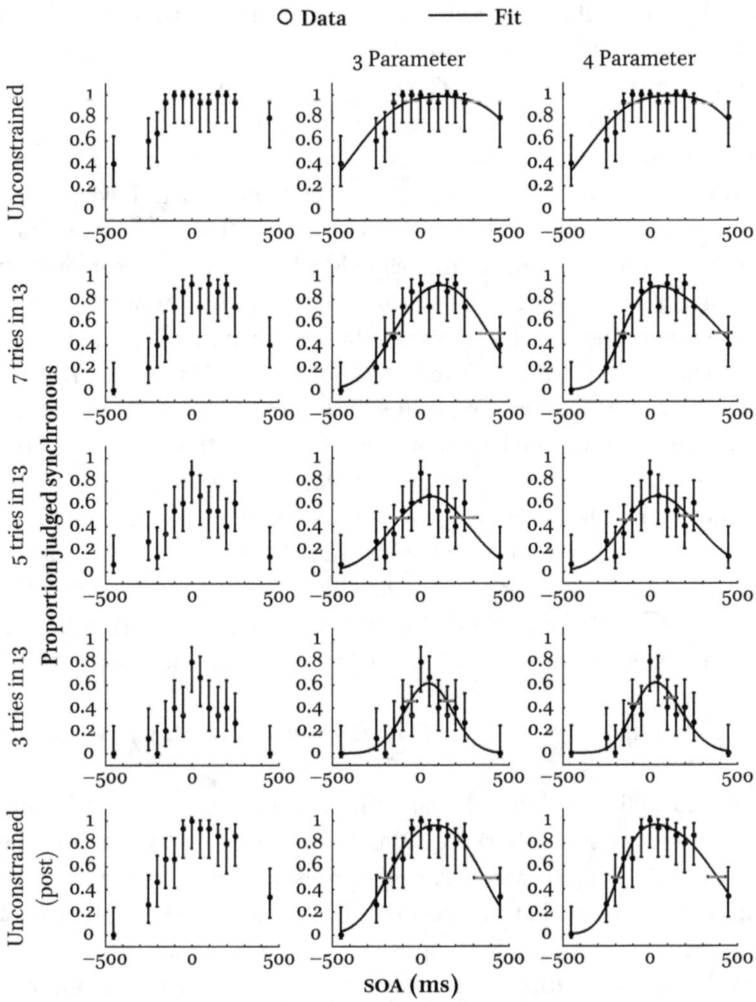


FIGURE 13.1 *Example SJ data. Each row shows data collected under different instructions (see main text). Left column: Data alone. Middle column: Data along with the predictions of a best-fitting three-parameter observer model. Right column: Data fitted with a four-parameter observer model, allowing asymmetry (i.e., a varied slope on each side of the psychometric function). Vertical error bars show 95% binomial confidence intervals on the data. Horizontal (grey) error bars surround estimates of two model parameters, and show 95% bootstrap confidence intervals. These two parameters represent the transition points from judgements of simultaneity to judgements of succession (or vice versa).*

pathway to a decision hub. The latency with which they do so is considered to be a random variable (i.e., to vary from trial to trial about some mean value according to a known distribution). The decision hub receives both signals, and, thus, has access to the subjective difference in arrival times between them ($\Delta t$), which is corrupted by their latency noise. Hence, $\Delta t$ is also a random variable, with a distribution that depends on its two contributors. If they are each distributed in a Normal/Gaussian way, $\Delta t$ is also Gaussian, with a variance equal to the sum of the two contributing signals' variances. The observer then interprets $\Delta t$ on each trial by placing decision criteria, typically one below and one above true synchrony, so that values that fall between them can be judged synchronous. These ideas are illustrated schematically in Figure 13.2. If, after reviewing it, you find that you are struggling with these concepts, I would suggest that you find out about the basics of signal detection theory, for example in Macmillan and Creelman (2005), before studying this chapter again.

If the two decision criteria that an observer uses to define simultaneity can be held perfectly constant across trials, this model predicts a psychometric function that is the *difference of two cumulative Gaussians*, each having the same variance but a different mean (Schneider & Bavelier, 2003). Hence both cumulative Gaussians can be described using just three parameters:

$$P\ \text{"simultaneous"} = \Phi\left(C_{High}, SOA, \sigma\right) - \Phi\left(C_{Low}, SOA, \sigma\right)^{3} \qquad (1)$$

where $\Phi$ is the normal cumulative density function. The two means represent the positions of the decision criteria ($C_{Low}$ and $C_{High}$) on the SOA axis, and the single standard deviation ($\sigma$) shared by both represents the variability in $\Delta t$.

What about if we doubt that our observer can hold their decision criteria perfectly constant across trials? If the positions of the two criteria are additionally considered to be Gaussian random variables (Yarrow, Jahn, Durant, & Arnold, 2011), the psychometric function becomes the difference of two cumulative Gaussians with different means and variances:

$$P\ \text{"sim."} = \Phi\left(C_{High}, SOA, \sigma_{High}\right) - \Phi\left(C_{Low}, SOA, \sigma_{Low}\right)^{4} \qquad (2)$$

3 <SimultaneityNoisyCriteria 12–22>.

4 <SimultaneityNoisyCriteria 23–33>. Note that this formula is an approximation and will break down if the criteria are close together and one is noisier than the other (because the cumulative Gaussians will overlap, producing negative predictions for judgements of simultaneity). One possible fix is to implement a simulation in such cases and assume that when noise in the criteria makes their order illogical, observers default to just using the less noisy criterion <SimultaneityNoisyCriteria 46>.

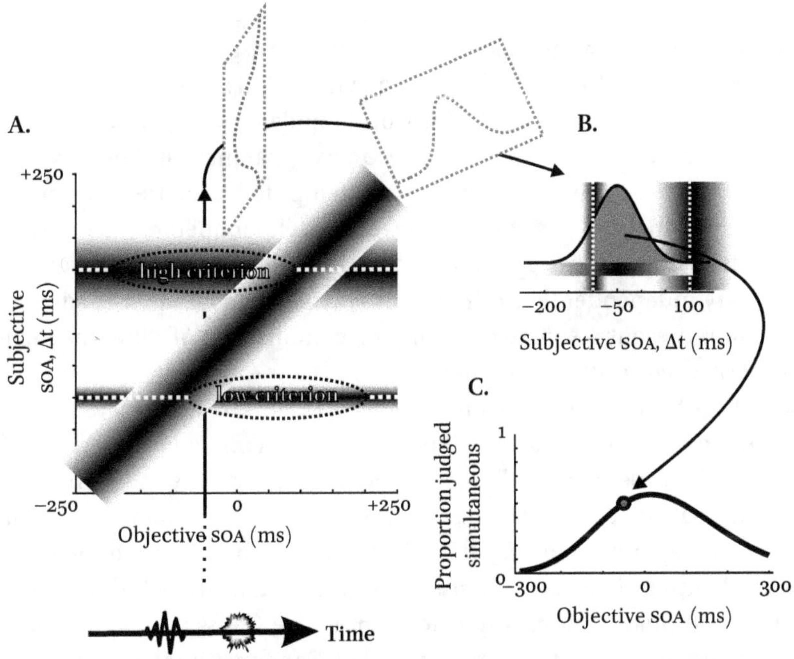


FIGURE 13.2 *Schematic of the four-parameter observer model outlined in the main text. A. Each SOA is presented many times. Each trial yields a noisy internal response (Δt, the subjective SOA). Hence the relationship between objective and subjective SOAs is depicted as linear (and in this case unbiased) but shading is used to denote the likelihood of each Δt value (darker shading denoting higher probabilities). Cutting vertically through this function for any given objective SOA yields the Gaussian distribution of resulting Δt values across trials. B. This probability density function (PDF) is shown for a -50 ms SOA. An observer will judge the trial synchronous when Δt falls between two decision criteria (shaded region between white dashed lines). The area under a PDF (to the left of some point) is given by a cumulative density function, so the shaded region is estimated as the difference of two cumulative Gaussians, one integrating all the way to the rightmost criterion, the other integrating only to the leftmost one. Shading around the criteria denotes additional criterion noise; criterion likelihood is highest where the shading is darkest. C. Resulting psychometric function (asymmetrical, due to differential criterion noise) with the point defined by the shaded region in Part B highlighted. Other points on the function are similarly obtained by integrating the region between the two criteria that is obtained at different SOAs.*

Hence, four parameters are needed to describe the predictions from this model. The means retain the same interpretation as before, corresponding to the mean positions of the decision criteria for synchrony, while the variance of each cumulative Gaussian now represents the sum of two sources of variability. Both cumulative Gaussians contain the variance in Δt, but each also uniquely contains the variance in the placement of the corresponding criterion. To expand slightly: Psychologically, we are now envisaging several contributing sources of noise – from variability in signal transmission times, and from variability in the placement of two decision boundaries. However, when we implement the model mathematically it becomes apparent that these psychological constructs are degenerate should we attempt to have a free parameter for each one.[5] To make it possible to recover model parameters (as outlined later, in Section 6) we must create composite parameters ($\sigma_{Low}$ and $\sigma_{High}$), which can vary independently from each other, and from all other model parameters, but represent a somewhat complex combination of different theoretical sources of sensory/decision noise.

These kinds of models have been developed by several authors for different tasks (Allan, 1975; Baron, 1969; Gibbon & Rutschmann, 1969; Schneider & Bavelier, 2003; Sternberg & Knoll, 1973; Ulrich, 1987; Yarrow et al., 2011). It is worth noting that the exact processes that lead $\Delta t$ to be a Gaussian random variable (i.e., independent *latency noise* in the two signals) can be incorrect without invalidating this whole approach. Consider that this kind of model also leads to the prediction of a (single) cumulative Gaussian psychometric function for TOJs. However, the cumulative Gaussian function is actually used very widely in psychophysics (whenever a judgement is made that effectively divides a continuous decision variable into two halves). This is because all that is really being assumed for this fit to be sensible is that the internal response that informs the decision (here $\Delta t$, but in other applications contrast, intensity, orientation, or whatever) has *in some way* accumulated Gaussian noise. The central limit theorem[6] of classical probability theory makes this a fairly appealing conjecture for many sensory domains, regardless of the exact processing steps that might precede a sensory judgement.

So far, I have described an observer model with two variants. The first, with three parameters, produces a symmetric psychometric function for SJs. The second uses four parameters and can additionally capture an asymmetry in the data. I will comment on formal methods for model selection in a later section. For now, I want to discuss the meaning of the model parameters, partly just to help make the models more interpretable, and partly in order to clarify what I

5 By "degenerate", I mean that such parameters could trade off perfectly with one another, such that different combinations lead to the exact same model prediction. This creates a problem in model fitting known as *non-identifiability*.

6 If you take a large number of random variables and add them together, the distribution of their sum will be Gaussian (even if the contributing variables are not).

think are some misconceptions that have arisen out of the common decision to fit an arbitrary function (the Gaussian) to SJ data.

To get us going, consider the first column of Figure 13.1. These data come from five blocks (of 195 trials each) of an SJ experiment, performed by a novice observer (using the method of constant stimuli, and evaluating synchrony between an LED flash and a 1000 Hz beep, both 10 ms in duration). In the first block, they were simply told to report simultaneity if that is what they perceived. In the second, third and fourth blocks, they were instead told to try and successfully guess the stimuli that were truly simultaneous, but given a maximum number of attempts (7, 5, and 3 in every set of 13 trials) on which they could make use of the "simultaneous" response option. In the fifth block, the original (standard) simultaneity instruction was repeated.

What effect did these altered instructions have on the psychometric function? When unconstrained, this participant, like many others I have tested, made extensive use of the synchronous response, so that they reported synchrony almost 100% of the time across a range of SOAs. If these data were fitted with a Gaussian, we might be tempted to interpret its standard deviation (or some linear transform of this value, such as the full width half height) as a measure of sensitivity to asynchrony. We might further be tempted to consider this parameter equivalent to the slope of the fitted function (or its inverse, the just noticeable difference) in a different task, like the TOJ.

However, consider what happens when instructions require the participant to be more conservative with their use of the synchronous response (Figure 13.1 rows 2–4). If they were simply insensitive across the range of SOAs that they originally reported as synchronous (see row 1) they would still perceive synchrony across this full range. Any constraint on the number of "synchronous" responses that could be made would yield a psychometric function with a flat plateau across this range, but with a ceiling at a proportion lower than 1.0 (because the limited responses would now have to be shared out at random across this region). This is not what occurs. Instead, the synchronous responses increasingly cluster close to true synchrony. The observer model I have outlined in this chapter describes these patterns of data quite naturally, as the result of *changing decision criteria*. Initially, the participant adopts quite loose criteria regarding what is synchronous, but the instructions force them to adaptively tighten them up in response to task requirements.[7] Fits are shown in the

7 It is easy to envisage other situations that might alter a participant's decision strategy, for example the speed with which they are expected to respond, or their beliefs about the proportion of stimuli that are actually simultaneous.

middle column for the three-parameter observer model I outlined earlier. The first two model parameters capture the position of the decision criteria.

If the width of the SJ function is a poor measure of sensitivity, what is a good measure? The answer is the *slope* of the function (on either side), which, under the three-parameter observer model, is determined directly by sensory noise in the $\Delta t$ distribution. This measure remains rather similar down the rows of Figure 13.1, consistent with our expectation that a change in instructions has not somehow radically adjusted the participant's levels of sensory precision. The $\sigma$ parameter of the SJ function, when implemented as I have described (as the difference of two cumulative Gaussians), is exactly equivalent to the $\sigma$ parameter of a sigmoidal psychometric function applied to TOJ data (when considered as the prediction of the same model). Note that fitting an arbitrary Gaussian provides no such way of dissociating the width of the SJ function from the steepness of the SJ function. The practical importance of this limitation is up for debate (it may, for example, be the case that slope and width of the SJ function are typically highly correlated, perhaps because noisy observers tend to choose liberal criteria; c.f. Magnotti, Ma, & Beauchamp, 2013). However, conceptually this distinction seems a sensible one to maintain, being closely related to the distinction between d-prime (d') and c made famous in classical signal detection theory (Green & Swets, 1966).

I have also discussed a four-parameter model, which is fitted in the right-hand column of Figure 13.1. Here, the asymmetry in the function arises from unequal variance in the placement of the two decision criteria (low and high) over the trials of the experiment. These sources of variance would sum with sensory noise in the $\Delta t$ distribution, uniquely at each decision boundary. If we believe that criterion noise may be present, it complicates our interpretation slightly, because we can never fully separate criterion noise from sensory noise in order to determine the magnitude of either one. All we can do is place an upper limit on sensory noise (being the smaller of the $\sigma^2$ estimates associated with the two sides of the SJ function). Note that this conflation of sensory noise and criterion noise applies equally to the interpretation of sigmoidal functions in TOJ and other (non-temporal) tasks: If both kinds of noise are assumed to be Gaussian, only their sum can be estimated from a psychometric function.

Before I conclude this section, it is important to consider a measure that is often derived in timing studies which I have not yet touched upon: The point of subjective simultaneity or PSS. Classically, this measure is estimated from TOJ tasks, being the SOA at which the two order responses are equally likely (implying maximum uncertainty about stimulus order). An analysis couched in terms of the kinds of observer model I have described here illustrates that this SOA represents the combination of a sensory bias (for example, if stimuli from one modality in an AV task must travel a shorter neural pathway than those from the other to reach the decision hub) and a decision bias (in placing a criterion to demarcate the two order responses; Sternberg & Knoll, 1973).

In the SJ task, a similar ambiguity is present (Yarrow et al., 2011). If a sensory bias exists, it seems plausible that the two criteria demarcating synchrony from asynchrony would simply be placed at equal distances from "subjective zero." In that case, we can simply average them to recover a single PSS with a purely sensory interpretation. Note, however, that such an "equal distance" assumption may not be applicable in many situations, particularly if the two stimuli are substantially different from one another. For example, two stimuli may persist to different extents within the brain, and this might influence how decision criteria are set (e.g., "I will call them simultaneous if I experience no gap between them"). A weaker claim would be that the PSS is very likely to lie *somewhere between* the two criteria. My personal preference is generally to report the two criteria, which may in any case provide greater insights about changes across conditions than a single inferred PSS (Yarrow et al., 2013; Yarrow et al., 2011). However, reviewers often request (quite reasonably) that the PSS also be reported for easier comparison with the previous literature. In this case, I would suggest that averaging the criteria is a good compromise.

## 6 Fitting Models to SJ Data

So far, I have alluded to the general notion of fitting observer models to SJ data in order to derive meaningful parameters, and described two variants of what I believe to be a sensible observer model for this purpose. If you are happy that the observer models I have suggested serve your experimental needs, then you may already have most of what you need from this chapter, because the Matlab code to fit these models is available.[8] With an intuitive understanding of the models, you can fit them and interpret their parameters (rather like having only an intuition about the maths underlying ANOVA is more than sufficient to apply this statistical tool). However, it is possible that you may want to fit variants of the models I have outlined (e.g., Yarrow, Minaei, & Arnold, 2015) or other models entirely, or that you are simply inquisitive about how models are fitted to data and wish for a deeper understanding of this process. In this section, I will provide a whistle-stop tour. Realistically, I can only touch on the topic of model fitting. If you would like to know more, I would highly

8 <SimultaneityDiffCumGaussMultistart> and <SimultaneityNoisyCriteriaMultistart>.

recommend that you take a look at Lewandowsky and Farrell (2010), an excellent and readable book on this topic from which much of what I will say has been gleaned. The Matlab functions provided as part of this chapter owe a large debt to the structure that Lewandowsky and Farrell introduce and the examples that they provide in their code snippets. Other very useful sources for what follows are the now classic papers by Wichmann and Hill (2001a, b) on fitting sigmoidal psychometric functions, and Myung's (2003) short tutorial on maximum likelihood estimation.

### 6.1 *Introduction to Model Fitting: Regression*

If you are attempting a chapter like this, I am going to assume that you are somewhat familiar with simple linear regression, so I will start there. You will recall that regression fits a straight-line model with two (or more) parameters to data. The parameters, for simple regression, are the slope ($s$) and intercept ($c$) of the line defined by the function $y = sx + c$. I am going to begin with an even simpler model, where $c$ is fixed to zero. Hence, the model I am working with has just one free parameter ($s$) and the model's predictions are captured in the equation $y = sx$.

Data for a regression-style problem come in the form of a vector of values of x ($\mathbf{x}$) and the associated vector of values of y ($\mathbf{y}$) so that $x_{1 \dots n}$ and $y_{1 \dots n}$ are matched pairs, for example the height and weight of a set of n participants. At this point I have a set of data and a parametric model with predictions defined by an equation. How should I go about finding the value of my parameter $s$, which maximises the fit of the model to the data? For this, I need to consider something called the discrepancy function. In the case of regression, the discrepancy function is based on summed squared error. If I pick a value of $s$, I can use my model equation $y = sx$ to find a prediction (about y) for each value of x in my data set. Then I can look at the actual value of y associated with each value of x in the dataset. Finally, I can subtract each predicted y from the corresponding y in the data, square this difference, and sum these values up to produce the summed squared error for the model (associated specifically with the particular value of $s$ that I have just picked and tested). If I were to repeat this process with many values of $s$ (for example stepping up from $s = 0$ to $s = 2$ in small increments) I could save each error value and plot out a discrepancy function, showing how discrepant the model predictions are from the data for different values of $s$. What I want to find is the minimum of this function, because that will be the value of $s$ that provides the best fit of model to data. These ideas are illustrated in Figure 13.3.

For my reduced regression problem (or indeed for much more realistic and complex linear regression problems) I wouldn't actually bother to generate a

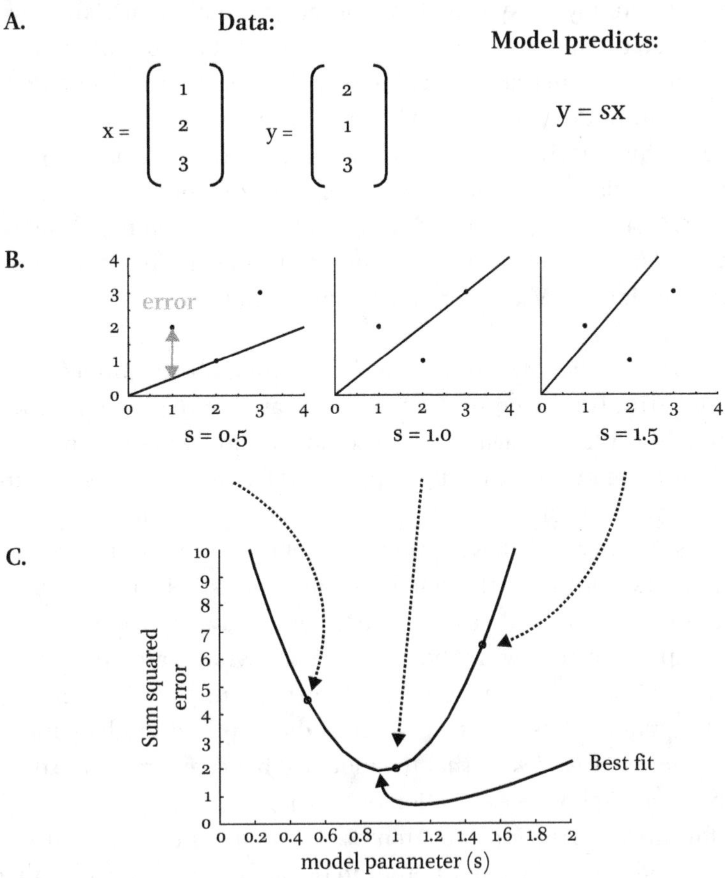


FIGURE 13.3 *Schematic of process for the generation of a discrepancy function. Here, a toy regression problem is illustrated. A. Three data points are shown, along with the equation that captures the model's predictions. B. The model's parameter,* s, *is varied. For each value of* s, *error is determined as the distance between the model prediction and each data point. C. To provide a metric of model fit, errors for each data point are summed and squared. For example, when* s = 0.5, *errors are 1.5, 0, and 1.5 (for the three data points), so squared errors are 2.25, 0 and 2.25, and the sum of squared error (*SSE*) = 4.5. Calculating* SSE *for all values of the model parameter* s *allows us to plot a discrepancy function. The best-fitting model parameter is the value of* s *that minimises this function.*

discrepancy function in this tedious iterative manner. For the toy example, I could just about crunch through the necessary maths, using calculus, in order to reach an analytic solution (by first deriving a formal expression for the discrepancy function, then differentiating it to find its slope, and finally setting this derivative to zero to find the minimum). For a more realistic regression problem, I could thank my lucky stars that competent mathematicians have already derived analytic solutions, and simply plug my data into those to find the best-fitting parameters in a single step. However, in the case of the SJ models that are our main interest here, we will actually end up doing something nearly as crude as the iterative search I have outlined above, because finding an expression for the discrepancy function in terms of the model parameters is not as trivial as just looking it up in a statistics textbook. Before getting to that, however, we need to touch on another important concept that is required when we move to fitting SJs: Maximum likelihood estimation.

### 6.2 *Maximum Likelihood Estimation and the Binomial Data Model*

Summed squared error (or the equivalent mean squared error) is an intuitive measure of model fit. We can clearly visualise how a model fits poorly if its predictions fall at a greater distance from the data. Furthermore, the squaring operation seems a sensible way to punish both positive and negative "prediction errors" (rather than having them cancel each other out when we sum over data points). However, this goodness-of-fit statistic is not generally applicable. Rather, it is a special case of a more generally applicable metric (with summed squared error giving the same answer when data are distributed normally and with equal variance at each level of prediction).[9] In general, to find best-fitting parameters, what we want to do is to find model parameters, which *maximise the likelihood* that the model at hand generated the data (known as maximum likelihood estimation or MLE).

Recall that for linear regression, we attempted to find parameters that minimised the summed squared error. In order to do so, we first had to be able to measure the summed squared error obtained with a particular parameter value. Analogously, in order to find a fit that maximises likelihood, we first need to able to measure the likelihood that a model generated our data given particular parameter values. With regression, we broke this process down by measuring error at each data point (and then squaring and summing them). With MLE, we can also begin at the level of a single data point.

In fact, we will begin by considering a single data point and a model with a single parameter. In doing so, we are (almost inadvertently) introducing an important concept in MLE fitting – the *data model*. The data model is our best guess about the statistical process that makes our data noisy. In the case of regression, the data model is Gaussian. We assume that our measurements are being corrupted by Gaussian noise. However, for synchrony judgements this would be the wrong data model. In an SJ experiment, the observer can

9 Which you will probably recognise as one of the assumptions for linear regression.

only select one of two options on any given trial. The observer model, which I described earlier generates predictions about the probability with which they will say "simultaneous" at each SOA. Hence, at each SOA, our experiment can be considered a *Bernoulli process* (like repeatedly flipping a coin, with a particular probability of coming up heads). When you sum the number of times one or other outcome is obtained from a Bernoulli process across a set number of trials, you get a *binomial distribution*. Hence, for an SJ experiment, at each SOA we expect our data to follow a binomial distribution, with a probability parameter that can be predicted by our observer model. For binomial data (denoted *X* here) the probability of getting exactly *k* successes (e.g. heads, or "synchronous" decisions) in *n* trials with a probability of success of *p* is:

$$p(X=k)=\binom{n}{k}p^{k}(1-p)^{n-k} \tag{3}$$ [10]

where:

$$\binom{n}{k}=\frac{n!}{k!(n-k)!} \tag{4}$$

Now we are ready to appreciate what it means to measure the likelihood that model *a* with parameter *p* generated data point X. Make model *a* a coin toss with parameter *p* = 0.5 (a fair coin). If data point X showed 7 heads out of n = 10, we can make a precise quantitative statement about how likely it is that this model generated those data. We do this by plugging the numbers into the formula for the binomial distribution above. The answer, as it happens, is 0.1172. If we adjusted the probability parameter *p* of our binomial distribution to 0.1, you will probably guess that the model with this parameter will not do such a good job of predicting our data, and indeed the calculation returns a value of only 0.00008. What this number is telling us is that given the data we observed, it is rather unlikely that it was generated from a binomial distribution with a probability parameter of 0.1. On the other hand, with a probability parameter of 0.7 (i.e. where the prediction looks very much like the data) we obtain a likelihood of 0.2668, because this combination of model and parameter value is much more likely.

What I have just described is exactly what we need to do at each data point (corresponding to each tested SOA) when we conduct an MLE fit to SJ data.

10 <BinomialLikelihood 9>.

We need to take the probability predicted by our model, and use it, along with the number of synchronous responses and the number of trials at that SOA, to obtain the likelihood of obtaining those data given binomially distributed data with the predicted probability. However, we still need to scale this calculation up in two ways. Firstly, we need to make this assessment for all data points, as I outline next, in order to generate the likelihood that the model generated the *complete* data set. Secondly, we need to perform this whole evaluation repeatedly, for the sets of probability values predicted by our observer model as we change that model's parameters. In this way, we can create a likelihood function that can be used as a discrepancy function. I outline that process in the next section.

How do we move from the likelihood that this predicted probability yielded this many synchronous judgements out of this many trials (i.e., a single prediction and a single data point) to the likelihood that a full set of predicted probabilities (one per SOA) gave rise to a full set of data points? We need to perform the calculation at each data point and then multiply the obtained probabilities together (because the probability of several independent events all occurring is simply their product). However, there are practical reasons for doing this in a slightly different way, not least the fact that when you multiply lots of probabilities together you soon end up with a very small number indeed, which can be tough for computers to represent. You may or may not recall that the logarithm of a product of one or more numbers equals the sum of the logarithm of each. Hence it is standard practice to calculate *log* probabilities, and *sum* them across data points.[11] We could then convert this back to a probability for the overall prediction, but given that we are looking for the maximum likelihood value, and log likelihood increases monotonically with increasing likelihood, it's more typical to simply carry on working with the log-likelihood values when we search for a best-fitting set of model parameters.[12]

---

11 E.g., <SimultaneityNoisyCriteriaWrapperForFMin 103–113>.

12 Actually, working with log likelihoods derived from the log of the equation for a binomial distribution imposes an unnecessary computational burden, because one of the terms ("N choose K") depends only on the data, not on the model's predictions, so will never vary as model parameters are changed. Hence it is not going to be relevant to finding the best-fitting parameters. In practice, we therefore tend to drop this term to speed things up. This is known as a "kernel" log-likelihood calculation. Using a kernel won't matter at all if you limit yourself to making comparisons between models fitted to the same data using the same kernel, but it will matter if you want to interpret the absolute value of log likelihood (or likelihood) for the best-fitting parameters. Fortunately, we don't generally need to do that. See e.g., <SimultaneityNoisyCriteriaWrapperForFmin 121–124>.

Before I move on to this process, I want to briefly touch upon one important feature of likelihood as a metric of goodness of fit: It is very sensitive to deviations from extreme predictions. What I mean by this is that if a model predicts probabilities of virtually zero or virtually 1 (as observer models often do) the likelihood of observing even a single trial at odds with this prediction is vanishingly small. What this means in practice is that a single lapse by an observer (say, pressing the wrong button by mistake) will have a disproportionately large effect on the resulting model fit (Wichmann & Hill, 2001a). For this reason, it is often worth considering incorporating a "lapse rate" parameter into our observer models. However, extra parameters are generally undesirable for various reasons, so a compromise position is to simply fix a small but reasonable lapse rate, which then adjusts model predictions at extreme SOAs. The code that accompanies this chapter incorporates a fixed lapse rate of 1%.[13] Essentially, model predictions are tweaked slightly to range from a very small probability of saying simultaneous to a very high probability of doing so without ever getting as low as 0 or as high as 1.

## 6.3 *Finding Best-Fitting Parameters*

So far, I have tried to explain how we determine the (log) likelihood that a set of model predictions (i.e. predicted probabilities at each SOA) generated a set of corresponding data. However, the probability of saying "synchronous" that an observer model predicts at each SOA depends on the parameters fed into the model. What we want is the set of parameters that generates the set of predicted probabilities that maximise the likelihood that the model generated the data. One very labour-intensive way to go about finding them would be to iteratively modify each parameter at all levels of the other parameters, determine log likelihood, and repeat to sample the entire parameter space. This approach, known as a grid search, is very similar to the one I outlined in my toy regression example. That model had just one parameter, and hence generated a discrepancy function that could be visualised in two dimensions (Figure 13.3). If my observer model had just one parameter, I could do something very similar and generate a log-likelihood function amenable to a 2D plot. The main difference would be that I would be looking for the maximum, rather than the minimum, of this function.

Even with a single parameter, this approach is slow, particularly if you want a high-resolution search and have little idea about the range within which your best-fitting parameter lies. However, with two parameters, it is necessary to iterate through all reasonable values of one parameter *at each* reasonable value

13 E.g., <SimultaneityNoisyCriteriaWrapperForFmin 93–94>.

of the other (i.e., all combinations of two parameters). I then end up with a discrepancy *surface* that must be visualised in 3D. With more parameters, my discrepancy function becomes very hard to visualise and, more importantly, the number of points that must be searched in a grid search grows exponentially. Hence a grid search is not very practical for the observer models I discuss here, with three and four parameters. Fortunately, many algorithms exist to help with searches of this kind. The most famous is the Nelder-Mead simplex search (Nelder & Mead, 1965).

The intuition for this approach is simple. Set a starting point (i.e. a reasonable guess for each parameter) and establish error of fit. Then, test a few more points in the vicinity to find their errors. Apply some geometric rules to try and figure out the slope of the discrepancy surface in this small region. Then, crawl down this slope, testing new parameter combinations as you go and re-applying the rules, in order to move towards a minimum. For a log-likelihood search, the approach needs to be tweaked slightly before the simplex algorithm will work, because we are seeking a maximum, not a minimum. However, simply inverting the obtained log-likelihoods is sufficient.[14] The set of rules embedded in the algorithm will then guide it to a best-fitting solution without having to sample the discrepancy surface exhaustively.[15]

You don't really need to know any more than that to perform a simplex search, as functions to implement it are readily available. However, you might want to find out a bit more in order to appropriately set the various options that these functions let you vary. One important fact to bear in mind is that a simplex search may struggle when the discrepancy surface is not smooth and well behaved. In particular, if there are *local minima* that vary from the *global minimum*, the simplex is likely to home in on the local minimum in the region where it started to search and get stuck there. A good sanity check for any search procedure is to use the model that is to be fitted to generate some data (based on a known combination of parameters) and then see if the fitting procedure recovers the model parameters successfully when initiated from various different start positions. In my experience, simplex searches often fail in this regard. In an attempt to overcome this issue, the code associated with this chapter actually combines grid-search and simplex-search approaches, by initiating a separate simplex search from each parameter combination defined by a grid search.[16]

14 E.g., <SimultaneityNoisyCriteriaWrapperForFmin 126 & 47>.

15 E.g., <SimultaneityNoisyCriteriaWrapperForFmin 44>.

16 E.g., <SimultaneityNoisyCriteriaMultistart 138–182>.

## 6.4 *Confidence Intervals around Model Parameters*

For comparisons involving groups of participants, recovering a set of best-fitting parameters for each participant in each condition is usually sufficient, and the standard error can then be computed across the sample (usually as an implicit part of common approaches to statistical inference like ANOVA). However, sometimes we wish to have an idea about how well parameters are being estimated *for each participant*. I will outline a couple of popular approaches.

Firstly, we can make use of a result from asymptotic statistical theory (i.e., theory that is true when our sample size in infinite), which (basically) tells us that there is a close relationship between the curvature of the log-likelihood surface at the point where we obtained the best-fitting parameters and the standard errors of those parameters. The intuition is that if changing a parameter by just a little bit makes the fit a lot worse, the parameter is tightly constrained and probably well estimated. In this situation, the point of best fit effectively sits in a steep-sided hole on the (negative) log-likelihood discrepancy surface (hence curvature is high). Formally, the curvature of the log-likelihood surface is captured by something called the Hessian matrix (a matrix of second-order partial derivatives). We can't work that out exactly without (at least) an analytic expression for the discrepancy function, and we don't have one for the kinds of observer model I have described here. However, we can approximate the Hessian using numerical methods (by measuring changes in log-likelihood in the best-fitting region via a series of small steps). Having done so, the inverse of the Hessian provides a covariance matrix for the model's best-fitting parameters, and the main diagonal values can, thus, be square rooted to estimate standard errors (which can then be straightforwardly converted to confidence intervals).[17]

It's questionable whether results from asymptotic statistical theory are actually going to hold for psychophysics experiments with fairly low numbers of data points and trials (Wichmann & Hill, 2001a). Hence a popular alternative approach is to estimate confidence intervals via bootstrapping. Bootstrapping theory, described in detail in Efron and Tibshirani (1994), tells us (roughly) that if we resample from a data set repeatedly *with replacement* to generate a new "bootstrap" data set of the same size, calculate a statistic of interest, and then repeat many times, the resulting distribution will allow us to make inferences about the standard error of that statistic. In the case of SJ models, a reasonable approach is to resample the data (known as non-parametric bootstrapping), fit the model to each resample, and record the parameters on each of around

17 E.g., <SimultaneityNoisyCriteriaWrapperForFmin 53–63> and <SimultaneityNoisyCriteriaMultistart 219–224>.

1999 such iterations to form parameter distributions. If these distributions are symmetric, we can pretty much just read values straight out of them to form confidence intervals (e.g., the 50th and 1950th values out of 1999 will give us a roughly 95% confidence interval). If they are not, we must do something more complicated, with the best choice being the bias-corrected and accelerated (BCa) approach. Because of the large number of fits that are required, bootstrapping is fairly slow. If the experiment contains many trials, the BCa method makes it even slower (because it incorporates additional "jackknife" resampling, implying one further fitting iteration for almost every trial).[18]

The code accompanying this chapter offers options to generate confidence intervals on fitted parameters. Confidence intervals sometimes imply statistical inference, as for example when they fail to overlap some value and thus imply that our statistic differs significantly from that value. However, in SJ experiments we are more likely to want to ask a question such as whether a particular parameter differs between two conditions for a single observer. To answer this kind of question, you will need to modify or develop the code. If we take the example of whether parameters vary across conditions, my recommendation would be to adopt a permutation test approach.

To do so, take the trials from both conditions and think of each trial as a card in a deck of cards. Making sure you keep each trial intact (i.e., without breaking the link between SOAs and responses) shuffle the trials and then deal them at random into two new piles, each representing a pseudo-condition. If your original conditions contained different numbers of trials, make sure the two pseudo-conditions match the size of the original conditions. For each pseudo-condition, perform a model fit. Now calculate the difference between model parameters in the two pseudo-conditions. This is the value you want to retain. Now repeat this whole process many times. What you are forming is a null distribution of the expected difference between model parameters that would occur just by chance. You can then compare the difference you actually obtained against this null distribution to generate a p value for your difference of interest.

## 7 Variants of SJ Observer Models

In this chapter, I have presented two variants of a latency-based observer model applied to the SJ task. Both assume that a single SOA will generate an internal response ($\Delta t$) that is a Gaussian random variable. Both assume a simple

18 E.g., <SimultaneityNoisyCriteriaMultistart 225–386>. Note that Matlab has inbuilt functions, which could have done most of this *if* you have the statistics toolbox extensions.

decision rule ("say synchronous if $\Delta t > C_{low}$ and $< C_{High}$", where C indicates decision criteria). The more complex variant also allows the two criteria to vary from trial to trial as Gaussian random variables. There are many variants of this kind of model that could be envisaged, some of which are considered in Sternberg and Knoll (1973) and Ulrich (1987).

This kind of model is generally presented as a consequence of two sensory signals travelling along independent pathways to a decision centre, with sensory noise reflecting variations in their latencies from trial to trial. However, the same predictions emerge if we assume the sensory noise accrues via some other process than latency variations (e.g., spike rate stochasticity) as long as the end result is a Gaussian $\Delta t$ distribution. This is an attractive feature, because in fitting our data, we may not want to commit to anything more than the fairly defensible position that noisy representations are quite likely to be Gaussian (a hallmark of classical signal detection theory).

If we stick closer to the process model in which sensory noise *is* latency noise, it is reasonable to argue that the Gaussian assumption must be a simplification. Latencies cannot be negative, so modelling them as Gaussian cannot be completely correct (although if the variance of the latency distribution is fairly small relative to the length of the neural pathway, the density below zero would be negligible). An alternative observer model based on the same basic principles has been developed by García-Pérez and Alcalá-Quintana (2012a, b, see also Chapter 12, this volume) who use exponential latency noise in place of Gaussian noise for each signal. The result is a four-parameter model, which can generate an asymmetric psychometric function for SJs and thus capture the same sorts of features as the four-parameter model presented here, but via the mechanism of an asymmetric $\Delta t$ distribution (rather than criterion noise). The authors are happy to provide fitting code for their model, which can also be scaled up to include extra parameters that deal with keying errors. They have a chapter in this volume.

Their model yields two noise parameters (one for each signal) and two further parameters, which seem distinct from the two criteria described here, but are in fact mathematically equivalent. García-Pérez and Alcalá-Quintana (2012a, b) describe $\tau$, a processing delay parameter, basically what most researchers think of as the PSS, and $\delta$, a resolution parameter, which defines the range of values judged synchronous. The two criteria I have described here map directly onto their parameters, being $\tau-\delta$ and $\tau+\delta$ (recall that I noted how a PSS could be recovered by averaging the positions of the two criteria). The differences in terminology seem to be driven by different theoretical positions. Whereas I view the decision criteria as being malleable components of the decision process, García-Pérez and Alcalá-Quintana (2012a, b) seem at least partly committed to a form of "low-threshold" or "triggered-moment" model where SOAs below some threshold cannot be recovered, so that an observer can only guess about order.[19] However, either kind of model can easily be fitted and interpreted from either theoretical perspective.

## 8 Choosing between Observer Models and Rejecting Participants

Two further reasonable questions one might ask are: 1) could my observer model have generated these data? and 2) does another observer model describe the data better? Model comparison is a large and complex topic, so once again, what I have to say here should be treated as a brief introduction rather than a comprehensive summary.

Let's begin by considering a metric I have not yet mentioned: *Deviance.* Deviance (sometimes called $G^2$) is a measure based on log likelihood, but which looks rather more like summed squared error, in that it is zero for a perfectly fitting model and large/positive for a poorly fitting model. Formally, deviance is two times the difference in log likelihood between the *saturated* model and the model with our current set of parameters. A saturated model is one that exactly predicts the data (which can always be accomplished by a model that has one parameter per data point). Hence it represents the situation with the maximum possible log-likelihood when predicting this particular set of data. Deviance is closely related to a simpler calculation ($-2 \times$ log likelihood) that forms the basis of a couple of well-known metrics for model comparison (the Akaike information criterion, AIC, and the Bayesian information criterion, BIC) and indeed is occasionally defined this way. That's because we are often only really interested in differences (in Deviance, or AIC, or BIC) between models, and the log-likelihood of the saturated model gets subtracted out in a comparison between two models (because it has contributed to the deviance in the same way for both) so calculating it is not necessary.

However, if you want to say something about the goodness of fit of a model *without* relating it to any other model, based on asymptotic statistical theory, you do need to calculate deviance properly. Asymptotically, it turns out that the deviance of a model fitted to data *when that model actually generated those data* follows a chi-square ($\chi^2$) distribution, with degrees of freedom equal to the number of data points minus the number of model parameters (note: for

19 García-Pérez and Alcalá-Quintana's commitment to this account is a little unclear, because they often let $\delta$ vary across experimental conditions, suggesting flexibility more akin to a criterion-based account. It may be that they believe a low-threshold exists, but that synchrony is often additionally reported beyond this hard limit.

data points, think of the number of SOAs tested, not the number of trials!) Hence, if we want to know if our model might have generated our data, we could check the best-fitting deviance against such a distribution to see how improbable this is. Unfortunately, it seems that this asymptotic result may not always be accurate for data sets of a size typical in psychophysics experiments (Wichmann & Hill, 2001a).

For this reason, Wichmann and Hill (2001a) suggest using Monte-Carlo simulation to assess whether a model is plausible. The idea is as follows. First, find the best-fitting set of model parameters. Second, create a set of data based on a simulation of the experiment in which that model generates the data. Third, find a fit to that data, and record the deviance.[20] Fourth, repeat steps two and three many times to generate a distribution of deviances that you would expect *when that model actually generated those sets of data.* Finally, look to see where the deviance of your actual fit sits on this distribution in order to assess if the model is likely to have generated the data. This approach is not implemented in the code accompanying this chapter, but should be feasible for you to implement yourself if you are interested in assessing whether your data are under or over- dispersed relative to what would be expected. However, although certainly informative, I find it a rather high bar to set if you are, for example, deciding whether to use a model or to include a participant. After all, even the most ardent defender of a particular observer model would be unlikely to argue that it really represents a complete characterisation of the psychological processes that are being modelled. I think that a model fit can be informative even if the model is a simplification of absolutely everything that observers do in experiments. To paraphrase George Box: All models are wrong, but that doesn't mean that they are not useful (Box, 1979).

With this in mind, my preference is to ask a slightly different question: Does this observer model seem to fit the data better than some other simpler account? This question is well aligned with what we generally do during statistical inference. For example, a simple (i.e., two parameter) regression is generally considered significant if it explains the data significantly better than an even simpler one-parameter model (i.e., just the mean).

What can we say about the deviance statistic as model complexity increases? Well, in general a complex model produces a better fit than a simple model whether it is correct or not, because more free parameters mean a greater ability to describe patterns that are actually just random noise

20 It doesn't actually have to be deviance. Log likelihood, or -2 × log likelihood would be fine too.

(at least for nested models).[21] Hence, simply finding a decrease in deviance for a more complex model is not enough to show that it is better. We need to instead show that the decrease in deviance is greater than that expected by chance. Although the asymptotic result I outlined above for expectations about *absolute* deviance may be unreliable with psychophysical data sets, another rather similar result may be more robust even when N is low. The *change* in deviance from a simpler to a more complex model also follows a $\chi^2$ distribution, but with degrees of freedom equal to the difference in free parameters between the models, as long as the models are nested.

The two observer models for SJs that I have discussed in this chapter are nested, so it's possible to make a decision about whether to use the more complicated one by comparing the deviances they each return. I have previously found the four parameter-variant to be justified for AV data with LED flashes and brief tones (Yarrow et al., 2011).

The code accompanying this chapter also includes an option, when fitting either of these models, to additionally fit a simpler model as a method of deciding whether to retain a participant as part of a group-level analysis. The logic here is that if a participant is simply guessing rather than taking the experiment seriously, they will be equally likely to say "synchronous" at any SOA, which can be captured by a straight horizontal line (effectively a model with just one parameter: their overall tendency to use one of the two keys). However, in SJ experiments we may also need to exclude participants who showed some ability to discriminate, but on only one side of the SJ function, implying that we failed to sample extreme enough SOAs to capture both of their transitions from synchrony to asynchrony. Although such an observer may have been concentrating well and following instructions, the model will return very poorly constrained and extreme parameter estimates. Hence, to look for this pattern, we should fit an intermediate model, a cumulative Gaussian, which can capture usable performance on one or other side of the SJ function, but not both. Only if the full SJ model provides a better fit relative to this partial performance model should the participant be retained (c.f. Yarrow et al., 2013).

I have now discussed what I believe is a reasonable approach to model comparison for nested models. I will finish this section by very briefly mentioning some possible approaches when models are not nested. Firstly, models can be compared using either AIC or BIC. Both of these statistics are equal

21 Two models are nested if (basically) the more complex model can generate all the same sets of predictions that the simpler model can generate, plus a bit more. For example, stepwise regression compares nested models. Strictly, this approach requires that models are nested *and* that one of them is correct.

to −2 times the log likelihood of the best-fitting model, but with a penalty applied to the model with more free parameters. For AIC the penalty is simply 2 per parameter, whereas for BIC it is (generally) slightly greater per parameter and depends on the number of data points in the fit. BIC is actually an approximation to the Bayes factor, an (arguably) more sophisticated form of model comparison in which model performance is considered across all parameter combinations, not just at the best-fitting values. A second tactic would be to develop a Monte-Carlo simulation approach similar to that outlined above in order to produce a distribution of expected deviance improvements if the more complex model is fitted to data generated by the simpler model. As mentioned earlier, model comparison is a substantial and complex field, and there are several other approaches that could be considered beyond those touched on here.

## 9 Alternative Approaches to Interpreting SJ Data

Fitting a model is a nice way to summarise a set of SJ data with a few meaningful parameters. However, those parameters are only likely to tell you something useful if the model is (at some level) correct. The fact is, there is no consensus about whether any given observer model is correct, or about how literally parameter values should be interpreted. These considerations might lead us to consider doing away with any kind of parametric fit. For example, we could analyse the data without a pre-processing step, so that proportion judged simultaneous at each SOA is the dependent variable, or we could attempt a non-parametric fit to derive summary measures.

The former approach is used occasionally, sometimes as a supplement to a parametric fit. For example, Zampini et al. (2005) simply applied an ANOVA to proportion simultaneous data, incorporating their set of SOAs as a second factor (the first factor being the two conditions they were comparing). Interactions and main effects can then be interpreted to explain differences between conditions, although it may be somewhat challenging to explain what is going on in a succinct manner, particularly when many conditions are tested. Another concern is the application of ANOVA in a situation where data are clearly non-normal. Proportion/percentage data are likely to be skewed (and less variable) at the extremes (i.e., where most participants report synchrony not at all or all of the time). It might be possible to address this concern using a more complex variant of the generalized linear (mixed) model with an appropriate link function (in place of an ANOVA), an approach that has been applied successfully for data yielding sigmoidal psychometric functions (Moscatelli, Mezzetti, & Lacquaniti, 2012).

If summary measures akin to thresholds and PSSs are desirable, an alternative to a parametric fit would be to simply draw straight lines (or use some form of spline interpolation) between data points and make some informal estimates on that basis (e.g., a window where the proportion judged simultaneous falls above 0.75, or the point at which the highest proportion of simultaneity judgements is reached). However, noisy data tend to make this problematic, as the psychometric function may then appear non-monotonic on one or both sides. More sophisticated non-parametric approaches have been developed, but mainly for the more common situation of a sigmoidal psychometric function (e.g., Miller & Ulrich, 2001; Zchaluk & Foster, 2009). In some cases, it is possible to adapt these procedures to the SJ task (Lee & Noppeney, 2011).

## 10 Ternary Data

Before the SJ reached its current level of popularity, several authors had considered expanding the TOJ to a ternary task in which the two order responses where supplemented with a "simultaneous" response option to indicate uncertainty about order. In fact, latency-based observer models for this situation are formally identical to those I have discussed for the SJ. In early analyses, the ternary task was typically considered to permit two binary divisions of the data, each yielding a sigmoidal psychometric function. In the first such division, the psychometric function was constructed by plotting the proportion of times that observers report either "simultaneous" or "A then B" (i.e., the proportion of times they said anything other than "B then A"). In the second division, it was constructed by plotting the proportion of times that observers report only "A then B." These two psychometric functions are displaced from one another along the SOA axis. Their difference represents the occasions when the observer responded synchronous. Note that this provides an intuitive link regarding why the SJ function can be described as the difference of two cumulative Gaussians.

In fact, we can fit observer models directly to these data without rearranging them into a binary format. The observer models make predictions directly about a ternary division, which equates to predicting two out of three probability values at each SOA (with the third being defined by the fact that probabilities sum to 1.0). The code accompanying this chapter includes options to perform such a fit based on the two models (i.e., the three and four parameter variants) that I described in Section 4. From a practical perspective, there is only one conceptually challenging point of difference. It is the data model (discussed for a binary fit in Section 6.2). Because there are three possible choices, the appropriate data model (applied at each SOA) is no longer the binomial distribution, but rather the multinomial distribution, which can provide an exact likelihood of obtaining any particular combination of probabilities that divide N choices into three bins when the actual probabilities of selecting each bin are known (or rather, for fitting purposes, predicted).[22]

## 11 Dual-Presentation SJ Data

Several authors have investigated the use of a dual-presentation SJ task in which two bimodal stimuli are presented (one after another) and compared, for example by reporting which one was (most) synchronous (Allan & Kristofferson, 1974; Powers, Hillock, & Wallace, 2009; Roseboom, Nishida, Fujisaki, & Arnold, 2011). This is a form of what would, in classical signal detection theory, be described as a two-alternative forced choice (specifically the two-interval forced choice variant). However, that designation is ambiguous (about whether there are two presentations or two response categories) and has been applied to cases where either or both of the possible qualifying conditions are met, which is probably why the dual-presentation SJ task has ended up being given a variety of names (e.g., temporal 2AFC; forced-choice successiveness discrimination; 2IFC SJ, where the classic SJ is referred to as 2AFC SJ in the same paper). I will label it the *2xSJ*.

The simplest form of the 2xSJ would have a synchronous standard on every trial along with a non-synchronous test pair. Based on the kind of observer models discussed in this chapter, the resulting psychometric function (plotting the probability of judging the standard more synchronous than the test against the test's SOA) is U-shaped and centred over the PSS. This approach represents a reasonable way to derive estimates of inverse precision (i.e., $\sigma_{\Delta t}$) but a fairly poor way to estimate the PSS, because having a synchronous standard on every trial provides feedback about objective synchrony. A simple solution is to also include a range of standards as well as a range of tests, in a roving standard design.

The observer model can be fitted to data even when both standard and test are non-zero, as described in detail by Yarrow et al. (2016; see also García-Pérez & Peli, 2014). To present all of the data, it is necessary to plot a function for each standard SOA (using several standard plots, or a single 3D plot), which is somewhat cumbersome, but not a major obstacle to using the task. A simple

22 <MultinomialLikelihood 9>.

observer model with three parameters captures PSS, sensory noise and an interval bias (i.e., a tendency to select one interval in preference to the other under uncertainty).

The 2xSJ task provides estimates that correlate fairly well with equivalent parameters estimated using TOJs, SJs, and ternary tasks. However, each trial takes longer than in those single-presentation tasks, which makes experiments more onerous. There are a few reasons why the roving-standard 2xSJ is still worth considering. Firstly, it asks about synchrony explicitly (unlike the TOJ) and by requiring relative judgements it reveals a point of maximal synchrony perception (whereas the SJ and ternary tasks often reveal a range of SOA values that are classified as synchronous). Secondly, it can be added in to a single-presentation task (as a follow-up question every two trials), which somewhat mitigates the burden of additional experimental time. Finally, a case can be made that it will be more resistant to some forms of decision-level bias (Morgan, Grant, Melmoth, & Solomon, 2015; Morgan, Melmoth, & Solomon, 2013). As with the other tasks I have described, code to fit data from the 2xSJ accompanies this chapter.[23] For further information, read the comments there and consult Yarrow et al. (2016).

## 12 Conclusion

In this chapter, I have outlined the benefits of fitting formal observer models to judgements about simultaneity, and described how this can be achieved using Matlab code (see book's GitHub repository). In doing so, I have presented one particular observer model in some detail, and highlighted the fundamentally subjective nature of the SJ task, which requires us to think carefully about how both the strategic decisions and perceptual sensitivity of a participant can affect their psychometric function. I have gone on to supply a brief overview of appropriate models for several closely related timing tasks. I hope I have also provided enough of a tutorial regarding bespoke model fitting and evaluation to allow the interested reader to go forward and explore their own models of perceived simultaneity. Modelling may seem intimidating, but in fact, a good understanding of just a few basic concepts (which is best gained through practical exploration) will take you a long way, providing tools to engage more fully with the timing literature. This is an endeavour I would very much encourage!

23 <TwoAFCSimultaneity_3PEq_Multistart_rawdata>.

## References

Allan, L.G. (1975). The relationship between judgments of successiveness and judgments of order. *Perception & Psychophysics, 18*(1), 29–36.

Allan, L.G., & A.B. Kristofferson (1974). Successiveness discrimination: Two models. *Perception & Psychophysics, 15*(1), 37–46.

Arnold, D.H., K. Petrie, R. Gallagher, & K. Yarrow (2015). An object-centered aftereffect of a latent material property: A squishiness visual aftereffect, not causality adaptation. *Journal of Vision, 15*(9), 4.

Baron, J. (1969). Temporal ROC curves and the psychological moment. *Psychological Science, 15*, 299–300.

Box, G.E.P. (1979). Robustness in the strategy of scientific model building. In Launer, R.L. & G.N. Wilkinson (Eds.), *Robustness in Statistics* (pp. 201–236). New York: Academic Press.

Efron, B., & R.J. Tibshirani (1994). *An introduction to the bootstrap.* CRC press.

García-Pérez, M.A. (2014). Adaptive psychophysical methods for nonmonotonic psychometric functions. *Attention, Perception, & Psychophysics, 76*(2), 621–641.

García-Pérez, M.A., & R. Alcalá-Quintana (2012a). On the discrepant results in synchrony judgment and temporal-order judgment tasks: A quantitative model. *Psychonomic Bulletin & Review, 19*(5), 820–846.

García-Pérez, M.A., & R. Alcalá-Quintana (2012b). Response errors explain the failure of independent-channels models of perception of temporal order. *Frontiers in Psychology, 3*, 94.

García-Pérez, M.A., & E. Peli (2014). The bisection point across variants of the task. *Attention, Perception, & Psychophysics, 76*(6), 1671–1697.

Gibbon, J., & R. Rutschmann (1969). Temporal order judgement and reaction time. *Science, 165*(891), 413–415.

Green, D.M., & J.A. Swets (1966). *Signal detection theory and psychophysics.* New York: Wiley.

Lee, H., & U. Noppeney (2011). Long-term music training tunes how the brain temporally binds signals from multiple senses. *Proceedings of the National Academy of Sciences of the United States of America, 108*(51), E1441–E1450.

Lewandowsky, S., & S. Farrell (2010). *Computational modeling in cognition: Principles and practice.* Sage.

Love, S.A., K. Petrini, A. Cheng, & F.E. Pollick (2013). A psychophysical investigation of differences between synchrony and temporal order judgments. *PloS One, 8*(1), e54798.

Macmillan, N.A., & C.D. Creelman (2005). *Detection theory: A user's guide* (2nd ed.). New York: Lawrence Erlbaum Associates.

Magnotti, J.F., W.J. Ma, & M.S. Beauchamp (2013). Causal inference of asynchronous audiovisual speech. *Frontiers in Psychology, 4,* 798.

Miller, J., & R. Ulrich (2001). On the analysis of psychometric functions: The spearman-kärber method. *Perception & Psychophysics, 63*(8), 1399–1420.

Miyazaki, M., S. Yamamoto, S. Uchida, & S. Kitazawa (2006). Bayesian calibration of simultaneity in tactile temporal order judgment. *Nature Neuroscience, 9*(7), 875–877.

Morgan, M., S. Grant, D. Melmoth, & J.A. Solomon (2015). Tilted frames of reference have similar effects on the perception of gravitational vertical and the planning of vertical saccadic eye movements. *Experimental Brain Research, 233*(7), 2115–2125.

Morgan, M., D. Melmoth, & J. Solomon (2013). Linking hypotheses underlying class A and class B methods. *Visual Neuroscience, 30*(5–6), 197–206.

Moscatelli, A., M. Mezzetti, & F. Lacquaniti (2012). Modeling psychophysical data at the population-level: The generalized linear mixed model. *Journal of Vision, 12*(11), 26.

Myung, I.J. (2003). Tutorial on maximum likelihood estimation. *Journal of Mathematical Psychology, 47*(1), 90–100.

Nelder, J.A., & R. Mead (1965). A simplex method for function minimization. *The Computer Journal, 7*(4), 308–313.

Powers, A.R., A.R. Hillock, & M.T. Wallace (2009). Perceptual training narrows the temporal window of multisensory binding. *The Journal of Neuroscience, 29*(39), 12265–12274.

Roseboom, W., S. Nishida, W. Fujisaki, & D.H. Arnold (2011). Audio-visual speech timing sensitivity is enhanced in cluttered conditions. *PloS One, 6*(4), e18309.

Rosenberger, W.F., & S.E. Grill (1997). A sequential design for psychophysical experiments: An application to estimating timing of sensory events. *Statistics in Medicine, 16*(19), 2245–2260.

Schneider, K.A., & D. Bavelier (2003). Components of visual prior entry. *Cognitive Psychology, 47*(4), 333–366.

Sternberg, S., & R.L. Knoll (1973). The perception of temporal order: Fundamental issues and a general model. In Kornblum, S. (Ed.), *Attention and performance IV* (pp. 629–686). London: Academic Press.

Stone, J.V., N.M. Hunkin, J. Porrill, R. Wood, V. Keeler, M. Beanland, M. Port, & N.R. Porter (2002). When is now? Perception of simultaneity. *Proceedings of the Royal Society of London Series B: Biological Sciences, 268,* 31–38.

Ulrich, R. (1987). Threshold models of temporal-order judgments evaluated by a ternary response task. *Perception & Psychophysics, 42*(3), 224–239.

Vroomen, J., & M. Keetels (2010). Perception of intersensory synchrony: A tutorial review. *Attention, Perception & Psychophysics 72*(4), 871–884.

Wichmann, F.A., & N.J. Hill (2001a). The psychometric function: I. Fitting, sampling, and goodness of fit. *Perception & Psychophysics, 63*(8), 1293–1313.

Wichmann, F.A., & N.J. Hill (2001b). The psychometric function: II. bootstrap-based confidence intervals and sampling. *Perception, & Psychophysics, 63*(8), 1314–1329.

Yarrow, K., I. Sverdrup-Stueland, W. Roseboom, & D.H. Arnold (2013). Sensorimotor temporal recalibration within and across limbs. *Journal of Experimental Psychology: Human Perception & Performance, 39*(6), 1678–1689.

Yarrow, K., N. Jahn, S. Durant, & D.H. Arnold (2011). Shifts of criteria or neural timing? The assumptions underlying timing perception studies. *Consciousness and Cognition, 20*, 1518–1531.

Yarrow, K., S.E. Martin, S. Di Costa, J.A. Solomon, & D.H. Arnold (2016). A roving dual-presentation simultaneity-judgment task to estimate the point of subjective simultaneity. *Frontiers in Neuroscience, 7*, 416.

Yarrow, K., S. Minaei, & D.H. Arnold (2015). A model-based comparison of three theories of audiovisual temporal recalibration. *Cognitive Psychology, 83*, 54–76.

Zampini, M., S. Guest, D.I. Shore, & C. Spence (2005). Audio-visual simultaneity judgments. *Perception & Psychophysics, 67*(3), 531–544.

Zchaluk, K., & D.H. Foster (2009). Model-free estimation of the psychometric function. *Attention, Perception, & Psychophysics, 71*(6), 1414–1425.

CHAPTER 14

# Using the Simon Effect in Simultaneity/Asynchrony Discrimination Tasks: Interest, Methods, and Limits

*Anne Giersch, Patrick E. Poncelet, Céline Z. Duval and Laurence Lalanne*

## 1 Introduction

In this chapter, we describe some methods to explore the automatic, unconscious processing of asynchronous events that are judged as being simultaneous. We first describe our conceptual framework, which was guided by clinical inquiries on patients with schizophrenia. We then describe how we applied the Simon effect, as a novel method of analyses. The code provided with the manuscript (see book's GitHub repository) will allow the reader to generate his/her own data. The aim of the chapter is to provide all the necessary information for the reader to train and calculate effects like the Simon effect by including some example data files (see Annex). It must be noted at the onset that these calculations are not meant to replace more standardized measures like the evaluation of the threshold. They are only meant to complement them. Also, it is not the aim of this manuscript to provide a set method. We rather advocate that some flexibility in our methods can sometimes provide additional and useful information on the processes we explore. In the final section of this chapter, we provide an example of how and to which aim other methods can be used when utilizing identical tasks.

The flow of our mental activity in time is unidirectional, from the past to the future. This directionality is an intrinsic property of all cognitive functions, which appears to be a given fact, and which we seldom reflect upon or include as a parameter of our experimental approaches. Its importance is evident in clinical populations for whom the continuous flow of thought is altered. For example, clinical experience suggests that patients with schizophrenia suffer from a disruption of the feeling of time continuity. This disruption has been described by many psychiatrists based on patients' reports (i.e., they are based on what the patients say within interviews with the psychiatrist; Andreasen, 1999; Fuchs, 2007; Minkowski, 1933; Vogeley & Kupke, 2007; Chapman, 1966). The drawback of these reports is that they are selected and interpreted subjectively by the psychiatrist. The impairments, thus, require objective experimentation and measurement. It is with the goal of objectively characterizing such a

DOI 10.1163/9789004280205_015

disruption that we devised new ways to look at temporal discrimination tasks. Although the aim of this chapter is to describe analysis methods and not the abnormalities of patients with schizophrenia, the development of these methods has been driven by their potential to be applied in our research.

Objectifying a disruption of the sense of time continuity requires an understanding of the mechanisms leading to the feeling of continuity to begin with. It is indeed far from clear whether our conscious access to sensory information is as continuous as experienced subjectively. Experimentally it is long known that two events are distinguished in time only if there is a minimal time asynchrony between them (Elliott & Giersch, 2016, for a review). For example, two visual events displayed in distinct spatial locations will have to be delayed by 30 to 50 ms to be distinguished in time (Brecher, 1932; Elliott, Shi, & Kelly, 2006; Pöppel, 1997; Wittmann, 2011). The fact that such an asynchrony is not zero leads to the concept of the temporal window, the interval of time within which the brain is hypothesized to process events as being co-temporal (Elliott et al., 2006; Elliott & Giersch, 2016; van Wassenhove, 2009; Wittmann, 2011). These results confirm that the conscious processing of information is not as continuous as experienced subjectively. As a matter of fact, if information is judged as being co-temporal within temporal windows, it is as if time stops for the period of the temporal window.

Other authors have also proposed that the processing of information is discrete rather than continuous (e.g., Pilz, Zimmermann, Scholz, & Herzog, 2013; Scharnowski et al., 2009; VanRullen, Zoefel, & Ilhan, 2014). In case of Pilz et al. and Scharnowski et al., this conclusion is derived from studies on visual integration rather than timing *per se*. These authors show that sensory stimuli require time to be processed. The consequence of this constraint in the processing time would be that information processing has an intrinsic inertia, which would not mirror external-event timing. Objects and scenes are indeed sensed and identified on the basis of multisensory signals, implying the need to integrate such sensory signals over space and sensory pathways. Even within a given modality like vision, bits of information have to be integrated before identification can occur (Boucart et al., 1994). This means that information is not processed instantaneously rather information is first decomposed and processed by specialized neurons. In the primary visual cortex, specialized pathways are used to process contour and surface information: neurons will respond selectively to orientation, code line-ends, and edges, or they will be activated by information related to color or texture. Whichever model is used to understand how information is bound together to identify forms and objects, time is required, because ambiguities regarding which information belongs to which object is often ambiguous (Grossberg & Pilly, 2008; Pugeault,

Wörgötter, & Krüger, 2010) and because it is also possible to mentally group or re-group objects (Pilz et al., 2013; Van Assche, Gos, & Giersch, 2008). A series of mechanisms are, thus, required to identify an object.

Finally, time is also needed to achieve neuronal synchronization. It is usually proposed that pools of neuron activations are formed to code all properties of objects. Such functional assemblies, however, require some time to be formed, as they rely on spikes synchronization (Varela, 1999). The perception of an object would not be available before such synchronization takes place, and, thus, accessing the experience of the object would need to be delayed. This may participate to inertia in information processing. The time required to experience each event, might, in turn, complicate the conscious temporal ordering of these events. In contrast, at initial stages, processing is parallel and accurate temporal properties of sensory information may be preserved at these initial, unconscious stages.

To check, therefore, whether temporal processing is more accurate at initial than at conscious stages, procedures are required to explore the implicit processing of information in time, in addition to explicit processing (see van Wassenhove, 2009, for additional arguments). 'Explicit' processing is defined here by the fact that, in a given task, participants are explicitly asked to make a conscious decision (e.g., decide whether or not stimuli are simultaneous; Coull & Nobre, 2008). Implicit processing is defined by no such required judgment by the participants. For example, implicit mechanisms play an important part in sensorimotor timing (Repp, 1999; van Wassenhove, 2009) and are not necessarily equated with explicit judgments (Martin, Giersch, Huron, & van Wassenhove, 2013; van Wassenhove, 2009). In other words, implicit processing relies on mechanisms that do not require conscious thought, but that nonetheless can influence what we experience.

## 2 Exploring Implicit Timing Mechanisms with the Simon Effect

There are several techniques that allow one to assess the properties of implicit mechanisms; we will describe only a subset of those below. This section merely describes new ways to explore implicit biases in participants' responses during simultaneity/asynchrony judgments. It should be noted that here, 'biases' refer to the fact that when participants press on a response key, they not only base their choice on a conscious decision.

The method we employed is based on the use of the Simon effect. The Simon effect reflects the participants' tendency to press the response button corresponding to the location where a stimulus is displayed on the screen.

Such a preferential response is generally present and it has been shown to be task independent (Hommel, 2011a). For example, if the task requires form discrimination, participants will tend to press on the side of the displayed image whatever its form. The precise mechanisms underlying this effect (i.e., motor or attentional) are still under debate (Hommel, 2011a, b; van der Lubbe & Abrahamse, 2011), but here we use the behavioural phenomenon as a tool to check the extent of implicit response biases during the simultaneity/asynchrony discrimination task. In this task, two stimuli are displayed on each trial and remain on the screen until the participant gives a response. The relative onset time of the two stimuli is manipulated and stimuli appear with a stimulus onset asynchrony (SOA) of 0 to 100 ms, in steps of 8 to 17 ms. The participants have to respond on whether the two stimuli are simultaneous or asynchronous by pressing a left or right response key. When the two stimuli are physically simultaneous, information is perfectly symmetrical on the two sides of the screen and there cannot be a Simon effect (i.e., there cannot be a bias to press to one side rather than the other due to the order of stimulus presentation). An asymmetry is present, however, in case of an asynchrony in the presentation of the two stimuli. Inasmuch the two stimuli do not disappear but they stay on the screen until a response is given, the only asymmetry is the onset asynchrony of the two stimuli. What can be determined in that case is whether this temporal asymmetry induces a tendency to press to the side of either the first or the second stimulus. The results of this test collected from healthy volunteers show that responses are more frequently given using the response key at the same side of the second stimulus (Lalanne, van Assche, & Giersch 2012a, Lalanne, van Assche, Wang, & Giersch 2012b). When the second stimulus is on the right, they will, thus, press more frequently on the right than the left response key, resulting in more asynchronous responses (when 'asynchronous' is on the right) and less simultaneous responses (since there is only two types of responses, if 'asynchronous' responses increase, simultaneous responses necessarily decrease). Conversely, when the second stimulus is on the left, participants tend to press more frequently the left response, and 'simultaneous' responses increase relative to 'asynchronous' responses. In healthy volunteers, this effect can be reformulated as asynchronous responses being more frequent in the left-right than in the right-left direction. This response imbalance suggests the presence of a direction bias, with asynchronies being more easily detected in one direction than the other. Such an explanation, however, does not explain the pattern of results obtained from patients with schizophrenia. Patients show this bias to the side of the second stimulus when the asynchrony is large enough, but they show an opposite bias when the asynchronies are sub-threshold (i.e., with small asynchronies patients have a bias to respond to the side of the first stimulus not the second). The reversal of the bias with short SOAs means that when the direction of the stimuli is left-right, patients tend to press more frequently on the left ('simultaneous') than on the right. Conversely, when the direction of the stimuli is right left, the patients tend to press on the right ('asynchronous'; Figure 14.1). By integrating the results of five experiments, we could demonstrate that patients show a bias to the side of the first stimulus even for asynchronies of only 8 ms (Giersch et al., 2015). These results lead to a larger amount of 'simultaneous' responses for left-right stimuli than for right-left stimuli, but only at the shortest SOAs. This precise effect has been replicated in five different groups of 18 to 20 patients (3 are published, Lalanne et al., 2012a, b). It can hardly be reduced to a global advantage for the right-left over left-right direction. Such a direction bias would have

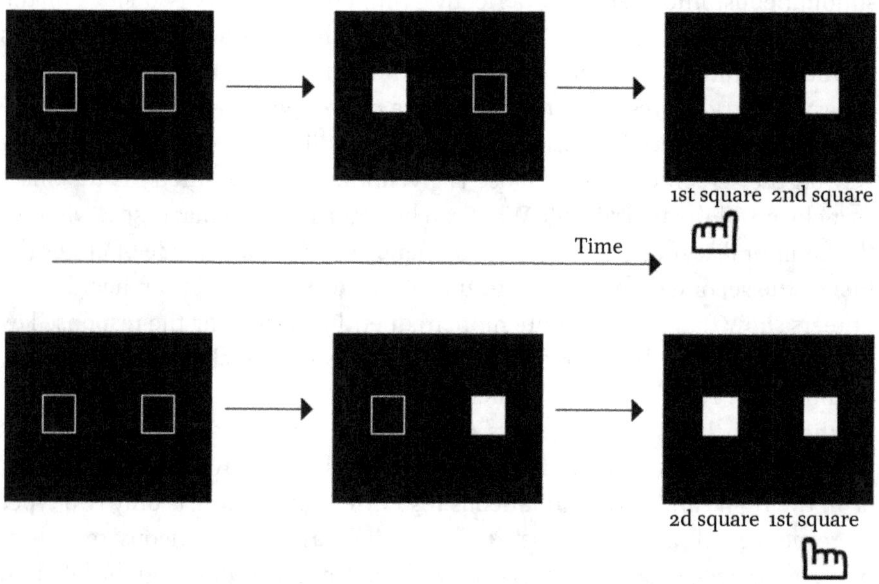


FIGURE 14.1 *Illustration of the Simon effect observed in patients affected by schizophrenia for SOAs of less than 20 ms, which is opposite in direction to the one found with larger asynchronies and in the healthy population. The task of the participants was to decide whether the two squares displayed on the screen are simultaneous or asynchronous, and to press one of the two response keys accordingly. The Simon effect in this situation results in a bias to answer using the button on the side of the first square in patients, which is opposite to what is found in other conditions. Such an effect leads to a higher number of responses on the left side when stimuli have been displayed in the left-right than in the right-left direction. In contrast the direction of the Simon effect in patients is in line with what is found in the healthy population for SOAs larger than 30 ms (i.e., there is a tendency to press the button to the side of the 2nd stimulus).*

been expected to be similar at all SOAs, which was not the case in patients. On the contrary, patients showed a bias to press to the side of the first stimulus at short SOAs, but a bias to press to the side of the second stimulus at long SOAs. The results, thus, suggest a bias to press to the side of the first stimulus when patients cannot perceive an asynchrony explicitly.

The readers can train themselves calculating the Simon effect with the example data provided with the manuscript (see book's GitHub repository: 'resultsRaw.txt').

Interestingly, for SOAs as short as 8 ms, the bias to the side of the first stimulus predicts the difficulties of the patients to explicitly detect asynchronies at larger SOAs. This correlation may suggest that the bias to the side of the 1st stimulus reveals an elementary mechanism at the core of our ability to predict and follow events over time (Giersch et al., 2015).

Difficulties at processing temporal information in patients with schizophrenia did not come as a surprise. As already emphasized, it has long been reported that the patients' thought flow is disrupted (Fuchs, 2007; Vogeley & Kupke, 2007). Psychiatrists had proposed that patients with schizophrenia have a difficulty to look forward in time (Minkowski, 1933). Importantly, these descriptions were made before the discovery of antipsychotics (in the 1950s), suggesting they are independent from treatment in schizophrenia. More recently, it has been proposed that patients display a cognitive dysmetria (Andreasen, 1999); that is, a difficulty to organize thoughts, especially in time. With this background in mind, the reversal of the Simon effect observed at small asynchronies could reveal how visual stimuli are processed in time, and might contribute to a more general difficulty at organizing information in time. We made an attempt at interpreting the biases to answer to the side of the first or second stimulus, and proposed that healthy participants follow the events in time, and, thereby, have their attention on the second stimulus by the time of their response. In contrast, patients would stay stuck with the first stimulus in case of undetectable asynchronies. The patients would have difficulties to predict and follow sequences of events over a few ms, and this might be at the origin of their difficulties at explicitly detecting asynchronies (Foucher et al., 2007; Lalanne et al., 2012a, b). These difficulties become huge when distractors are present, or in case of multisensory signals (Giersch et al., 2009; Martin et al., 2013; Schmidt et al., 2011).

Alternative explanations (eye movements, difficulty to relate stimuli in space, interhemispheric transfer, basal Simon effect; reviewed in Giersch, Lalanne, van Assche, & Elliott, 2013) were ruled out: eye movements had been controlled for by having participants fixating at the center of the screen, effects have been recorded with intra- as well as with inter-hemispheric presentation, the typical Simon effect was not impaired per se in patients. Despite this, it was necessary to test our hypothesis, i.e., to verify that participants can automatically follow stimuli over time, even when unable to tell these stimuli apart in time. We have done this in healthy volunteers by using a priming paradigm (Poncelet & Giersch, 2015), which is described in the following section.

## 3 Testing Implicit Timing Mechanisms with a Priming Task

To test whether the processing of visual stimuli over short intervals occurred automatically, we devised a new task, derived from earlier priming tasks (Elliott, Shi, & Sürer, 2007). In our task, priming stimuli were not related to the participants' task, but nonetheless influenced performance. They were two empty frames displayed on the screen, one on the left and one on the right of the screen center. These two frames were subsequently filled in by the target stimulus that elicited the response. In a first task, only one frame was filled in with a white square. Participants were instructed to press on a response key to the side of the target, as quickly as possible (Figure 14.2). The main manipulation regarded the frames that preceded the target as the frames' onsets were either simultaneous or asynchronous, but the asynchrony was short enough not to be perceived by participants (this was checked after the experiment). Participants were slightly (but significantly) faster in conditions where the target was displayed to the side of the second frame rather than to the side of the first one, when there was at least 75 to 100 ms between the frames and the target (Poncelet & Giersch, 2015).

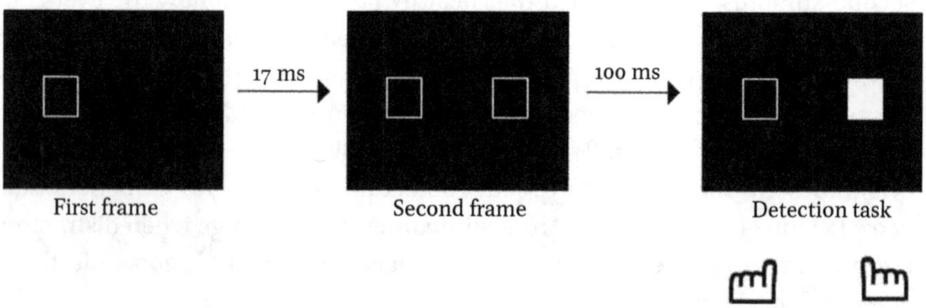


FIGURE 14.2 *Illustration of the priming task used to explore how participants follow events in time. Two frames are displayed either simultaneously or with an asynchrony too short to be perceived (17 ms). The task of the participant was to detect the target, which corresponds to the filling-in of one of the two frames. Participants are typically faster when the target is to the side of the 2nd than the 1st frame (Poncelet & Giersch, 2015).*

We had similar effects when both frames were filled in with an SOA of 100 ms and participants decided which frame was last filled in (Figure 14.3). With such an arrangement, we had to take into account whether frames and targets were displayed in the same direction or in the opposite direction. We showed that participants were faster when targets were displayed in the opposite direction relative to the primes. This was observed independent of whether participants had to press to the side of the first or second stimulus. We interpreted these effects as reflecting a shift of attention towards the second prime (Poncelet & Giersch, 2015). As a matter of fact, it is known that cueing the location of the first stimulus in a sequence of two facilitates temporal order judgment (Spence & Parise, 2010, for a review). The shift of attention towards the second prime would displace attention on the location of the first target when primes and targets are displayed in opposite directions. This cueing effect of the second prime would explain that participants are faster at making temporal judgments when the second prime is the location of the first target (i.e., when primes and targets are shown in opposite directions).

These results are important inasmuch they seem to be independent of direction perception. Else responses should have been faster when both primes and targets were displayed in the same direction. This was not the case, suggesting the results are related to time rather than direction (i.e., spatial perception). Second, we checked that the effects did not depend on the side of the response. We showed priming effects to be identical when participants had to decide about the side of the first target and when they had to decide about the side of the second one. These results, thus, suggest that participants are biased to shift their attention to the second stimulus and answer using the button on

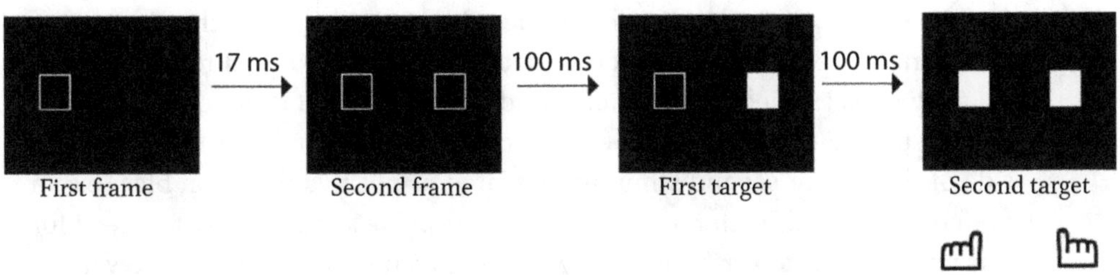


FIGURE 14.3 *Illustration of the priming paradigm used to explore the influence of asynchronous priming stimuli (i.e., frames with an undetectable SOA of 17 ms) in a temporal order judgment task. Whether participants had to press to the side of the first or second target did not change the results. Response times were always faster when the first target had been to the side of the second frame. The move of attention towards the second frame would prime the location of the first target and facilitate performance, similar to what happens with prior entry effects.*

its side. Such tendency is present even in tasks that do not require a temporal judgment (the detection task). Hence, it seems sensible to imply that the same phenomenon of attention shift occurs during the simultaneity/asynchrony discrimination task. In the simultaneity/asynchrony discrimination task, there is no priming stimulus (no frame) but stimuli are also displayed one after another, and attention can be expected to shift towards the second one after both stimuli have been displayed. This might explain the bias to the side of the second stimulus in healthy volunteers during simultaneity/asynchrony discrimination tasks. Conversely, the impairment of this effect at short asynchronies in patients with schizophrenia suggests that patients with schizophrenia have difficulties to shift their attention from the first to the second visual stimulus in a sequence of two, as if stuck with the first one. This would be consistent with the clinical hypotheses that they have difficulties at following and/or predicting stimuli.

The Simon effect, thus, represents a way to investigate the implicit processing of visual stimuli over time. However, there are several limits that should be kept in mind when using these procedures. First, the equipment should be checked carefully, and it should be verified with photocells that the delays between stimuli are as expected. Several parameters of the stimuli may influence the results, like the eccentricity of the stimuli, or their luminance. The typical Simon effect, which represents the bias to press on the side of a single stimulus, may also affect the results: if patients and controls differ on the ability to be biased by the location of the stimuli, then it may induce a group difference that would have nothing to do with a shift of attention in time. An impaired typical Simon effect should, thus, be controlled for. Since there are always two stimuli in the simultaneity/asynchrony discrimination task, the best control for the typical Simon effect is to present two stimuli on the right or two stimuli on the left side of the screen (instead of one on the left and one on the right). In this way, the typical effect should be a bias to respond on the side of the two stimuli, even when the task is to decide about their asynchrony. The critical analysis is to check that patients and controls do not differ on this typical Simon effect. Finally, it should be verified whether the tendency to press to the side of the first or second stimulus is related or not to a direction bias (e.g., a preference for a direction over the other one). The simplest way to check for this possibility is to verify that the pattern of results is the same whatever the response side (i.e., right response key for 'simultaneous' responses and left key for 'asynchronous' responses vs. the reverse).

It should be noted that a potential difficulty is related to eye movements. In the results described above on simultaneity/asynchrony discrimination, the participants were required to look towards the center of the screen and this was monitored continuously via an eyetracker. Whenever a saccade or fixation was recorded outside the center of the screen, the trial was discarded and run again at the end of the experiment. It would be useful to use at least a chin and a headrest to minimize the variance in the results. Overall, the Simon effect can be used as a valuable tool as long as one keeps in mind its limitations and interprets the results with some caution. Nonetheless, the Simon effect is a tool that can be adapted to many questions. In the following section, we provide some examples of questions that can be addressed with similar methods.

## 4 Analyzing the Responses in the Simultaneity/Asynchrony Discrimination Task in Innovative Ways

We have described how the Simon effect can be applied in the simultaneity/asynchrony discrimination task to reveal nonconscious mechanisms of information processing in time. We hope that the readers see that there are many innovative ways of analyzing data in addition to the standard methods, provided one keeps in mind the limits of each method (i.e., the need to check for all possible confounding factors). We think that other methods might be applied to address additional questions.

In Poncelet and Giersch (2015), we proposed that patients with schizophrenia do not only have difficulties to automatically shift attention in time but also to predict the stimuli. We reasoned that prediction mechanisms should be involved to enable attention to be shifted from the first to the second stimulus, especially when the stimulus onsets are separated by very short delays. It must be reminded here that the results detailed above have shown that in healthy volunteers attention moves only after 75 to 100 ms towards the second stimulus in a sequence of two. This is consistent with the attention literature (see Poncelet & Giersch, 2015, for a thorough discussion on this point). This late attention shift implies that after the first stimulus onset, attention moves only with a delay towards the second stimulus. If the second stimulus is displayed for less than 20 ms after the first one, it means that attention moves only 80 ms after the second stimulus onset. Meanwhile attention is probably on the first stimulus to facilitate its processing. For attention to shift towards the second stimulus, it would, thus, help if the participant anticipated this second stimulus. This means that the participants would not simply follow events one after another but predict sequences of events, and prepare for their processing. This hypothesis could be checked using a trial-by-trial analysis (that is, an analysis of performance that verifies whether performance on a given trial depends of the properties of the trial presented previously). It has indeed been repeatedly shown that participants can better prepare a response if they expect it at a precise point in time. This preparatory improvement can be observed when a cue indicates a delay (Correa, Lupiáñez, & Tudela, 2006), but also when looking at the effects from one trial to the next trial. In the latter case, the experience of a delay during trial N-1 primes the participants to expect the same delay at trial N (Capizzi, Correa, Wojtowicz, & Rafal, 2015; Schröter et al., 2015). On the reverse, adaptation effects might also occur as in multisensory integration (Van der Burg, Alais, & Cass, 2013). Van der Burg et al. have indeed shown that when participants are exposed to an audiovisual asynchrony, they recalibrate very fast sensory processing in order to process information as being synchronized. They have also shown that such adaptation does not occur in case of unimodal information (visual only or auditory only; Harvey, Alais, & van der Burg, 2014). However, the Simon effect allowed us to show an unexpected sensitivity to short asynchronies. This led us to wonder whether SOAs are not expected in a more accurate way than previously known.

This hypothesis can be evaluated by measuring the ability to detect asynchronies and taking into account what occurred during the trial before, not only the order of the stimuli (both trials left-right or right-left, or trials with an opposite stimuli direction) but also the relative asynchrony of the consecutive trials (trial N with a shorter, equal, or larger SOA than trial N-1). These hypotheses can be understood within the general framework of predictive coding, whereby information is constantly predicted. Real sensory information is compared to predictions, enabling prediction errors to be detected (Friston, 2008). It has been proposed that 'predictive coding' might apply generally to brain functioning (Friston, 2008). In the context of the simultaneity/asynchrony discrimination task, trial N can be predicted on the basis of trial N-1. When stimuli are displayed on trial N, the similarity or discordance with the previous trial should, thus, be automatically detected (see Di Luca & Rhodes, 2016, for an application of predictive coding and Bayesian models to expectations in time, and Tschacher, Giersch, & Friston, 2017, for applications in mental health). In the simultaneity/asynchrony discrimination task, the detection of a difference between successive trials may incite participants to adapt their predictions, and predictions themselves could influence the response of the participants in two ways. The most trivial effect would be a facilitated detection of asynchronies in case of a similarity between trials N-1 and N (i.e., more asynchronous responses). Conversely, there would be a deterioration of performance, i.e., more 'simultaneous' responses, in case of a difference between trials N-1 and N. A deterioration of performance would be akin to an adaptation effect, but related to relative SOA and not only direction (see e.g., the duration channel model for sub-second durations; Heron et al., 2012). Other effects may also be observed, related to the Simon effect. As discussed above, the prediction is especially important for the perception of the second stimulus, so that it is not missed. If the second stimulus is predicted on the basis of the previous trial, it might bias the response on this trial in a way that could be revealed by a modulation of the ability to detect stimuli in this location. The outcome of this influence would depend on whether the direction, the SOA, or both are predicted and compared from trial to trial. The main point here is that it is perfectly feasible to explore all possibilities, and this could help to check further to which extent patients with schizophrenia have difficulties at predicting sequences of stimuli in time (Giersch, Isope, & Lalanne, 2016). The Simon effect is, thus, by no means a unique method, but only one example among the many things that can be done to improve our understanding of how we process information in time.

## 5 Conclusion

We used the Simon effect to derive hypotheses regarding how participants follow visual stimuli in time at short time scales and how this is impaired in patients with schizophrenia. The same kind of method can be further adapted, e.g., to check for the implicit ability to predict very short SOAs in time. In other words, the Simon effect and related methods, are tools that can extend standard data analysis methods. As for any behavioral outcomes, the Simon effect does not provide direct insight into the mechanisms at play during a task, but it offers the possibility to make an inference. A major limitation of this type of inference is that participant' responses are influenced by multiple aspects of the stimuli presented. Hence, the interpretation of the effects described above should be considered with caution. As it has been described at the beginning of the paper, the analysis of the Simon effect should be performed using multiple ways of testing, e.g., the priming task to validate the interpretation. Despite the limitations mentioned, these methods of data analysis can be used to complement standard RT and error analyses, and are powerful enough to yield heuristic hypotheses concerning both processing in healthy volunteers and impairments in patients.

## Appendix: List of the Files Provided with the Manuscript (See Book's GitHub Repository)

'SyncTest' is the Matlab code for the experiment with synchronous vs. asynchronous stimuli. Some explanation on how the experiment is conducted, the instructions, and computer requirements, are provided in the header of the code. Questions can be addressed to the first author of this chapter.

'resultsRaw.txt' corresponds to the raw data generated with the 'SyncTest' program.

'Results_With_variable_Names.xls' is the same file as the 'resultsRaw.txt' in excel but with variable names.

Both .txt and .xls data files can be used by importing them in a statistics software package to calculate percent errors by taking into account the order of the stimulus (left-right versus right-left).

## References

Andreasen, N.C. (1999). A unitary model of schizophrenia: Bleuler's "fragmented phrene" as schizencephaly. *Archives of General Psychiatry, 56*, 781–787.

Boucart, M., S. Delord, & A. Giersch (1994). The computation of contour information in complex objects. *Perception, 23*, 399–409.

Brecher, G.A. (1932). Die Entstehung und biologische Bedeutung der subjectktiven Zeiteinheit – des Momentes. *Zeitschrift für Vergleichende Physiologie, 18*, 204–243.

Capizzi, M., A. Correa, A. Wojtowicz, & R.D. Rafal (2015). Foreperiod priming in temporal preparation: testing current models of sequential effects. *Cognition, 134*, 39–49.

Chapman, J. (1966). The early symptoms of schizophrenia. *British Journal of Psychiatry, 112*, 225–51.

Correa, A., J. Lupiáñez, & P. Tudela (2006). The attentional mechanism of temporal orienting: determinants and attributes. *Experimental Brain Research, 169*, 58–68.

Coull, J., & A. Nobre (2008). Dissociating explicit timing from temporal expectation with fMRI. *Current Opinion in Neurobiology, 18*(2), 137–144.

Di Luca, M., & D. Rhodes (2016). Optimal perceived timing: Integrating sensory information with dynamically updated expectations. *Scientific Reports, 6*, 28563.

Elliott, M.A., & A. Giersch (2016). What happens in a moment? *Frontiers in Psychology, 6*, 1905.

Elliott, M.A., Z. Shi, & S.D. Kelly (2006). A moment to reflect upon perceptual synchrony. *Journal of Cognitive Neuroscience, 18*, 1663–1665.

Elliott, M.A., Z. Shi, & F. Sürer (2007). The effects of subthreshold synchrony on the perception of simultaneity. *Psychological Res*earch, 71, 687–693.

Foucher, J.R., M. Lacambre, B.T. Pham, A. Giersch, & M.A. Elliott (2007). Poorer time resolution in schizophrenia: Longer windows of simultaneity for visual, auditory and bimodal stimuli. *Schizophrenia Research, 97*, 118–127.

Friston, K. (2008). Hierarchical models in the brain. *PLoS Computational Biology*, 4, e1000209.

Fuchs, T. (2007). The temporal structure of intentionality and its disturbance in schizophrenia. *Psychopathology, 40*, 229–35.

Giersch, A., P. Isope, & L. Lalanne (2016). Implicit timing as the missing link between neurobiological and self disorders in schizophrenia? *Frontiers in Human Neuroscience, 10*, 303.

Giersch, A., L. Lalanne, C. Corves, J. Seubert, S. Zhuanghua, J. Foucher, & M.A. Elliott (2009). Extended visual simultaneity thresholds in patients with schizophrenia. *Schizophrenia Bulletin, 35*, 816–825.

Giersch, A., L. Lalanne, M. Assche, & M.E. Elliott (2013). On disturbed time continuity in schizophrenia: An elementary impairment in visual perception? *Frontiers in Psychology 4*, 281.

Giersch, A., P. Poncelet, R.L. Capa, B. Martin, C.Z. Duval, M. Curzietti, M. Hoonacker, M. Assche, & L. Lalanne (2015). Disruption of information processing in schizophrenia: the time perspective. *Schizophrenia Research, 2*, 78–83.

Grossberg, S., & H.K. Pilly (2008). Temporal dynamics of decision-making during amotion perception in the visual cortex. *Vision Research, 48*, 1345–73.

Harvey, C., E. Van der Burg, & D. Alais (2014). Rapid temporal recalibration occurs crossmodally without stimulus specificity but is absent unimodally. *Brain Research, 1585*, 120–30.

Heron, J., C. Aaen-Stockdale, J. Hotchkiss, N.W. Roach, P.V. McGraw, & D. Whitaker (2012). Duration channels mediate human time perception. *Proceedings of the Biological Sciences, 279*(1729), 690–8.

Hommel, B. (2011a). The Simon effect as a tool and heuristic. *Acta Psychologica, 136*, 189–202.

Hommel, B. (2011b). Attention and spatial stimulus coding in the Simon task: A rejoinder to van der Lubbe and Abrahamse (2010). *Acta Psychologica, 136*, 265–268.

Lalanne, L., Van Assche, M., & Giersch, A., (2012a). When predictive mechanisms go wrong: disordered visual synchrony thresholds in schizophrenia. *Schizophrenia Bulletin, 38*, 506–513.

Lalanne, L., Van Assche, M., Wang, W., & Giersch, A., (2012c). Looking forward: an impaired ability in patients with schizophrenia? *Neuropsychologia, 50*, 2736–2744.

Martin, M., A. Giersch, C. Huron, & V. Wassenhove (2013). Temporal event structure and timing in schizophrenia: preserved binding in a longer "now". *Neuropsychologia, 51*, 358–371.

Minkowski, E. (1933). *Le temps vécu*. Presses Universitaires de France: Paris.

Pilz, K.S., C. Zimmermann, J. Scholz, & M.H. Herzog (2013). Long-lasting visual integration of form, motion, and color as revealed by visual masking. *Journal of Vision, 13*, 1–11.

Poncelet, P.E., & A. Giersch (2015). Tracking visual events in time in the absence of time perception: Implicit processing at the ms level. *PLoS One, 10*(6), e0127106.

Pöppel, E. (1997). A hierarchical model of temporal perception. *Trends in Cognitive Science, 1*, 56–61.

Pugeault, N., F. Wörgötter, & N. Krüger (2010). Disambiguating multi-modal scene representation using perceptual grouping constraints. *PLoS One, 5(6)*, e10663.

Repp, B.H. (1999). Control of expressive and metronomic timing in pianists. *Journal of Motor Behavior, 31*, 145–164.

Scharnowski, F., J. Rüter, J. Jolij, F. Hermens, T. Kammer, & M.H. Herzog (2009). Long-lasting modulation of feature integration by transcranial magnetic stimulation. *Journal of Vision, 9*, 1–10.

Schmidt, H., J. McFarland, M. Ahmed, C. McDonald, & M.A. Elliott (2011). Low-level temporal coding impairments in psychosis: Preliminary findings and recommendations for further studies. *Journal of Abnormal Psychology, 120*, 476–482.

Schröter, H., T. Birngruber, D. Bratzke, J. Miller, & Ulrich, R. (2015). Task predictability influences the variable foreperiod effect: Evidence of task-specific temporal preparation. *Psychological Research, 79*, 230–237.

Spence, C., & C. Parise (2010). Prior-entry: A review. *Consciousness & Cognition, 19*, 364–379.

Tschacher, W., A. Giersch, & K. Friston (in press). Embodiment and schizophrenia: A review of implications and applications. *Schizophrenia Bulletin*.

Van Assche, M., P. Gos, & A. Giersch (2008). Perceptual organization across hemispheres. *Visual Cognition, 16*, 1135–1139.

Van der Burg, E., D. Alais, & J. Cass (2013). Rapid recalibration to audiovisual asynchrony. *Journal of Neuroscience, 33(37)*, 14633–7.

Van der Lubbe, R.H.J., & E.L. Abrahamse (2011). The premotor theory of attention and the Simon effect. *Acta Psychologica, 136*, 259–264.

VanRullen, E., B. Zoefel, & B. Ilhan (2014). On the cyclic nature of perception in vision versus audition. *Philosophical Transactions of the Royal. Society of London B Biological Sciences, 369*, 20130214.

van Wassenhove, V. (2009). Minding time in an amodal representational space. *Philosophical Transactions of the Royal Society of London B Biological Sciences, 364*, 1815–30.

Varela, F.J. (1999). The specious present: A neurophenomenology of time consciousness. In Petitot, J., F.J. Varela, B. Pachoud, J.M. Roy (eds.), *Naturalizing phenomenology. Issues in contemporary phenomenology and cognitive science*. Stanford: Stanford University Press.

Vogeley, K., & K. Kupke (2007). Disturbances of time consciousness from a phenomenological and neuroscientific perspective. *Schizophrenia Bulletin, 33*, 142–56.

Wittmann, M. (2011). Moments in time. *Frontiers in Integrative Neuroscience, 5*, 1–9.

CHAPTER 15

# Tracking Time in the Infant Brain

*Franziska Kopp*

## 1 Introduction

From early on, infants learn to detect, predict, and adjust to internal and external events through interaction with their environment. Research shows that aspects of timing inherent in these events seem to modulate and facilitate infants' perceptual and cognitive processing. In addition to allowing for inferences about developmental trajectories and dynamics in human ontogeny, the investigation of these timing mechanisms in infants also allows the study of the structures and functions that build the foundation of the later, mature cognitive system. That is, identifying these fundamental processes provides insight into how timing processes in adults are basically organized. Infancy research has developed a number of useful methodological approaches that made these assessments possible. In this chapter, I will review some of the most established research approaches for this very young age group. The review is also intended for researchers who are interested in the investigation of timing mechanisms early in development and who are unfamiliar with this methodology.

## 2 Experimental Research in Infancy

In the context of experimental research conducted with adults, experimenters can expect at least to some extent that their participants exhibit the behavior that the experiment is intended to elicit; that they are able to give relatively precise responses, such as reaction time data or verbalizations; and that they exhibit a certain level of cooperation during testing. In contrast, research methodology used with infants often relies on the monitoring of indicators that allow only indirect conclusions about the internal processes under investigation. Infancy research implies accepting the fact that usually no precise task instructions are possible; that unambiguous, overt responses cannot necessarily be expected; and that the researcher often is not able to predict to what extent infants respond to the setting and the stimuli as they are expected

 | DOI 10.1163/9789004280205_016

to do. In other words, the variability in interpreting acquired infant data is typically high (Aslin 2007).

More than in any other age group, infants' willingness to cooperate in experimental testing and subsequent data evaluation depends on a variety of factors. First, very young children have only a short attention span (Aslin & Fiser 2005). That is, experimenters need to plan their study in a way that it allows recording data within a short period of time. Second, the content of the experimental stimuli is typically confined to material that holds at least a minimum of the infant's interest and that attracts attention to the intended place. This factor is particularly important for data collection that requires repeated presentation of the same stimuli (e.g., in event-related brain potentials designs, see below). On the other hand, the experimental stimuli should fulfill the demands of being standardized and controlled enough in order to allow for appropriate scientific inference.

Third, high levels of distractibility to both internal and external variables can generally be observed in infant populations. Therefore, visual and acoustic shielding is of particular relevance. Where not explicitly assessed, behavior is recommended to be video-recorded. Doing so guarantees that the researcher analyzes data from the experimental phases in which the infants were or were not attentive, depending on the research question. The increased distractibility also includes increased dependency in social situations, such as a caregiver's presence or absence as well as the number and experience of the experimenters. Fourth, in addition to attentional restrictions, very young children can be expected to display disproportionally high rates of random behavior, such as increased motor activity (de Haan 2007). This confounded variable may result in data contamination, data loss, high fluctuation, and high attrition rates, particularly in the context of physiological data assessment.

Fifth, following up on the finding of short attention spans, results obtained in infant experimental paradigms can be assumed to interact with infant development in other domains. For example, cognitive development early in infancy is very closely associated with social and motor development (e.g., Kopp & Lindenberger 2011, 2012). In other words, the experimental design, settings, and parameters need to take into account the respective developmental stage in several domains. Moreover, some of the experimental paradigms applied in infancy research (e.g., on perceptual processes) interact with memory. Hence, a potential confound with memory-related specificities, such as the reliability of recall and the temporal extent of memory (Bauer 2006), need to be considered in the interpretation of data thought to reflect the perceptual and cognitive processes of interest. Furthermore, overt responses in experimental situations may interfere with premature levels of motor control and of planning and executive processes (de Haan 2007).

Sixth, in general, infant data reveal a high degree of intra- and inter-individual variability (e.g., Gilmore & Thomas 2002). This constraint is particularly significant with respect to inferential statistics. Due to high attrition rates, researchers are often required to test a large number of participants in order to identify genuine, underlying psychological mechanisms observed in reliable statistical effects.

## 3 Assessment of Infants' Neural Activity

Given the limitations of the interpretability of behavioral data in this young age group, it seems particularly helpful to assess physiological parameters or some other online data. In the last couple of years, neuroscience has made substantial progress in the development and application of promising methods in the infancy domain. I will discuss a few of the common questions regarding the neuroscientific approaches used with infants.

Brain undergoes considerable developmental changes both in terms of structure and function. Experimental approaches have to take such changes into account. Comparability between the mature (adult) and the immature (infant) brain activity may not always be evident. Early in ontogeny, brain maturation processes, such as synaptogenesis or the beginning of synaptic pruning, play a major role. Pronounced lifespan changes in synaptic density have been demonstrated (Huttenlocher & de Courten 1987), showing a major increase of the number of synapses after birth, while synaptic pruning is initiated a few weeks after birth and continues over the lifespan. These mechanisms are associated with progressive specialization and differentiation, both in the behavioral and the neural domain. Johnson and Munakata (2005) described the structural changes as specialization, dissociation, and structural integration. In particular, development includes processes of narrowing, increased specialization, increased localization, and enhanced focal activation. In addition to these structural changes, some empirical findings point to developmental changes in connectivity (Eiselt et al. 2001; Grieve et al. 2004; Thatcher, Walker, & Giudice, 1987). For example, according to one line of research (Thatcher 1992), changes in connectedness—indicated by changes in coherence measures of electroencephalography (EEG)—can be observed especially in the first four years of life. In the left hemisphere, sequential lengthening of intracortical connections takes place, whereas in the right hemisphere, sequential contractions of intra-cortical connections can be observed. Furthermore, connectedness was shown to be modulated by experience-dependent variables in infants, such as motor behavior (Bell & Fox 1996) or specific cognitive capabilities (Bell & Fox 1992).

To date, EEG is one of the most widely used assessment methods for infants' neural signatures. One implication of the immaturity of brain activity early in life is the differences in scalp recordings compared to adult EEG signals. First, the functional equivalent of adult EEG frequency bands can be found in lower frequencies in small children. For example, while the frequency assigned as alpha is typically in the range of about 8–12 Hz in the adult EEG, it is in the range of about 6–9 Hz in young infants (e.g., Bell 2002; Marshall, Bar-Haim, & Fox, 2002). As a consequence, studies addressing brain's oscillatory activity have to take this issue into account. Second, developmental specificities can also be observed in event-related brain potentials (ERP). ERP components are significant deflections of neural responses time-locked to an internal or external event. Typically, they are operationalized as averaged neural discharges across a sufficient number of experimental trials. Modulations of these components are thought to reflect corresponding psychological states and processes. Infant ERP components usually differ from adult ERP components in terms of amplitude, latency, or polarity (Jing & Benasich 2006; Kushnerenko et al. 2002; Little, Thomas, & Letterman, 1999; McIsaac & Polich 1992; Wunderlich, Cone-Wesson, & Shepherd, 2006). Infant ERP data show higher inter-individual variability with respect to amplitudes and latencies (Thomas et al. 1997) and higher interference with movement artifacts or random noise. Short-term variability elicited by repeated presentation of the same or similar stimuli may be high (e.g., Thomas & Lykins 1995; Wiebe et al. 2006).

Both these and other factors require an adjustment of experimental EEG designs and parameters for infants. As for data acquisition, one has to keep in mind the overall high levels of movement artifacts, potential inattention, and fussiness resulting in high dropout rates. Moreover, for data analysis, some parameters, such as amplitude criteria for artifact control, need to be adjusted as well.

For the evaluation of experimental methodology, it is important to note that results from one source of data may dissociate from results obtained from other sources. This finding can be more pronounced than in adult research, given that adult participants may provide more precise behavioral measures than infants. In some cases, data may complement each other fruitfully. For example, de Haan and Nelson (1997) investigated infants' face perception and found ERP differences between the recognition of their mother's and a dissimilar stranger's face; however, there was no indication that the infants recognized the mother's face in the behavioral results.

## 4 Timing in Early Communication

Why should researchers investigate timing early in human ontogeny? The high relevance of temporal perception and action mechanisms becomes evident in the way infants learn about the world and, in particular, in the way they communicate with others. Infants learn through interaction with events in their environment, primarily from and through interaction with other people (e.g., Kopp & Lindenberger 2011, 2012). As such, perceptual and cognitive systems mature in interaction with the social communicative processes infants are engaged in. Research shows that much of the communicative content of early interaction is conveyed through the specific timing of interactive parameters (Feldman & Greenbaum 1997).

For this type of investigation, infants are usually placed in situations of free play or structured play with other people, and behavior is coded with respect to the temporal parameters of interest. Another frequently used experimental setup for the study of timing in early interaction is a double-screen setup (each individual perceives audio-visual responses of his/her interaction partner via screen). It allows the manipulation of interaction dynamics, such as social contingency or temporal contingency, while essential features of face-to-face interactions are maintained (e.g., Nadel et al. 1999).

From these investigations, we have learnt that infants learn to expect contingent responses from their environment, both in terms of content and timing, and that they are able to detect discrepancies in these interactional patterns from a very young age (e.g., Nadel et al. 1999; Striano, Henning, & Stahl, 2005, 2006). They use temporal information from an ongoing social interaction and may express the timing dynamics either in the same sensory modality (e.g., imitation; Meltzoff & Moore 1977) or internally transfer it to another sensory modality (e.g., affect attunement; Jonsson & Clinton 2006). In line with this finding, coordinated temporal interaction has been observed between infants' gazing behavior and adults' vocalization (Crown et al. 2002).

Time-series analyses of mother-infant interactions revealed that temporal coordination and synchrony play important roles in affect transfer and are related to cognitive competencies later in development (Feldman 2007; Feldman & Greenbaum 1997; Jaffe et al. 2001; Kaye & Fogel 1980). Moreover, infants' temporal interaction dynamics seem to be closely related to specific temporal patterns in the speech of their adult interaction partners (Condon & Sander 1974). Even preverbal infants engage in coordinated mutual vocalizing with adults, for example, by establishing tonal synchrony of the pitch of their utterances (van Puyvelde et al. 2015; van Puyvelde et al. 2010). On the other hand, disruptions in the temporal parameters of the reciprocal exchange between adults and infants may be associated with clinical conditions, such as depression (e.g., Beebe et al. 2008; Field, Healy, Goldstein, & Guthertz, 1990), or with infant risk conditions (e.g., Lester, Hoffman, & Brazelton, 1985).

Why does time play such an important role in early communication? It has been repeatedly hypothesized that the tendency and the capacity to engage in interactions that are temporally coordinated with other people may be associated with biological rhythms, such as sleep-wake cyclicity or cardiac vagal tone (Feldman 2006), and that internal rhythms may be determinants of social interactions (Feldman 2007). This assumption has been corroborated by empirical findings on mother-infant synchrony obtained through dynamical systems modeling demonstrating both self-regulation dynamics and interpersonal coupling effects (Zentall, Boker, & Braingart-Rieker, 2006).

Based on their internal rhythmicity, children continuously learn to develop a sense of timing of their behavior in interaction with the timing of external events. Accumulated experience in this exchange with the environment enables them to develop their perceptual capacities and predictions about the timing, which in turn allows infants to develop and adjust their actions in the world.

## 5 Timing Processes as Seen through Behavioral Data

Next to the macro perspective of communication processes, researchers have been demonstrating temporal processing in infancy on a micro level. Most of this research uses observed overt infant behavior, such as eye gaze, in order to make inferences regarding the hypothesized corresponding internal states. This approach often leaves room for ambiguity and variance.

One of the most commonly applied experimental paradigms takes advantage of the phenomenon that repeated presentation of the same stimulus results in habituation to this stimulus (operationalized as decreased looking time) and the presentation of a novel stimulus in subsequent dishabituation (operationalized as increased looking time; Colombo & Mitchell 2009; Fantz 1964). The procedure may include a habituation sequence of fixed trial presentation or infant-controlled habituation (relying on the real looking time toward a stimulus). This approach is helpful in assessing the capacity of detecting differences in stimuli in preverbal infants. However, the understanding of the processes underlying habituation is a subject of debate (Sirois & Mareschal 2002; Turk-Browne, Scholl, & Chun, 2008). Regarding the specific investigation of timing processes, one has to consider that the repeated presentation of stimuli in itself contains a temporal dimension.

A second behavioral approach frequently used in experimental infancy research is the inference about psychological states via the assessment of visual preference. Typically, two or more stimuli are presented simultaneously or successively, and the proportional duration of gazing toward one of the stimuli is assumed to inform the internal representation and processing of this stimulus as compared to the other stimulus/stimuli. In other words, conclusions about the internal stimulus relation are drawn from the external stimulus relation. Looking times are usually compared against chance level. The procedure may include a familiarization phase to a specific stimulus, after which novel stimuli are introduced for comparison; otherwise, infants utilize their experience and knowledge without a specific familiarization phase.

Experimental approaches using infant gaze as the dependent variables are subject to a number of limitations. For example, the validity and reliability of the results and the conclusions that can be drawn may be challenged (see factors influencing infant data evaluation described above). Moreover, the question is how comparable different looking times really are, for example, with respect to different age groups or the possible confound of memory interference effects (Houston-Price & Nakai 2004). Furthermore, the interpretability of results in the absence of a statistically reliable effect or the evaluation of familiarity versus novelty preference are debatable (for detailed discussions, see Aslin 2007; Aslin & Fiser 2005). Nevertheless, statistically significant results may provide reliable indicators of internal mental states, particularly detection and discrimination capabilities.

These behavioral approaches have been used to study infants' cognitive capacities to perceive temporal patterns and relations. Sensitivity to temporal phenomena, such as tempo, duration, rhythm, velocity, or synchrony between sensory modalities in multisensory events, has been demonstrated at early ages (Bahrick 2001; Byrne & Horowitz 1984; Dannemiller & Freedland 1989; Lewkowicz, Leo, & Simion, 2010; Pickens & Bahrick 1997; Spelke 1979). These capacities undergo developmental changes in terms of precision and complexity during the first months and years of life (Bahrick 2001; Bahrick & Lickliter 2004; Lewkowicz 2000a; Pickens et al. 1994). Infants' capacity to perceive intersensory synchrony, which is assumed to precede responsiveness to duration, rate, and rhythm (Lewkowicz 2000b), seems to be of particular relevance to their early experience of learning about and interacting with the world. In other words, being able to relate two sensory stimulus components as belonging to each other based on their temporal coincidence helps infants to extract and ascribe meaning to the world around them. Very young infants can already detect asynchrony between audition and vision in audiovisual stimuli (Bahrick 1983; Dodd 1979; Lewkowicz 1996, 2010). Furthermore, young children use synchrony between auditory and visual stimuli for rhythm discrimination (Bahrick & Lickliter 2000), affect discrimination (Flom & Bahrick 2007), speech processing (Hollich, Newman, & Jusczyk, 2005), and word learning (Jesse and Johnson 2016). Perceptual experience of audiovisual synchrony relations and active experience with the timing of audiovisual events (e.g., drumming experience) may, in turn, increase infants' sensitivity to discern between audiovisual synchrony and asynchrony (Gerson et al. 2015; Pons et al. 2012).

## 6 Eye Movements

Global eye gaze data can provide empirical evidence to a number of research questions. In the domain of timing processes, however, it may be useful to rely on measures and paradigms that allow for the possibility of tracking temporal dynamics. Based on the idea that looking behavior may suggest what is in an infant's mind, internal timing dynamics can be made visible with higher precision and validity.

Eye tracking has been increasingly used in infant populations. The availability and application of this method have improved substantially, allowing for the assessment of precise spatial and temporal information regarding infants' eye gaze (e.g., Aslin & McMurray 2004; Gredebäck, Johnson, & von Hofsten, 2010).

These experimental techniques increase the possible spectrum of insight into psychological processes in infants and allow for inferences about temporal aspects of selective attention (Lewkowicz & Hansen-Tift 2012), action perception (Van Elk et al. 2008), categorization (McMurray & Aslin 2004), scanning dynamics (Hunnius & Geuze 2004), attentional disengagement (Hunnius, Geuze, & van Geert, 2006), anticipatory processes (Hunnius & Bekkering 2010; McMurray & Aslin 2004), predictive changes (such as the representation of temporarily occluded objects; Gredebäck & von Hofsten, 2004), integration of audiovisual speech information (Guiraud et al. 2012), the role of audiovisual temporal synchrony in infants' attention to a talker's face (Hillairet de Boisferon et al. 2016), and the timing of responses to multisensory stimuli as compared to unisensory stimuli (Neil et al. 2006).

For example, to assess specific aspects of social interactions, such as speech perception, it may be helpful to gain information about infants' precise gaze direction. It is known that, in the second half of the first year of life, infants shift their attention from a talker's eyes to a talker's mouth suggesting a developmental shift in the use of available speaker information. Using eye-tracking methodology, Hillairet de Boisferon et al. (2016) were able to demonstrate developmental differences of attentional indicators of the direction and the duration of infant gaze as a function of audio-visual speech coherence. Thus, high spatial and temporal resolution allow for greater insight into the dynamics of social and cognitive development.

## 7 Neural Dynamics

Neurophysiological data may also provide insight into infants' internal states relatively independent of the child's overt behavior. In many cases, neurophysiology complements behavioral observation. While some neurophysiological imaging techniques, such as near-infrared spectroscopy (NIRS) and magnetic resonance imaging (MRI), are increasingly used in the infancy domain (e.g., Aslin & Mehler 2005; De Vita et al. 2006; Dehaene-Lambertz, Dehaene, & Hertz-Pannier, 2002; Emberson, Richards, & Aslin, 2015; Prastawa, Gilmore, Lin, & Gerig, 2005), other techniques, such as EEG, are already well established. EEG measures allow monitoring of neural activity with high temporal resolution, which makes them a preferred method for investigating timing dynamics in the very young brain. Typically, neural activity is studied as recorded in either a continuous state or behavior, such as rest or play, or time-locked to a specific external or internal event. As discussed above, the human EEG undergoes pronounced developmental changes (e.g., Picton & Taylor 2007) and thus the interpretation of EEG data elicited as responses to an experimental manipulation has to consider these changes.

Oscillatory activity in neural signals provides information about temporal fluctuations in the frequency domain of the EEG. Spectral analysis has been successfully used for several years in the investigation of spontaneous EEG (e.g., Bell & Fox 1992, 1996). However, the acquisition of infants' time-locked oscillatory responses to experimental stimulus manipulation is still a sparse field of research. Here, due to the specificities of the infant EEG (see above), some methodological questions are not yet sufficiently resolved. Furthermore, the correlation of neural activity in certain frequency bands to perceptual or cognitive states and processes is often not as clear as it is in comparable adult research. Yet, initial studies have provided promising findings regarding the temporal processing in the infant brain. Of particular interest have been infants' neural responses to action observation processes.

Differences in event-related synchronization of alpha/mu band activity were observed in infants' observation of ongoing goal-directed versus non-goal-directed movements (Nyström et al. 2010) and in observation versus execution of goal-directed action (Marshall, Young, & Meltzoff, 2011). Moreover, temporal neural dynamics associated with the observation of movements were significantly related to the infants' own motor experience (van Elk et al. 2008). Analysis of oscillatory activity also has the potential to provide information about the time function of internal states during the processing of ongoing actions. For example, alpha band activity was found to be attenuated not only during the observation of a grasping action, but also prior to the event when the stimulus allowed for anticipation of the occurrence of this action (Southgate, Johnson, Osborne, & Csibra, 2009). When actions are temporally occluded and the timing of movements is manipulated through introduction of continuous versus non-continuous movement, EEG signatures showed that attention- and memory-related processes (revealed in alpha and theta oscillations) support infants' tracking and internally representing observed movement (Bache et al. 2015).

A more widely established measure in infancy research is event-related brain potentials (ERP, see above). One experimental procedure that has produced a large body of literature is the measurement of the mismatch negativity (MMN), an ERP component elicited and modulated by deviant acoustic stimuli in a continuous stream of homogeneous stimuli (e.g., Cheour 2006; Jing & Benasich 2006). A major advantage of the MMN is that it can be assessed already at the beginning of life and also while infants are asleep (Cheour et al., 2002a; Cheour et al., 2002b). MMN modulations reflect preattentive processing and can be regarded as stable and reliable indicators of the temporal dynamics of auditory sensory memory. Using MMN assessments, very early neural responsiveness to several dimensions of timing were revealed, including stimulus duration (e.g., Cheour et al., 2002a, 2002b), interval timing (Brannon et al. 2004), ratio of occurrence of different inter-stimulus intervals (Brannon et al. 2008), temporal resolution operationalized as gap detection thresholds (Trainor et al. 2001), and variations in the frequency spectrum of sounds (Kushnerenko et al. 2007), among others.

Apart from mismatch responses, ERP correlates of temporal processing in infants were identified both for unisensory (e.g., Kushnerenko et al. 2001; Purhonen et al. 2005; Rosander et al. 2007) and multisensory stimuli. As described earlier, infants rely preferably on temporal coincidence of sensory information to make sense of the world around them. Hence, the study of infants' capacities to bind information from two sensory modalities together, such as audition and vision, has received increased interest. Using ERP, early processing differences were shown for congruence versus incongruence of concurrently presented auditory and visual information (Bristow et al. 2008). Moreover, in line with findings in the adult brain, in very young infants visual stimuli modulate auditory ERP responses when presented simultaneously with acustic stimuli (Hyde et al. 2010). ERP modulations were also observed in response to asynchronous versus synchronous presentation of a static face and of speech sounds as well as in response to congruence versus incongruence of dynamic visual and auditory speech streams (Hyde et al. 2011).

Neural correlates of audiovisual synchrony relations—independent of identity, congruence, or static versus dynamic presentation—were examined in two studies using non-speech stimuli (Kopp 2014; Kopp & Dietrich 2013). Infants saw and heard a person clapping her hands at a fixed time interval. In an infant-controlled habituation paradigm, they did not detect a temporal discrepancy of 200 ms between audition and vision behaviorally (Kopp 2014), but showed dishabituation to a 400-ms asynchrony (Kopp & Dietrich 2013). In contrast, neural activity differentiated not only between synchrony and asynchrony (400 ms) perception, but also between synchrony and the subliminal temporal discrepancy (200 ms). Although the experimental manipulation included a temporal shift only in the visual modality, auditory ERP activity was significantly modulated relative to the synchrony conditions in both experiments. Moreover, results demonstrated that infants predictively adjusted their ongoing neural activity very early after stimulus onset, resulting in an asynchronous (400 ms) or temporally fused (200 ms) multisensory percept, respectively. In other words, depending on temporal synchrony relations between vision and audition and on how they were perceived behaviorally, brain signatures showed significantly different activity modulations that followed expectancy processes.

These two latter studies are examples of the potential of the assessment of online physiological measures and of the usefulness of collecting complementary measures (in this case, behavior and EEG) in the infancy domain. The overview in this chapter has demonstrated that the investigation of timing and temporal perception early in human development is still a developing research area. Some progress has been made both in the development of appropriate experimental methodology in infancy and in the understanding of timing mechanisms. Timing plays an important role both on a macro level, as seen in social interactions, and on a micro level, as observed in individual perceptual and cognitive processes. It seems fair to assume that research methods providing fine-tuned, online behavioral, and physiological measures are increasingly used. They make it possible to address the correlation between infants' psychological states and the processing characteristics of internal and external events in high temporal resolution, which is particularly useful in the field of timing phenomena. For present and future research, the combination of expertise in infancy research, neuroscience, cognitive psychology, and other related disciplines seems the most promising in terms of increasing insight into this fascinating field of research.

## References

Aslin, R.N. (2007). What's in a look? *Developmental Science, 10*, 48–53.

Aslin, R.N., & J. Fiser (2005). Methodological challenges for understanding cognitive development in infants. *Trends in Cognitive Sciences, 9*, 92–98.

Aslin, R.N., & B. McMurray (2004). Automated corneal-reflection eye tracking in infancy: Methodological developments and applications to cognition. *Infancy, 6*, 155–163.

Aslin, R.N., & J. Mehler (2005). Near-infrared spectroscopy for functional studies of brain activity in human infants: Promise, prospects, and challenges. *Journal of Biomedical Optics, 10*, 011009.

Bache, C., F. Kopp, A. Springer, W. Stadler, U. Lindenberger, & M. Werkle-Bergner (2015). Rhythmic neural activity indicates the contribution of attention and memory to the processing of occluded movements in 10-month-old infants. *International Journal of Psychophysiology, 98*, 201–212.

Bahrick, L.E. (1983). Infants' perception of substance and temporal synchrony in multimodal events. *Infant Behavior and Development, 6*, 429–451.

Bahrick, L.E. (2001). Increasing specificity in perceptual development: Infants' detection of nested levels of multimodal stimulation. *Journal of Experimental Child Psychology, 79*, 253–270.

Bahrick, L.E., & R. Lickliter (2000). Intersensory redundancy guides attentional selectivity and perceptual learning in infancy. *Developmental Psychology, 36*, 190–201.

Bahrick, L.E., & R. Lickliter (2004). Infants' perception of rhythm and tempo in unimodal and multimodal stimulation: A developmental test of the intersensory redundancy hypothesis. *Cognitive, Affective, & Behavioral Neuroscience, 4*, 137–147.

Bauer, P.J. (2006). Constructing a past in infancy: A neuro-developmental account. *Trends in Cognitive Sciences, 10*, 175–181.

Beebe, B., J. Jaffe, K. Buck, H. Chen, P. Cohen, S. Feldstein, & H. Andrews (2008). Six-week postpartum maternal depressive symptoms and 4-month mother-infant self- and interactive contingency. *Infant Mental Health Journal, 29*, 442–471.

Bell, M.A. (2002). Power changes in infant EEG frequency bands during a spatial working memory task. *Psychophysiology, 39*, 450–458.

Bell, M.A., & N.A. Fox (1992). The relations between frontal brain electrical activity and cognitive development during infancy. *Child Development, 63*, 1142–1163.

Bell, M.A., & N.A. Fox (1996). Crawling experience is related to changes in cortical organization during infancy: Evidence from EEG coherence. *Developmental Psychobiology, 29*, 551–561.

Brannon, E.M., M.E. Libertus, W.H. Meck, & M.G. Woldorff (2008). Electrophysiological measures of time processing in infant and adult brains: Weber's law holds. *Journal of Cognitive Neuroscience, 20*, 193–203.

Brannon, E.M., L. Wolfe Roussel, W.H. Meck, & M. Woldorff (2004). Timing in the baby brain. *Cognitive Brain Research, 21*, 227–233.

Bristow, D., G. Dehaene-Lambertz, J. Mattout, C. Soares, T. Gliga, S. Baillet, & J.-F. Mangin (2008). Hearing faces: How the infant brain matches the face it sees with the speech it hears. *Journal of Cognitive Neuroscience, 21*, 905–921.

Byrne, J.M., & F.D. Horowitz (1984). The perception of stimulus shape: The influence of velocity of stimulus movement. *Child Development, 55*, 1625–1629.

Cheour, M. (2006). Development of mismatch negativity (MMN) during infancy. In M. de Haan (Ed.), *Infant EEG and event-related potentials* (pp. 171–178). Hove: Psychology Press.

Cheour, M., R. Ceponiene, P. Leppänen, K. Alho, T. Kujala, M. Renlund, V. Fellman, & R. Näätänen (2002a). The auditory sensory memory trace decays rapidly in newborns. *Scandinavian Journal of Psychology, 43*, 33–39.

Cheour, M., E. Kushnerenko, R. Ceponiene, V. Fellman, & R. Näätänen (2002b). Electric brain responses obtained from newborn infants to changes in duration in complex harmonic tones. *Developmental Neuropsychology, 22*, 471–479.

Colombo, J., & D.W. Mitchell (2009) Infant visual habituation. *Neurobiology of Learning and Memory, 92*, 225–234.

Condon, W.S., & L.W. Sander (1974). Neonate movement is synchronized with adult speech: Interactional participation and language acquisition. *Science, 183*, 99–101.

Crown, C.L., S. Feldstein, M.D. Jasnow, B. Beebe, & J. Jaffe (2002). The cross-modal coordination of interpersonal timing: Six-week-olds infants' gaze with adults' vocal behavior. *Journal of Psycholinguistic research, 31*, 1–23.

Dannemiller, J.L., & R.L. Freedland (1989). The detection of slow stimulus movement in 2- to 5-month-olds. *Journal of Experimental Child Psychology, 47*, 337–355.

De Haan, M. (Ed.) (2007). *Infant EEG and event-related potentials*. Hove: Psychology Press.

De Haan, M., & C.A. Nelson (1997). Recognition of mother's face by six-month-old infants: A neurobehavioral study. *Child Development, 68*, 187–210.

De Vita, E., A. Bainbridge, J.L.Y. Cheong, C. Hagmann, R. Lombard, W.K. Chong, J.S. Wyatt, E.B. Cady, R.J. Ordidge, & N.J. Robertson (2006). Magnetic resonance imaging of neonatal encephalopathy at 4.7 Tesla: Initial experiences. *Pediatrics, 118*, 1812–1821.

Dehaene-Lambertz, G., S. Dehaene, & L. Hertz-Pannier (2002). Functional neuroimaging of speech perception in infants. *Science, 298*, 2013–2015.

Dodd, B., (1979). Lip reading in infants: Attention to speech presented in- and out-of-synchrony. *Cognitive Psychology, 11*, 478–484.

Eiselt, M., J. Schindler, M. Arnold, H. Witte, U. Zwiener, & J. Frenzel (2001). Functional interactions within the newborn brain investigated by adaptive coherence analysis of EEG. *Clinical Neurophysiology, 31*, 104–113.

Emberson, L.L., J.E. Richards, & R.N. Aslin (2015). Top-down modulation in the infant brain: Learning-induced expectations rapidly affect the sensory cortex at 6 months. *pnas, 112*, 9585–9590.

Fantz, R.L. (1964). Visual experience in infants: Decreased attention to familiar patterns relative to novel ones. *Science, 146*, 668–670.

Feldman, R. (2006). From biological rhythms to social rhythms: Physiological precursors of mother-infant synchrony. *Developmental Psychology, 42*, 175–188.

Feldman, R. (2007). Parent-infant synchrony and the construction of shared timing; physiological precursors, developmental outcomes, and risk conditions. *Journal of Child Psychology and Psychiatry, 48*, 329–354.

Feldman, R., & C.W. Greenbaum (1997). Affect regulation and synchrony in mother-infant play as precursors to the development of symbolic competence. *Infant Mental Health Journal, 18*, 4–23.

Field, T., B. Healy, S. Goldstein, & M. Guthertz (1990). Behavior-state matching and synchrony in mother-infant interactions of nondepressed versus depressed dyads. *Developmental Psychology, 26*, 7–14.

Flom, R., & L.E. Bahrick (2007). The development of infant discrimination of affect in multimodal and unimodal stimulation: The role of intersensory redundancy. *Developmental Psychology, 43*, 238–252.

Gerson, S.A., A. Schiavio, R. Timmers, & S. Hunnius (2015). Active drumming experience increases infants' sensitivity to audiovisual synchrony during observed drumming actions. *PLoS ONE, 10*, e0130960.

Gilmore, R.O., & H. Thomas (2002). Examining individual differences in infants' habituation patterns using objective quantitative techniques. *Infant Behavior & Development, 25*, 399–412.

Gredebäck, G., & C. Hofsten (2004). Infants' evolving representation of moving objects between 6 and 12 months of age. *Infancy, 6*, 165–184.

Gredebäck, G., S. Johnson, & C. Hofsten (2010). Eye tracking in infancy research. *Developmental Neuropsychology, 35*, 1–19.

Grieve, P.G., R.G. Emerson, J.R. Isler, & R.I. Stark (2004). Quantitative analysis of spatial sampling error in the infant and adult electroencephalogram. *NeuroImage, 21*, 1260–1274.

Guiraud, J.A., P. Tomalski, E. Kushnerenko, H. Ribeiro, K. Davies, T. Charman, M. Elsabbagh, & M.H. Johnson (2012). Atypical audiovisual speech integration in infants at risk for autism. *PLoS ONE, 7*, e36428.

Hillairet de Boisferon, A., A.H. Tift, N.J. Minar, & D.J. Lewkowicz (2016). Selective attention to a talker's mouth in infancy: Role of audiovisual temporal synchrony and linguistic experience. *Developmental Science, 20*. e12381.

Hollich, G., R.S. Newman, & P.W. Jusczyk (2005). Infants' use of synchronized visual information to separate streams of speech. *Child Development, 76*, 598–613.

Houston-Price, C., & S. Nakai (2004). Distinguishing novelty and familiarity effects in infant preference procedures. *Infant and Child Development, 13*, 341–348.

Hunnius, S., & H. Bekkering (2010). The early development of object knowledge: A study of infants' visual anticipations during action observation. *Developmental Psychology, 46*, 446–454.

Hunnius, S., & R.H. Geuze (2004). Developmental changes in visual scanning of dynamic faces and abstract stimuli in infants: A longitudinal study. *Infancy, 6*, 231–255.

Hunnius, S., R.H. Geuze, & P. Geert (2006). Associations between the developmental trajectories of visual scanning and disengagement of attention in infants. *Infant Behavior and Development, 29*, 108–125.

Huttenlocher, P.R., & C. Courten (1987). The development of synapses in striate cortex of man. *Human Neurobiology, 6*, 1–9.

Hyde, D.C., B.L. Jones, R. Flom, & C.L. Porter (2011). Neural signatures of face-voice synchrony in 5-month-old human infants. *Developmental Psychobiology, 53*, 359–370.

Hyde, D.C., B.L. Jones, C.L. Porter, & C. Flom (2010). Visual stimulation enhances auditory processing in 3-month-old infants and adults. *Developmental Psychobiology, 52*, 181–189.

Jaffe, J., B. Beebe, S. Feldstein, C.L. Crown, M.D. Jasnow, P. Rochat, & D.S. Stern (2001). Rhythms of dialogue in infancy: Coordinated timing in development. *Monographs of the Society for Research in Child Development, 66*, 1–132.

Jesse, A., & E.K. Johnson (2016). Audiovisual alignment of co-speech gestures to speech supports word learning in 2-year-olds. *Journal of Experimental Child Psychology, 145*, 1–10.

Jing, H., & A.A. Benasich (2006). Brain responses to tonal changes in the first two years of life. *Brain & Development, 28*, 247–256.

Johnson, M.H., & Y. Munakata (2005). Processes of change in brain and cognitive development. *Trends in Cognitive Sciences, 9*, 152–158.

Jonsson, C.-O., & D. Clinton (2006). What do mothers attune to during interactions with their infants? *Infant and Child Development, 15*, 387–402.

Kaye, K., & A. Fogel (1980). The temporal structure of face-to-face communication between mothers and infants. *Developmental Psychology, 16*, 454–464.

Kopp, F. (2014). Audiovisual temporal fusion in 6-month-old infants. *Developmental Cognitive Neuroscience, 9*, 56–67.

Kopp, F., & C. Dietrich (2013). Neural dynamics of audiovisual synchrony and asynchrony perception in 6-month-old infants. *Frontiers in Psychology, 4*.

Kopp, F., & U. Lindenberger (2011). Effects of joint attention on long-term memory in 9-month-old infants: An event-related potentials study. *Developmental Science, 14*, 660–672.

Kopp, F., & U. Lindenberger (2012). Social cues at encoding affect memory in 4-month-old infants. *Social Neuroscience, 7*, 458–472.

Kushnerenko, E., R. Ceponiene Balan, P., V. Fellman, M. Huotilainen, & R. Näätänen (2002). Maturation of auditory event-related potentials during the first year of life. *Cognitive Neuroscience and Neuropsychology, 13*, 47–51.

Kushnerenko, E., R. Ceponiene, V. Fellman, M. Huotilainen, & I. Winkler (2001). Event-related potential correlates of sound duration: Similar pattern from birth to adulthood. *NeuroReport, 12*, 3777–3781.

Kushnerenko, E., I. Winkler, J. Horváth, R. Näätänen, I. Pavlov, V. Fellman, & M. Huotilainen (2007). Processing acoustic change and novelty in newborn infants. *European Journal of Neuroscience, 26*, 265–274.

Lester, B.M., J. Hoffman, & T.B. Brazelton (1985). The rhythmic structure of mother-infant interaction in term and preterm infants. *Child Development, 56*, 15–27.

Lewkowicz, D.J. (1996). Perception of auditory-visual temporal synchrony in human infants. *Journal of Experimental Psychology: Human Perception and Performance, 22*, 1094–1106.

Lewkowicz, D.J. (2000a). Infants' perception of the audible, visible, and bimodal attributes of multimodal syllables. *Child Development, 71*, 1241–1257.

Lewkowicz, D.J. (2000b). The development of intersensory temporal perception: An epigenetic systems/limitations view. *Psychological Bulletin, 126*, 281–308.

Lewkowicz, D.J. (2010). Infant perception of audio-visual speech synchrony. *Developmental Psychology, 46*, 66–77.

Lewkowicz, D.J., & A.M. Hansen-Tift (2012). Infants deploy selective attention to the mouth of a talking face when learning speech. *PNAS, 109*, 1431–1436.

Lewkowicz, D.J., I. Leo, & F. Simion (2010). Intersensory perception at birth: Newborns match nonhuman primate faces and voices. *Infancy, 15*, 46–60.

Little, V.M., D.G. Thomas, & M.R. Letterman (1999). Single-trial analyses of developmental trends in infant auditory event-related potentials. *Developmental Neuropsychology, 16*, 455–478.

Marshall, P.J., T. Young, & A.N. Meltzoff (2011). Neural correlates of action observation and execution in 14-month-old infants: An event-related EEG desynchronization study. *Developmental Science, 14*, 474–480.

Marshall, P.J., Y. Bar-Haim, & N.A. Fox (2002). Development of the EEG from 5 months to 4 years of age. *Clinical Neurophysiology, 113*, 1199–1208.

McIsaac, H., & J. Polich (1992). Comparison of infant and adult P300 from auditory stimuli. *Journal of Experimental Child Psychology, 53*, 115–128.

McMurray, B., & R.N. Aslin (2004). Anticipatory eye movements reveal infants' auditory and visual categories. *Infancy, 6*, 203–229.

Meltzoff, A., & K. Moore (1977). Imitation of facial and manual gestures by human neonates. *Science, 19*, 75–78.

Nadel, J., I. Carchon, C. Kervella, D. Marcelli, & D. Réserbat-Plantey (1999). Expectancies for social contingency in 2-month-olds. *Developmental Science, 2*, 164–173.

Neil, P.A., C. Chee-Ruiter, C. Scheier, D.J. Lewkowicz, & S. Shimojo (2006). Development of multisensory spatial integration and perception in humans. *Developmental Science, 9*, 454–464.

Nyström, P., T. Ljunghammar, K. Rosander, & C. Hofsten (2010). Using mu rhythm desynchronization to measure mirror neuron activity in infants. *Developmental Science, 14*, 327–335.

Pickens, J., & L.E. Bahrick (1997). Do infants perceive invariant tempo and rhythm in auditory-visual events. *Infant Behavior and Development, 20*, 349–357.

Pickens, J., T. Field, T. Nawrocki, A. Martinez, D. Soutullo, & J. Gonzalez (1994). Full-term and preterm infants' perception of face-voice synchrony. *Infant Behavior and Development, 17*, 447–455.

Picton, T.W., & M.J. Taylor (2007). Electrophysiological evaluation of human brain development. *Developmental Neuropsychology, 31*, 249–278.

Pons, F., M. Teixidó, J. Garcia-Morera, & J. Navarra (2012). Short-term experience increases infants' sensitivity to audiovisual asynchrony. *Infant Behavior and Development, 35*, 815–818.

Prastawa, M., J.H. Gilmore, W. Lin, & G. Gerig (2005). Automatic segmentation of MR images of the developing newborn brain. *Medical Image Analysis, 9*, 457–466.

Purhonen, M., R. Kilpeläinen-Lees, M. Valkonen-Korhonen, J. Karhu, & J. Lehtonen (2005). Four-month-old infants process own mother's voice faster than unfamiliar voices – Electrical signs of sensitization in infant brain. *Cognitive Brain Research, 24*, 627–633.

Rosander, K., P. Nyström, G. Gredebäck, C. Hofsten (2007). Cortical processing of visual motion in young infants. *Vision Research, 47*, 1614–1623.

Sirois, S., & D. Mareschal (2002). Models of habituation in infancy. *Trends in Cognitive Sciences, 6*, 293–298.

Southgate, V., M.H. Johnson, T. Osborne, & G. Csibra (2009). Predictive motor activation during action observation in human infants. *Biology Letters, 5*, 769–772.

Spelke, E.S. (1979). Perceiving bimodally specified events in infancy. *Developmental Psychology, 15*, 626–636.

Striano, T., A. Henning, & D. Stahl (2005). Sensitivity to social contingencies between 1 and 3 months of age. *Developmental Science, 8*, 509–518.

Striano, T., A. Henning, & D. Stahl (2006). Sensitivity to interpersonal timing at 3 and 6 months of age. *Interaction Studies, 7*, 251–271.

Thatcher, R.W. (1992). Cyclic cortical reorganization during early childhood. *Brain and Cognition, 20*, 24–50.

Thatcher, R.W., R.A. Walker, & S. Giudice (1987). Human cerebral hemispheres develop at different rates and ages. *Science, 236*, 1110–1113.

Thomas, D.G., & M.S. Lykins (1995). Event-related potential measures of 24-hour retention in 5-month-old infant. *Developmental Psychology, 31*, 946–957.

Thomas, D.G., E. Whitaker, C.D. Crow, V. Little, L. Love, M.S. Lykins, & M. Letterman (1997). Event-related potential variability as a measure of information storage in infant development. *Developmental Neuropsychology, 13*, 205–232.

Trainor, L.J., S.S. Samuel, R.N. Desjardin, & R.R. Sonnadara (2001). Measuring temporal resolution in infants using mismatch negativity. *NeuroReport, 12*, 2443–2448.

Turk-Browne, N.B., B.J. Scholl, & M.M. Chun (2008). Babies and brains: Habituation in infant cognition and functional neuroimaging. *Frontiers in Human Neuroscience, 2*, 16.

Van Elk, M., H.T. Schie, S. Hunnius, C. Vesper, & H. Bekkering (2008). You'll never crawl alone: Neurophysiological evidence for experience-dependent motor resonance in infancy. *NeuroImage, 43*, 808–814.

Van Puyvelde, M., G. Loots, L. Gillisjans, N. Pattyn, & C. Quintana (2015). A cross-cultural comparison of tonal synchrony and pitch imitation in the vocal dialogs of Belgian Flemish-speaking and Mexican Spanish-speaking mother-infant dyads. *Infant Behavior and Development, 40*, 41–53.

Van Puyvelde, M., P. Vanfleteren, G. Loots, S. Deschuyffeleer, B. Vinck, W. Jacquet, & W. Verhelst (2010). Tonal synchrony in mother-infant interaction based on harmonic and pentatonic series. *Infant Behavior and Development, 33*, 387–400.

Wiebe, S.A., C.L. Cheatham, A.F. Lukowski, J.C. Haight, A.J. Muehleck, & P.J. Bauer (2006). Infants' ERP responses to novel and familiar stimuli change over time: Implications for novelty detection and memory. *Infancy, 9*, 21–44.

Wunderlich, J.L., B.K. Cone-Wesson, & R. Shepherd (2006). Maturation of cortical auditory evoked potentials in infants and young children. *Hearing Research, 212*, 185–202.

Zentall, S.R., S.M. Boker, & J.M. Braungart-Rieker (2006). Mother-infant synchrony: A dynamical systems approach. Retrieved from: https://www.researchgate.net/publication/252672051_Mother-Infant_synchrony_A_dynamical_systems_approach.

www.ingramcontent.com/pod-product-compliance
Lightning Source LLC
LaVergne TN
LVHW010624300926
849162LV00001B/51